Applied Probability and Statistics

BAILEY · The Elements of Stochastic Processes with Applications to the Natural Sciences

BARTHOLOMEW · Stochastic Models for Social Processes, *Second Edition*

BENNETT and FRANKLIN · Statistical Analysis in Chemistry and the Chemical Industry

BHAT · Elements of Applied Stochastic Processes

BLOOMFIELD · Fourier Analysis of Time Series: An Introduction

BOX and DRAPER · Evolutionary Operation: A Statistical Method for Process Improvement

BROWNLEE · Statistical Theory and Methodology in Science and Engineering, *Second Edition*

BURY · Statistical Models in Applied Science

CHERNOFF and MOSES · Elementary Decision Theory

CHOW · Analysis and Control of Dynamic Economic Systems

CLELLAND, deCANI, BROWN, BURSK, and MURRAY · Basic Statistics with Business Applications, *Second Edition*

COCHRAN · Sampling Techniques, *Second Edition*

COCHRAN and COX · Experimental Designs, *Second Edition*

COX · Planning of Experiments

COX and MILLER · The Theory of Stochastic Processes, *Second Edition*

DANIEL · Application of Statistics to Industrial Experimentation

DANIEL and WOOD · Fitting Equations to Data

DAVID · Order Statistics

DEMING · Sample Design in Business Research

DODGE and ROMIG · Sampling Inspection Tables, *Second Edition*

DRAPER and SMITH · Applied Regression Analysis

DUNN and CLARK · Applied Statistics: Analysis of Variance and Regression

ELANDT-JOHNSON · Probability Models and Statistical Methods in Genetics

FLEISS · Statistical Methods for Rates and Proportions

GNANADESIKAN · Methods for Statistical Data Analysis of Multivariate Observations

GOLDBERGER · Econometric Theory

GROSS and CLARK · Survival Distributions

GROSS and HARRIS · Fundamentals of Queueing Theory

GUTTMAN, WILKS and HUNTER · Introductory Engineering Statistics, *Second Edition*

HAHN and SHAPIRO · Statistical Models in Engineering

HALD · Statistical Tables and Formulas

HALD · Statistical Theory with Engineering Applications

HARTIGAN · Clustering Algorithms

HILDEBRAND, LAING, and ROSENTHAL · Prediction Analysis of Cross Classifications

HOEL · Elementary Statistics, *Third Edition*

continued on back

PREDICTION ANALYSIS OF CROSS CLASSIFICATIONS

PREDICTION ANALYSIS OF CROSS CLASSIFICATIONS

DAVID K. HILDEBRAND

University of Pennsylvania

JAMES D. LAING

University of Pennsylvania

HOWARD ROSENTHAL

Carnegie-Mellon University

John Wiley & Sons

New York · London · Sydney · Toronto

Library of Congress Cataloging in Publication Data:

Hildebrand, David K   1940-
  Prediction analysis of cross classifications.

  Bibliography: p.
  Includes index.
  1. Prediction theory.   I.   Laing, James D., 1935-
joint author.   II.   Rosenthal, Howard, 1939-   joint
author.   III.   Title.

QA279.2.H54   519.5'4   76-25575

ISBN 0-471-39575-7

Printed in the United States of America

10 9 8 7 6 5 4 3 2 1

To Our Parents

FRANK and JOYCE HILDEBRAND

MELVIN and FAY LAING

ARNOLD and ELINOR ROSENTHAL

Whose Intent Was
to Keep Us Out of
the
Error Cells

PREFACE

This book presents a general approach to the analysis of cross-classified data. Those analyzing cross tabs, as cross classifications are commonly known, are frequently puzzled as to which, if any, of the large set of current statistical techniques is really appropriate to their research problems. Aimed toward creating a more direct link between research models and data analysis models, this book provides a general approach for using data to evaluate or select propositions that predict states of one variable from one or more other variables.

Our approach to prediction analysis is based on a formal language that, although simple in form, permits a precise statement of predictions derived from either verbal or mathematical statements of scientific models. A statistical measure evaluates the prediction success of each distinct statement in the language.

The prediction language and associated statistical measures represent a natural, simple, and unified method that ought to furnish a more coherent perspective to what currently appears as a fragmented area of data analysis. In fact, we show that many current measures of association can be interpreted as special applications of our method. On the other hand, our goal is to develop prediction analysis rather than to detail all earlier statistics. Selecting for attention only items of the prior literature relevant to our purposes, we do not include every known measure of statistical association and method of modeling probabilities. Consequently this is not a cookbook. Successful data analysis requires thought, planning, and decisions on the part of the researcher. Critical thinking about the basic issues of substantive theory and its empirical evaluation is an essential complement to the methods presented here.

We illustrate applications of our methods with a large number of cross classifications of real data from published research. These are intended for our primary target audience, people now or about to be involved in the substantive analysis of data. Although our own interests have led us to emphasize examples from sociology and political science, those fields by no means delineate the disciplinary boundaries for these methods.

Many examples pertain to the evaluation of game theory and coalition theories. In part, this emphasis again reflects our own substantive interests. More importantly, game-theoretic examples advantageously illustrate three of our primary analytic goals: evaluating predictions that are stated *a priori* rather than estimated from the data; assessing propositions whose predicted set of events may include more than one outcome; analyzing cross classifications that sometimes are too large for visual inspection. However game theory is not a prerequisite. This book discusses the evaluation of a wide variety of theory, whether simple or complex, *a priori* or developed from the data.

Writing for a diverse audience, we have kept the mathematics simple, banishing to appendixes matters demanding more than high school algebra and elementary probability theory. Our presentation has been directly influenced by the experience gained in using draft versions of the book in an undergraduate statistics course for psychologists at the University of Pennsylvania; in undergraduate political science courses at the University of California, Berkeley, and Stanford University; in an undergraduate Social Science Research Institute sponsored by the Roper Public Opinion Research Center at Williams College; and in graduate courses at Stanford and Carnegie-Mellon University. Although the basic text should not pose problems for anyone who has taken a one-semester course in probability and statistics, the concepts and methods introduced should interest readers who are highly sophisticated in methodology; we have attempted to make the presentation sufficiently brief for the tastes of the mathematically skilled.

Many readers with statistical training or prior experience in data analysis will be primarily oriented by the regression and correlation concepts of the linear model. The concepts of prediction analysis are both analogous to and distinct from the linear model, as indicated in Sections 3.2.7 and 7.5. Readers without prior exposure to the linear model may wish to skip those sections. Similarly readers not previously exposed to the analysis of cross classifications may initially wish to proceed directly to Chapter 3 after reading Chapter 1 and Section 2.1.

The first three chapters form the common trunk of the book. They, particularly Chapter 3, should be grasped firmly before proceeding further. In contrast, any of the remaining chapters may be read independently. For example, readers who have identified a research problem they can analyze with the methods of Chapter 3 may turn directly to Chapter 6 for the statistical procedures needed to make inferences about populations from sample data. Similarly one can turn to the multivariate analysis presented in Chapter 7 without having covered the applications of the bivariate method found in Chapters 4 and 5.

Although this book presents a unified approach to prediction analysis, it is only a beginning. Particularly in the later chapters, we seek to identify some open and challenging questions. The practitioner who looks to this, or any, book for a "one best way" to data analysis will quickly lose an illusion. In fact, our discussion of real examples emphasizes the need for alternative evaluations in prediction analysis. We present this book not as dogma but as inquiry.

DAVID K. HILDEBRAND
JAMES D. LAING
HOWARD ROSENTHAL

Philadelphia, Pennsylvania
Pittsburgh, Pennsylvania
June 1976

ACKNOWLEDGMENTS

We are indebted to many people. Melvin Hinich, statistician and interdisciplinary social scientist par excellence, initiated our collaboration. From the beginning Richard Morrison has bombarded us with highly constructive critiques. Some of his specific contributions are acknowledged in the text. Frank Sim helped in the initial identification of the problem addressed in this book. The reader engaged by John Wiley & Sons provided remarkably precise and insightful advice. We have also benefitted from discussions with numerous colleagues and students. Although we cannot name them all, we would especially like to mention Richard Bartlett, Kenneth Friend, Robert Lucas, Richard McKelvey, Scott Olmsted, Peter Ordeshook, Samuel Popkin, and Larry Rose. As Dean of the Graduate School of Industrial Administration at Carnegie-Mellon University, Richard Cyert made it possible for us to spend a summer developing the methods found in this book. This work has also been supported by Grants GS-33754 and SOC72-05245 A01 from the National Science Foundation. The patience and understanding of our secretaries, Connie Paris and Marjorie Farinelli, allowed us an unmentionable number of revisions of the original draft. Finally, we owe profound thanks to our families. At last we can give an acceptable answer to their frequent query, "Isn't that book done yet?"

Reprint permission was granted by Basic Books, Inc. for Figure 3-3 and Figure 3-4 from Chapter 3, "Differentiation," from *The Structure of Organizations*, by Peter M. Blau and Richard A. Schoenherr, copyright 1971 by Basic Books, Inc., Publishers, New York. Our Table 1.2 was adapted from *Amateurs and Professionals in British Politics, 1918–1959*, by Philip W. Buck, by permission of The University of Chicago Press, copyright 1963. We thank the Human Relations Area Files Press for permission to adapt our Table 3.2 from *Organization of Work*, 2nd printing, by Stanley H. Udy, Jr., HRAF Press, New Haven, copyright 1967.

Figure 5 from our article "Prediction Logic: A Method for Empirical Analysis of Formal Theory," *Journal of Mathematical Sociology*, has been reprinted with permission of Gordon and Breach. We acknowledge the permission to reprint, in condensation, Table 7.4 from *Parliament, Parties, and Society in France, 1946–58*, by Duncan MacRae, published by St. Martin's Press, Inc. We thank the American Political Science Association for permission to reprint material from several articles appearing in *The American Political Science Review*; the American Sociological Association and the author, H. Edward Ransford, for permission to reprint data from Table 1 of "Blue Collar Anger," *American Sociological Review*, Vol. 37, 1972; and the American Statistical Association and the authors, Leo A. Goodman and William Kruskal, for permission to reprint tables appearing on p. 745 of "Measures of Association for Cross Classifications," *Journal of the American Statistical Association*, Vol. 49, 1954.

                                                                  D.K.H.
                                                                  J.D.L.
                                                                  H.R.

CONTENTS

PREDICTION ANALYSIS OF CROSS CLASSIFICATIONS

SCIENTIFIC PREDICTION

...laws serve to explain events and theories to explain laws; a good law allows us to predict new facts and a good theory new laws. At any rate, the success of prediction... adds credibility to the beliefs which led to it, and a corresponding force to the explanations they provide. ...

...if we can predict successfully on the basis of a certain explanation we have good reason, and perhaps the best sort of reason, for accepting the explanation [Kaplan, 1964, pp. 346, 350].

1.1 INTRODUCTION

Prediction plays a fundamental role in the development, evaluation, and application of scientific theory. An advancing science produces theory whose ability to predict empirical events grows in reliability and precision. In order to formulate and evaluate such theory, scientists require both (1) language sufficiently rich and precise that relevant predictions can be stated exactly and (2) methods for accurately gauging the correspondence between observations and predictions.

The prediction language must permit exact specification of predicted relations. The relations may include not only one-to-one relations, such as "each son's socioeconomic status tends to be the same as his father's" or "the Republicans' presidential vote in the next election will be two-thirds of their vote in the previous election," but also one-to-many relations, such as "the Republicans' presidential vote will be one- to three-quarters of their previous vote." In the latter case any outcome between one-quarter and three-quarters is consistent with the prediction, whereas in the one-to-one case only the specific outcome two-thirds is consistent.

1

Whatever the form of prediction, the analytic methods should permit measuring the degree to which observed events conform to the specified prediction. Techniques for specifying and evaluating scientific predictions have been developed by statisticians and other methodologists, such as econometricians, psychometricians, and biometricians. This book supplements that research technology.

In the following pages we present a formal language called prediction logic for expressing certain types of predictions and describe methods for using data to evaluate the success achieved by any proposition stated in this language. The prediction logic, based on a simple adaptation of elementary logic, is an easy formalization of natural language frequently used to discuss relations among variables. The methods for data analysis also are readily understood and applied. Despite its apparent simplicity, the approach developed here can be used to state and evaluate highly complex scientific propositions.

1.1.1 The Squared Error Approach to Quantitative Variables

The most extensively developed prediction languages and associated methods of data analysis apply to theories predicting one-to-one (functional) relations among variables measured at the *interval* or *ratio* levels. For example, heat increases the length of a steel bar. In physical science, the increase is predicted to be a specific function of the bar's original length and the change in temperature. This one-to-one proposition predicts just one value of the predicted variable (final length) for each distinct combination of predictor variables (original length and change in temperature). Any *ratio* variable, such as length, is measured on a scale that has a natural origin (absolute zero) and some arbitrary unit of measurement. Thus whether length is measured in inches or centimeters, 0 in. and 0 cm both indicate zero length, but 10 in. is not the same as 10 cm. Similarly the Kelvin temperature scale, which is based on an absolute zero, is a ratio scale. In contrast, both origin and unit of measurement are arbitrary in an *interval* scale. For example, although zero on the centigrade (Celsius) temperature scale is set at the temperature at which water freezes at sea level, this corresponds to 32° Fahrenheit. The units of measurement also are selected arbitrarily in these scales so that the lowest temperature at which water boils is 100°C and 180°F higher than the freezing point. Differences between values of an interval scale satisfy the conditions of ratio scale measurement, but the values themselves do not. For convenience we refer to variables measured by ratio or interval scales as *quantitative* variables.

Quantitative variables such as length, temperature, money, and votes have a number of advantages. Predictions relating quantitative variables can be expressed mathematically in a variety of functional forms (such as linear, exponential, hyperbolic). Values of ratio variables (and differences in values of interval variables) can be manipulated as real numbers; therefore concepts such as average value and squared difference between predicted and observed values can be used to evaluate predictions. For example, if a steel bar is predicted to be 400 cm long but its true length is 398 cm, then the *squared prediction error* is $(400-398)^2 = 4$ cm^2. Currently the predominant forms of reasonably sophisticated methods of prediction analysis are based on using squared errors to evaluate functional predictions of quantitative variables.

1.1.2 Qualitative Variables and the Inapplicability of Methods for Quantitative Variables

These mathematical concepts, so important to physical science, with its emphasis on quantitative variables such as velocity, acceleration, and voltage, cannot be applied in many research contexts frequently encountered in the social and life sciences. For example, one cannot calculate the average religious affiliation in a group of two Protestants, one Catholic, and one Jew. Nor can prediction error be measured by squaring the "distance" between predicted and observed religious affiliation, "(Catholic–Buddhist)^2."

Religious affiliation is a *qualitative* variable. Such variables are represented by nominal or ordinal scales. A *nominal* scale describes objects or events by just one element of a set of mutually exclusive classes or states: for example, {Protestant, Catholic, Jew, Other, None}.[1] (Throughout this book we use braces to enclose the list of elements defining a set. Although sets are always assumed to be finite, most of the methods in the book would readily apply to variables with an infinite number of classes.) The order in which nominal classes are listed is arbitrary; therefore the variable for gender can be expressed as {female, male}, or {male, female}. Frequently a qualitative variable is expressed symbolically. A capital letter represents the variable and a lower-case letter with numeric indices denotes the variable's states. For example, let X denote a variable with C states $\{x_1, x_2, \ldots, x_j, \ldots, x_C\}$. Although the states of a nominal variable can be indexed in this way with numeric "names," no more significance can be attributed to these numbers than to any other names.

An *ordinal* scale orders or ranks nominal classes in terms of the

---

[1] Any measurement scale satisfies the conditions for a nominal scale; hence any data analysis technique designed for nominal scales can also be used for other measurement scales, although such a use does not take advantage of additional information provided by these scales.

"amount" of the characteristic being measured.[2] We follow the standard convention of listing ordinal states in order of decreasing amounts or degrees of the variable. Thus x_1 is a greater X-value than x_2, x_2 is greater than x_3, and so on. For example, a respondent's approval of the President's performance in office might be described by one of the mutually exclusive states in the set {strongly approves, approves, disapproves, strongly disapproves} or, symbolically, $\{x_1, x_2, x_3, x_4\}$.

As the religious affiliation example illustrated, neither the prediction languages nor the analytic methods designed for quantitative variables are generally suitable for analyzing predictions of such qualitative variables as occupation, preference ordering of three candidates, or ordinally measured approval of the present incumbent. Unfortunately no research technology of appeal comparable to squared error methods has emerged for qualitative variables.

Yet prediction analysis methods for qualitative variables are needed in many scientific contexts. It is not feasible, in practice or principle, to measure some variables with quantitative scales. Although psychometricians and other measurement specialists have made important progress in quantification by developing scaling procedures for qualitative information, concepts such as gender, religion, occupation, and political party continue to be subject to qualitative analysis.

1.1.3 Qualitative versus Quantitative Theory

Even if the appropriate quantification procedures can be specified, they may require more restrictive theoretical assumptions than those needed for a qualitative assessment. A good example of this is the analysis of individual preferences over a set of alternatives, such as whether the main course at dinner is steak, lobster, or Peking duck. Suppose that an individual's preferences are ordered and transitive (for example, Peking duck is preferred to lobster, lobster is preferred to steak, and, by transitivity, Peking duck to steak). Then the rank of any alternative in the preference ordering may be described by one of the classes in the ordinal variable {first choice (rank = 1), in a two-way tie for first (rank = 1.5), second choice or in a three-way tie (rank = 2), in a two-way tie for last (rank = 2.5), third choice (rank = 3)}.[3] (Note that the numbers 1.5 and 2.5 are used as convenient

---

[2] Various subtypes of ordinal scales can be identified [examples include ordinal scales with a natural origin (zero) and scales in which the *distance* between ordinal classes can be ordered], but these distinctions are not used in this book. For extensive treatments of measurement, see Torgerson (1958) and Coombs (1964).

[3] Knowing the rank of each main course, we can identify an individual's preference ordering. For example, if steak and lobster are 1.5 and Peking duck is 3, then the individual is indifferent between steak and lobster and prefers both to the duck. Thus no information is lost by our particular rank assignments.

labels only and not to assign interval values to rank.) If we can assume that, in making decisions, the individual also satisfies certain additional, hence more restrictive, assumptions (such as choosing in specified ways when faced with choices among alternative gambles on what the main dish will be), then utility theory can be applied to measure the preferences by an interval (utility) scale (Luce and Raiffa, 1957, Ch. 2). Yet a theory requiring these restrictive assumptions must be less general than a similar theory requiring only ordinal measurement.

A recent trend in social science is to develop theory based on the less restrictive assumptions of qualitative measurement.[4] To test such theory, we need to develop appropriate methods for data analysis that do not require the stronger assumptions associated with quantitative measurement.

1.1.4 One-to-Many Predictions

In addition to being restricted to quantitative variables, the focus of the most appealing classical procedures for data analysis is limited to one-to-one predictions. Whether measurement is qualitative or quantitative, previous data analysis methods are generally inappropriate for evaluating one-to-many predictions. Such predictions occur frequently in daily life and science: (a) It will rain or snow. (b) Given a patient's symptoms and the most recent theory, the physician may only be able to narrow the diagnosis to several alternative maladies. (c) In analyzing the control of corporations with small numbers of stockholders, one might apply Riker's (1962) theory of minimal winning coalitions to predict that the directors will represent only coalitions of stockholders whose combined votes are just large enough that the coalition will fail to have a majority if any stockholder leaves the coalition. For example, if Able holds 48% of the votes (stock), Betty and Charlie each own 24%, and Daphne controls 4%, then the theory could be used to predict that directors would represent any one of four blocks {(A and B), (A and C), (A and D), (B, C, and D)}. On the other hand, if Able were to control more than half of the votes, then the theory would predict the single event that the directors represent only Able. (d) Historians of the 1972 Democratic presidential primary election might be willing to make the one-to-one statement for Pennsylvania that "Suburban lower-income whites voted for Wallace," but only the one-to-many statement that "Urban middle-income blacks voted for Humphrey,

---

[4] For example, theory of social choice (Arrow, 1963) and general equilibrium theory in economics (Arrow and Hahn, 1971) can be developed by representing each person's preferences only ordinally rather than by a cardinal utility function. Our arguments in the preceding text disagree with Wilson's (1970) statement that ordinal variables are inadequate for complex models in social science.

Muskie, McGovern, or Jackson." (*e*) Von Neumann and Morgenstern's (1947) solution concept for *n*-person games may be used to predict that the division of payoffs within the coalition will be any one of the many alternatives within a particular subset of feasible payoff divisions.

1.1.5 An Overview

This book presents methods for both one-to-many and one-to-one predictions about qualitative variables. We present an approach for specifying and evaluating a broad class of alternative predictions concerning qualitative data. Although in developing this approach we occasionally relate our viewpoint to predictions with quantitative variables as well,[5] the book's primary focus is on prediction analysis of cross classifications of qualitative variables.

   Continuing this chapter's introduction to prediction analysis, Section 1.2 identifies some important characteristics for distinguishing among alternative forms of predictions. Section 1.3 considers the traditional model of "nonrelation" between two variables—statistical independence. There it is argued that much previous research with qualitative variables has been concerned with measuring the degree to which data depart from this "nonrelation" (whatever the "direction") rather than, as seems more appropriate, with evaluating the prediction success of the investigator's particular theory. Section 1.4 briefly overviews conventional types of predictions relating ordinal variables. (This topic is considered extensively in Chapter 5.) Next, in Section 1.5, a nontechnical discussion of the linear model for relating quantitative variables identifies advantages of this widely used technique that we wish to incorporate in our approach, and limitations that we want to avoid. Finally Section 1.6 identifies some critical methodological issues, and Section 1.7 defines the language of prediction logic for stating propositions relating two qualitative variables.

   The book's subsequent argument proceeds as follows. Chapter 2 identifies criteria we think should be satisfied by a measure of the prediction success of any proposition stated in the logic, and also demonstrates that currently established statistical methods fail to satisfy these criteria. Chapter 3 presents a general model for defining a specific measure, $\nabla_{\mathcal{P}}$, for each distinct prediction logic proposition relating two variables. Chapters 4 and 5 develop research applications and further elaborations of the bivariate model. The methods developed to this point are for analyzing prediction success when the data describe an entire population or represent

---

[5] The methods presented here can be applied rather directly to quantitative variables with discrete value states or to continuous variables whose unconditional density functions are known. More effort is required before the methods can be applied to continuous variables whose unconditional densities are not known.

a sufficiently large sample that sampling considerations may be ignored. Chapter 6 considers bivariate statistical inference when only a (perhaps small) sample of data is available for evaluating prediction success in the larger population. Methods are presented there for testing hypotheses and constructing confidence intervals for bivariate prediction logic propositions stated *a priori*, and for evaluating propositions selected from inspection of the data. The first six chapters focus on the special case of bivariate propositions. Chapter 7 then extends the prediction logic, $\nabla_{\mathscr{P}}$ measure, and statistical theory for multivariate prediction analysis.

1.2 STYLES OF PREDICTION

Scientific predictions come in many styles. This section makes some preliminary distinctions among alternative prediction styles in order to lay foundations for developing our approach and delimiting its domain.

1.2.1 Number of Independent, Dependent, and Intervening Variables

First, predictions differ in the number of variables they assign to various roles. Let *independent variable* denote a (predictor) variable that is used to predict another variable and let *dependent variable* refer to a predicted variable. These two roles are not mutually exclusive; an *intervening variable* plays both roles, being an independent variable in one component of the prediction and a dependent variable in another. For example, in the typical prediction relating a son's occupation to his education and his father's occupation, the father's occupation is the independent variable, the son's education is intervening, and the son's occupation is the dependent variable. These three descriptions of variables identify roles assigned (perhaps arbitrarily) in the prediction; they connote nothing about causality.

This book develops methods for evaluating the success achieved by a proposition in predicting a single dependent variable. When a theory involves more than one dependent variable, these methods can be used to evaluate the theory's success with respect to each dependent variable separately, and with respect to composite variables (discussed later) formed by combining dependent variables.

The central portion of this book is devoted to propositions relating two variables, one treated as independent, the other as dependent. Bivariate prediction analysis is emphasized if only because it is prerequisite to later elaborations of the model. For example, in multivariate analysis the measure of prediction success attributed to one independent variable when other independent variables are controlled is very similar to the bivariate measure.

As the research applications considered in Chapters 3 to 5 demonstrate, a rich variety of complex scientific propositions can be addressed with the bivariate methods. This breadth partly results because it is possible to combine qualitative variables into a single composite variable.

This device may be used to transform a multivariate proposition into equivalent bivariate form, thus bringing it within the scope of bivariate methods. For example, Sawyer and MacRae's (1962) game-theoretic model of campaign strategy discussed in Chapter 3 predicts whether there will be one, two, or three Democratic candidates in each of the three-member legislative districts in Illinois. It also predicts the number of Republican candidates. These two dependent variables can be combined into a single composite variable with nine categories: {(1 Democrat, 1 Republican), (1 Democrat, 2 Republicans),...,(3 Democrats, 2 Republicans), (3 Democrats, 3 Republicans)}. In a more complex example discussed in Chapter 4, Laing and Morrison's (1974) model of coalition behavior in a laboratory situation uses a composite independent variable identifying 293 distinct bargaining situations to predict a composite dependent variable describing 39 coalition outcomes.

1.2.2   *A priori* versus *ex post facto* Predictions

A second distinction is whether the prediction is stated *a priori* or selected *ex post*. *Ex post facto* propositions are selected after analysis of the data. Specifically, any prediction that cannot be made without some information about the data's observed distribution on the dependent variable is *ex post*. For example, the Illinois predictions would be rendered *ex post* if the predictions used knowledge that the Democrats nominated two candidates more frequently than either one or three candidates. In contrast, *a priori* predictions are stated before the data are analyzed. An *a priori* prediction for each case can be made knowing only that case's location on the independent variable(s) [and not knowing its location on the dependent variable]. Although a temporal ordering is not necessary for *a priori* prediction, any proposition is *a priori* if it can be applied before the occurrence of the event described by the dependent variable.

Science typically advances by alternating between *ex post* and *a priori* prediction. A postulated law is modified or selected *ex post* after analyzing one set of data, then is used *a priori* to predict new data, and so on. The methods for both *a priori* and *ex post* prediction needed in research should facilitate the interplay between these perspectives. The methods developed in this book satisfy these requirements. However, investigators should attempt to specify predictions *a priori*, if only because (as shown in Chapter 6) fewer observations are needed to test an *a priori* prediction adequately than to select an *ex post* proposition with confidence.

1.2.3 Event Predictions

Third, we distinguish propositions that *predict each case's state on the dependent variable* (denoted as *event predictions*) from those that do not. Event predictions permit prediction error to be calculated for each observed case, one at a time. (We amplify the meaning of *case* later in discussing a prediction's "degree.") For example, a proposition that predicts an individual legislator's vote on a bill is an event prediction: "Liberals will vote in favor, conservatives will vote against or abstain." On the other hand, "Representatives of farm districts are more likely than others to vote against the bill" is not an event prediction for individual legislators. Clearly, "X and Y are not unrelated" is not an event prediction, nor is "Y is a linear function of X." However "Individuals with x dollars of gross income will have less than $0.1x$ dollars of deductions for charitable contributions" makes a prediction about each case's dependent variable (Y) state, given its value for the independent variable. This book investigates only event predictions, whether they predict a single state for each case (one-to-one prediction) or predict that the case lies in a specified set of dependent variable states (one-to-many prediction).

1.2.4 Absolute versus Actuarial Predictions

Fourth, following Clausen (1950, pp. 573–574), we distinguish between

> ..."absolute" prediction (i.e., involving unequivocal specification of performance for the individual case...) and actuarial prediction (involving specification of the probability that individuals of stated characteristics will behave in a given way).

An *absolute* prediction always predicts the same set of dependent variable states for all cases that have identical independent variable states. For example, "If the legislator is a liberal from an urban district, then predict that person will vote in favor of this bill." Once the vote is cast the investigator can determine for any "liberal from an urban district" whether the prediction was correct. Borrowing a term from game theory, an absolute event prediction is a *pure strategy prediction*: if this, then (always) predict that.[6]

Now consider the statement "The chance of rain during each day in July is $1/3$." By definition this statement is an *actuarial* prediction since it specifies the probabilities of various dependent variable states: rain with probability $1/3$, no rain with probability $2/3$. In evaluating an actuarial proposition the investigator has several prominent options. One is to

---

[6] We prefer the terms *pure strategy* and *absolute* to *deterministic*, in part in order to avoid connotations related to causality. This book focuses on *prediction*, leaving the choice of causal interpretations (if any) to the investigator.

evaluate the extent to which observed proportions match predicted probabilities. By this approach the prediction in the example fits the data perfectly if it rains exactly 1/3 of the days in July. Of course such perfect fits to proportions do not always indicate successful event predictions. If we know that a coin is fair, we can accurately guess the proportion of heads in a billion tosses yet be entirely incapable of predicting the events represented by single tosses of the coin.

A variety of techniques are available for assessing the fit of an actuarial proposition to proportions observed in the data,[7] including the chi-square goodness-of-fit test discussed in Chapter 2. However, as mentioned previously, we seek here to evaluate the extent to which such propositions succeed in *predicting events* described by the dependent variable, rather than to measure *goodness of fit*.

To this end consider the following optional ways in which an actuarial proposition may be applied to each case as an event prediction. The first option is to predict (always) that any case has the dependent variable state asserted by the actuarial proposition to be most likely.[8] Since the most likely event according to the proposition in the example is "no rain" (probability 2/3), no rain is predicted for every day in July. A potentially disadvantageous feature of this approach is that it translates different actuarial propositions into the same pure strategy prediction: the prediction is "no rain" for any proposition assigning a probability greater than 1/2 to this event—whether the event is supposed to occur with probability 2/3, 3/5, or, say, 9/10. The next approach avoids this.

An actuarial proposition may be regarded as a probability mixture of alternative absolute predictions. In applying the proposition to any case, one of these alternatives can be selected in accordance with the probability assigned to it by the actuarial proposition. By this approach the proposition in the example is interpreted to say "With probability 1/3, select the prediction 'rain'; if 'rain' is not selected, predict 'no rain'." For example, a fair die is rolled once for each day in July. If a 5 or 6 is rolled, then rain is predicted (and otherwise no rain is predicted) for that day. Drawing another analogy from game theory, we use the term *mixed strategy prediction* to denote any actuarial proposition that is applied probabilistically to yield an absolute event prediction for each case. Mixed strategy predictions represent actuarial propositions in a form that can be evaluated with methods developed in Chapter 4. Until then absolute predictions of events will be our main focus.

---

[7] For example, see Hays (1973), Theil (1972), and Hoel, Port, and Stone (1971).

[8] Another variation is to predict that the dependent variable state will *not* be that deemed to be "least likely" by the actuarial proposition.

1.2.5 Observations and the Degree of a Prediction

We have not yet been explicit as to the meaning of the term *observation*. An observation is a recorded occurrence of an event. Events are described in terms of one or more variables in the proposition being analyzed. Thus the unit observation (frequently called the *unit of analysis*) depends on the investigator's choice of variables and analytic domain within the research context. Consider, for example, research on the relation between amount of formal education and interest shown in politics, as demonstrated by watching a particular televised debate of presidential candidates. One study might use each person interviewed as an observation ("Respondents with at least some college tended to watch the debate, whereas those with less education did not"); another might treat each household as an observation ("The TV is more likely to be tuned to the debate if the head of household is a college graduate"); and a third might establish the census tract as the unit of analysis ("The proportion of households watching the debate increases with the average level of education in the tract").[9]

Given the unit observation as specified in the proposition, the next distinction among alternative styles of prediction concerns whether the proposition makes a prediction for each observation singly or, rather, predicts only the result of relative comparisons of two or more observations on the dependent variable.

> A *degree-1* proposition makes a prediction for each observation taken one at a time (for example, predicts "watched the debate" for each college graduate interviewed).

> A *degree-2* proposition makes predictions about the relative comparison of two observations on the dependent variable (for example, if the first person in a pair of observations has more formal education than the other, predict that the first watched more of the debate).

In degree-2 prediction, then, a *pair* of observations constitutes a single "case." Similarly a degree-3 proposition pertains to relative comparisons among three observations, and so on.

---

[9] Robinson (1950) has shown that care must be taken in making inferences from one level of aggregation to another. For example, high viewership of a televised debate in regions with high levels of education is not sufficient evidence for inferring that educated persons tended to watch the debate. There is an extensive literature on aggregation (Christ, 1968; Hannan, 1971; Rosenthal, 1973). A good nontechnical discussion is found in Eulau (1969). In this book we do not consider aggregation problems.

Often, particularly in the social sciences, predictions are stated in the form "Y tends to increase with X" (for example, "Time devoted to watching the political debate increases with years of formal education"). Such a proposition does not make an event prediction for each observation (degree-1). However it implies an event prediction of degree-2 ("The more educated person in the pair watched more of the debate"). Thus propositions of the form "the more the X, the greater the Y" can be represented as event predictions of degree-2, thus bringing them within the domain of the methods developed in this book.

1.2.6 Error Evaluation

Finally, scientific prediction is rarely perfect; usually some prediction error can be expected. A goal of science is to reduce prediction error to tolerable levels, but theory need not eliminate error altogether to be useful. We distinguish such scientific prediction from assertions of logical tautologies and postulated universal truths. There is no reason for evaluating tautologies with data—they are established by logical proof, regardless of observed events. Postulated universal truths are empirical statements about the existence or nonexistence of a critical observation: one black swan falsifies the universal statement "All swans are white" and establishes the proposition "There exists at least one black swan." *This book focuses on methods for using data to evaluate scientific propositions that cannot be established or disproved with just one observed case.*

In sum, as indicated earlier in discussing alternative styles of scientific prediction, we present in this book methods for evaluating a bivariate or multivariate proposition's success in predicting events described by the dependent variable(s) for each case (single observation or relative comparison of two or more observations), whether the proposition is given *a priori* or *ex post*, and whether it predicts just one or several events for each state of the independent variable(s) or represents a pure or mixed strategy of prediction. The broad range of research examples considered in this book suggests that the domain of the methods, as delimited earlier, includes a rich variety of scientific applications.

Scientific propositions also differ in terms of the level at which variables are measured. The next sections consider prediction with variables defined at various levels of measurement, beginning with nominal variables, then moving, successively, to ordinal and quantitative scales.

1.3 PREDICTION FOR NOMINAL VARIABLES: STATISTICAL INDEPENDENCE AND "UNRELATEDNESS"

In the analysis of nominal variables the concept of statistical independence has been the traditional criterion for investigating whether two qualitative variables are said to be "related."

In order to review this concept we first explain some notation used throughout this book. Let the variable X denote the set of C mutually exclusive states (categories or classes) $\{x_1, x_2, \ldots, x_j, \ldots, x_C\}$. To illustrate, the following example describes a person's occupation by just one of the classes in the set {professional, business, manual worker, other}.[10] Similarly, define the variable Y as the set of R states $\{y_1, y_2, \ldots, y_i, \ldots, y_R\}$. The cross classification of Y and X denotes the set of RC ways in which each state of one variable can be paired with each state of the other: $\{(y_1, x_1), (y_1, x_2), \ldots, (y_2, x_1), \ldots, (y_i, x_j), \ldots, (y_R, x_C)\}$. Formally we express this set as the Cartesian product $Y \times X$.

As an example, consider a classic bivariate relation explored in political sociology: that between occupation and political affiliation. Let X be the occupation variable just described. Let Y be the set of political party affiliation categories {Labor, Conservative, Other}. Buck (1963) categorized British ministers and parliamentary private secretaries from 1918 to 1955 on these variables. We treat his $N = 597$ observations as a *population*. Of the 597 observations, 61 were Conservative party members from "business" occupations and are cross-classified as the pair (Conservative, business).

The ratio of the number of these observations to the total population size, $61/597 = .102$, is the probability that a member randomly chosen from the population would be classified as Conservative, business.[11] (Throughout the book decimals are rounded to three places, although trailing zeroes may be omitted.)

1.3.1   General Representation of a Cross Classification

In general, a $Y \times X$ cross classification may be represented as the contingency table with R rows and C columns shown in Table 1.1. The RC cells of Table 1.1 contain probabilities P_{ij} that any observation in the population is located in states y_i and x_j, where $0 \leqslant P_{ij} \leqslant 1$.

Since the .102 probability of (Conservative, business) occurs at the intersection of the second row and second column in Table 1.2A, we have $P_{22} = .102$ for the example. The probabilities shown in the margins of Table 1.1 for each row and column (often called *marginal probabilities* or *marginals*) indicate the probability that an observation lies in the respective row

---

[10] Since the states of a nominal variable must be mutually exclusive, only one class may be used to describe a given person. Thus if double occupations such as "business and professional" are encountered, we must exclude such persons from the analytic domain, define the variable as describing a person's "primary" occupation, or modify the original set of classes so that each person can be described by just one variable state.

[11] In fact, Buck obtained data for only 597 of the 695 members of the true population. We deal with these 597 individuals as a population. Some investigators would want to regard the 597 (or 695) individuals as one sample of an infinite population of possible historical realizations.

Table 1.1   Bivariate contingency table

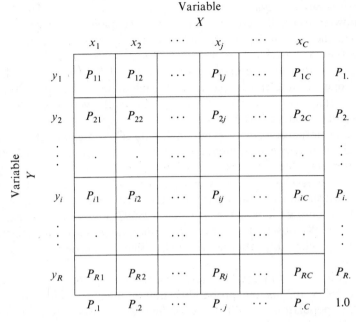

For all $i \in \{1, 2, \ldots, R\}$ and for all $j \in \{1, 2, \ldots, C\}$,

$$0 \leqslant P_{ij} \leqslant 1$$

$$\sum_j P_{ij} = P_{i.}$$

$$\sum_i P_{ij} = P_{.j}$$

$$\sum_i \sum_j P_{ij} = \sum_i P_{i.} = \sum_j P_{.j} = 1$$

or column. Thus, for example, $P_{1.}$ is the (unconditional) probability that an observation lies in the state y_1; the "dot" in the symbol $P_{1.}$ indicates that the variable X whose index j is omitted has been ignored. Note that the marginal is equal to the sum of its row or column entries and that both sets of marginals must sum to one, as is required of a probability measure. Note also that the sum of the RC cell entries $\Sigma_i \Sigma_j P_{ij}$ equals $P_{11} + P_{12} + \cdots + P_{ij} + \cdots + P_{RC} = 1.0$.

Table 1.2 Measuring departures from statistical independence

(A)   Actual Population Probabilities *Not* Statistically Independent

		Occupation				
		Profes-sional	Busi-ness	Manual worker	Other	
Political party affiliation	Labor	.122	.017	.164	.067	.370
	Conservative	.355	.102	.000	.075	.533
	Other	.064	.020	.000	.013	.097
		.541	.139	.164	.156	1.000
						($N = 597$)

(B)   Population Probabilities if the Variables were Statistically Independent, Given Marginal Probabilities in (A)

		Occupation				
		Profes-sional	Busi-ness	Manual worker	Other	
Political party affiliation	Labor	.200	.051	.061	.058	.370
	Conservative	.288	.074	.087	.083	.533
	Other	.053	.014	.016	.015	.097
		.541	.139	.164	.156	1.000

(C)   Deviations from Statistical Independence in (A)

		Occupation			
		Profes-sional	Busi-ness	Manual worker	Other
Political party affiliation	Labor	− .078	− .035	+ .103	+ .009
	Conservative	+ .067	+ .028	− .087	− .008
	Other	+ .011	+ .007	− .016	− .002

Each entry $= P_{ij} - P_{i.}P_{.j}$

Source: (A) is adapted from Buck (1963, p. 122).

1.3.2   Statistical Independence

Let us now return to the example and consider the use of occupation in predicting party. Suppose an investigator who has no theory specifying a particular relation between these two variables simply predicts that the variables are related. In analyzing this prediction we need to indicate what constitutes evidence that occupation and party are *not* related. Typically the standard of nonrelation is statistical independence. *Two variables are statistically independent in a population if and only if $P_{ij} = P_{i.} P_{.j}$ for all i and j.* That is, the cell probabilities equal the product of the corresponding marginal probabilities. For the actual population probabilities represented in Table 1.2A, occupation and party are not independent. In a hypothetical population that has the marginal probabilities of Table 1.2A but is statistically independent, the probability of observing a Conservative manual worker, for example, is $.087 = .533 \times .164$. In fact the actual population probability is zero. Table 1.2B shows the cell probabilities if statistical independence were to hold for the marginal probabilities of Table 1.2A. The actual deviations from statistical independence are shown in Table 1.2C which displays the terms $\{ P_{ij} - P_{i.} P_{.j} \}$, the differences between the entries in (A) and (B).

More generally, *if X and Y are statistically independent,* then the probability of any observation falling in category y_i given that it is in x_j, that is, the conditional probability of y_i given x_j, is

$$P_{y_i | x_j} \equiv \frac{P_{ij}}{P_{.j}} = \frac{P_{i.} P_{.j}}{P_{.j}} = P_{i.}$$

The equation shows that under statistical independence the conditional probability of y_i does not depend on x_j. For example, in the hypothetical population of Table 1.2B, the conditional probability of Labor among professionals is $.200/.541 = .370$, the marginal probability of Labor. Thus if the variables are statistically independent it is not useful to know a person's occupation for *any* prediction about party affiliation. Therein lies the importance of the statistical independence concept. It is an indication of complete "nonrelation" between X and Y.

Most present statistical techniques evaluate the degree of departure from statistical independence or some alternative "nonrelation" criterion that includes, but is not restricted to, statistical independence.

As shown in greater detail in Chapter 2, each of these techniques typically provides a single summary number for the cross classification. If the number is a test statistic, it can provide only the doubly negative testimony that the variables are "not unrelated." If the number is a measure of prediction success, it evaluates only whether a single, data-determined prediction makes relatively few errors. However these methods provide little help in measuring the success achieved by a prediction specified by the investigator. For example, we could predict that "Manual workers are Labor, businessmen are Conservatives, and professionals are Other." Knowing either that occupation and party are not independent or that some other prediction succeeds does not establish the success of this particular prediction. In Chapter 3 we develop a general model that yields a specific measure of the success achieved by a particular proposition. Soon we must make the meaning of "prediction success" explicit. Now, however, it is useful to continue examining prediction methods associated with various levels of measurement. We next consider prediction for ordinal variables before discussing the powerful prediction languages and analytic methods available for quantitative variables.

1.4 PREDICTION FOR ORDINAL VARIABLES: RELATIVE COMPARISONS OF OBSERVATIONS

The previous comments about prediction for nominal variables also apply to ordinal variables, since an ordinal scale is, first of all, a set of nominal classes. For example, suppose that, on the basis of the highest academic degree awarded, we describe the level of formal education completed by any voter in a population by assigning just one class in the ordered set $X : \{ x_1$ (master's degree or PhD), x_2 (college degree), x_3 (some college work but no college degree), x_4 (high school diploma but no college work), x_5 (neither high school diploma nor any college work)}. Also assume that each person's support for civil liberties is described, on the basis of preferences among alternative public policies, by the variable $Y : \{ y_1$ (very strong), y_2 (moderately strong), y_3 (moderately weak), y_4 (very weak)}. A variety of degree-1 event predictions can be made about the 4 (rows) $\times$ 5 (columns) cross classification of these two variables. For one example, let the proposition $\mathcal{P}$ predict y_1 if the voter has an advanced degree (has $X = x_1$), y_1 or y_2 if $X = (x_2$ or $x_3)$, and $y_1, y_2,$ or y_3 if $X = x_4$. (Usually we omit the prediction for any state of the independent variable for which the predicted set includes all dependent variable states. Thus the preceding prediction omits the prediction when $X = x_5$.) One of the many alternatives to this one-to-many pure strategy prediction is the proposition $\mathcal{P}'$: "Predict y_1 if x_1, y_2 if x_2, y_3 if $x_3,$ and y_4 if x_4 or x_5."

The proposition $\mathcal{P}'$ is but one example of many degree-1, event predictions consistent with the proposition $\mathcal{P}''$: "Civil libertarianism tends to increase with formal education." $\mathcal{P}''$ predicts that "Y tends to increase with X," but it does not supply enough information to predict the support offered for civil libertarian policies by any voter whose X-state (level of education) is given; that is, $\mathcal{P}''$ makes no event prediction for any voter.

Indeed researchers typically use this type of verbal statement with measures that assess predictions concerning observation pairs. Two of the most widely applied are due to Goodman and Kruskal (1954) and Somers (1962). In the Goodman and Kruskal approach, pairs tied on either variable are excluded from the domain of the prediction. Within the domain the prediction has the form, "If Able has more formal education than Betty, then Able is predicted to offer stronger support than Betty for civil liberties." Somers' approach makes a similar prediction but ties are not excluded.

We examine both approaches in greater detail in Chapter 5. For the present it suffices to emphasize that neither approach permits the analysis of the many other possible predictions about pairs. Of still greater importance neither approach applies to such degree-1 predictions as, "If Able has received no academic degree, then Able's support for civil liberties is *not* very strong, but if Able has earned a bachelor's or advanced degree, then Able's support for civil liberties is (moderately or very) strong."

The distinction between degree-1 predictions concerning single observations and the more common predictions of order relationship for observation pairs is important. In an example discussed in Chapter 5, Goldberg (1969) used various ordinal variables as independent variables with a dichotomous dependent variable indicating whether a son defected from or was loyal to his father's party identification. For many purposes it is clearly more interesting to predict whether a given individual defects given his independent variable state than to predict how he behaves relative to some other individual given their relative positions on the independent variable. It is not our aim, however, to make a case for degree-1 rather than degree-2 predictions. On the contrary, the methods developed later in the book provide the flexibility needed to handle both types.

1.5   PREDICTION FOR QUANTITATIVE VARIABLES: LINEAR MODELS

In moving from qualitative to quantitative variables we encounter a dramatic increase in the precision and flexibility of available methods for stating and analyzing predictions. Since values of quantitative variables can be manipulated as real numbers, predictions can be stated in the precise language of algebraic functions. Thus a bivariate prediction can be stated in the form $y = f(x)$, where f is a function: as in the earlier example,

"Republican presidential vote in 1976 = 2/3 Republican vote in 1972," a single value of Y is predicted for any observation at a given location on the independent variable.

1.5.1 *A priori* and *ex post* Linear Predictions

Of the various functions, statistical methods are most extensively developed for a particular type—the linear function. In the bivariate case the linear function can be written as

$$y = \beta_0 + \beta_1 x \tag{1}$$

where β_0 and β_1 are coefficients. In other words, if $X = x$, predict that Y has the value $y = \beta_0 + \beta_1 x$.

When predictions stated as linear functions are assessed in terms of squared errors, one can readily evaluate the extent to which data support the predictions and, even with relatively "small" samples of data, obtain "reasonably" accurate estimates of any unknown coefficients. Moreover the methods extend nicely to problems involving several variables. In Sections 3.2.7 and 7.6 we provide a brief discussion of these features of the linear model and make comparisons with methods for prediction analysis of cross classifications.

Our concern in this section, however, is to emphasize the great flexibility that exists for stating predictions as linear functions. Even with this flexibility, linear functions are restricted to one-to-one predictions, a limitation we seek to avoid in developing a prediction language for qualitative variables. To illustrate these points this section focuses on the statement of linear predictions. We defer discussing how to evaluate linear predictions with data until Section 3.2.7.

Linear functions can be used to state event predictions about quantitative variables either *a priori* or *ex post*. In *a priori* linear prediction the values of the parameters (β_0 and β_1) are specified before the data are analyzed, so that the linear function predicts each observation's value for the dependent variable, given its value for the independent variable. Thus in the presidential voting example $\beta_0 = 0$ and $\beta_1 = \frac{2}{3}$ were given *a priori*. In *ex post* linear prediction the investigator specifies only the general linear form (1) of the relation before analyzing the data, treating the parameters β_0 and β_1 as unknown quantities. The data are analyzed and those values for β_0 and β_1 are selected that provide the "best fit" (as discussed in Section 3.2.7) of (1) to the data. Only after these best-fitting parameter values have been specified can the *ex post* linear proposition be evaluated as an event prediction for the dependent variable.

The flexibility of linear functions may be illustrated with an example that is based on Blau and Schoenherr's (1971) research into the organizational structure of bureaucracy. Let the independent variable SIZE be the number of employees in the bureau and LEVELS be the total number of levels in the bureau's official hierarchy. Figure 1.1 displays hypothetical data describing a (fictitious) population of bureaus in terms of these two variables. Each point represents a bureau. Plotting all bureaus yields a "cloud" of points rather than the one-to-one relation that would result if SIZE could be used to predict LEVELS without error. The squared prediction error for an observation is measured as the square of the *vertical* distance between the observation and the prediction line. Considering all observations forming the cloud, the least amount of total squared error results when the linear function is as plotted in Figure 1.1.

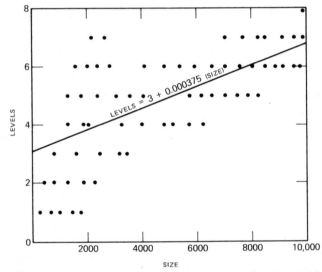

Figure 1.1 Predicting number of hierarchial LEVELS as linear function of a bureau's SIZE, showing hypothetical data.

1.5.2 Transformed Variables

Note that so far our prediction represents LEVELS as a linear function of the independent variable. But what if the observed relationship is better described by a function that is not linear? In order to use the powerful analytic methods associated with linear prediction, the variables must be transformed so that the relationship among the new variables is linear. A

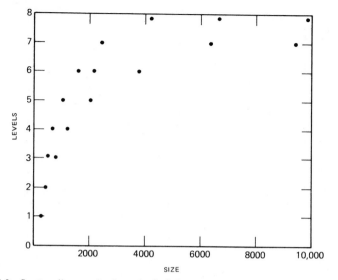

Figure 1.2 Scatter diagram for hypothetical data in which the relationship of LEVELS and SIZE is nonlinear.

wide variety of transformations (such as logarithmic, exponential, arc sin, logit) is used to increase the domain of linear models. Blau and Schoenherr report that a logarithmic model is superior to a linear model for describing their data. Suppose, for example, that the data on bureaus had the distribution plotted in Figure 1.2. In this cloud it is apparent that as SIZE increases, LEVELS increases at a diminishing rate. However we can create a new variable, LSIZE $= \log_{10}(\text{SIZE})$, to transform a function that is nonlinear in SIZE,

$$\text{LEVELS} = \beta_0 + \beta_1 \left[\log_{10}(\text{SIZE}) \right]$$

to one that is linear in the transformed variable,

$$\text{LEVELS} = \beta_0 + \beta_1 (\text{LSIZE})$$

as plotted in Figure 1.3.

1.5.3   Nonlinear One-to-One Predictions

Some functions, however, cannot be transformed in a direct manner. If the observed values of Y must be integers, for example, continuous functions such as LSIZE will generally lead to some observations being predicted to have nonintegral values of Y. In this case any linear prediction misspecifies

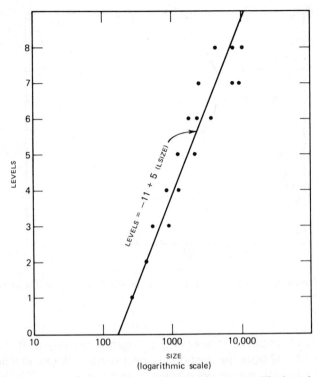

Figure 1.3 LEVELS as a a linear function of the logarithm of SIZE. The hypothetical data of Figure 1.2 are replotted here using a logarithmic scale for SIZE.

the bivariate relation and is guaranteed to lead to errors.[12] For example, although SIZE is "approximately" continuous in Blau and Schoenherr's data—the number of employees ranges across approximately 8000 integers in their data—the dependent variable, LEVELS, is limited to only five integral values. If, indeed, the event prediction is to yield an integral value for the dependent variable, then the appropriate function must exhibit discontinuities as in the step function displayed in Figure 1.4 (where a logarithmic X-scale has been used for clarity). Letting $0 < \alpha_0 < \alpha_1 < \alpha_2 < \alpha_3$, this function predicts that the number of

$$
\begin{aligned}
\text{LEVELS} &= 3 && \text{if } 0 < \text{SIZE} \leqslant \alpha_0 \\
&= 4 && \text{if } \alpha_0 < \text{SIZE} \leqslant \alpha_1 \\
&= 5 && \text{if } \alpha_1 < \text{SIZE} \leqslant \alpha_2
\end{aligned}
$$

[12] For a more complete discussion of this point see McKelvey and Zavoina (1975).

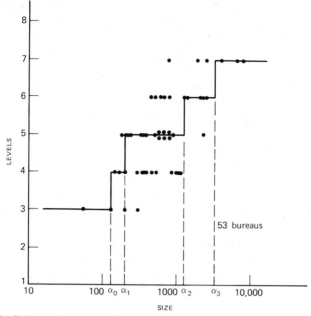

Figure 1.4 Predicting an integer number of LEVELS as a (one-to-one) function of bureau SIZE. Source: Blau and Schoenherr (1971, p. 69).

$$
\begin{aligned}
\text{LEVELS} &= 6 && \text{if } \alpha_2 < \text{SIZE} \leqslant \alpha_3 \\
&= 7 && \text{if } \alpha_3 < \text{SIZE}
\end{aligned}
$$

Figure 1.4 shows Blau and Schoenherr's scatter diagram for 53 bureaus and the values for the α parameters that we selected by visual inspection of these data. [The fitting or estimation of values for these parameters that minimize squared prediction error is a complex problem; we know of only a cumbersome solution procedure that, by trial and error, searches over all possible values for the α.]

1.5.4 One-to-Many Predictions

Notice that the step function shown in Figure 1.4 is accurate only for extreme values of SIZE. The model is without error in these data when it predicts three or seven hierarchical levels, but, for example, when it predicts LEVELS = 5, the observed values range from three to seven. The pattern of these errors suggest that the data would be described more accurately by a one-to-many (nonfunctional) relation. Predictions specifying one-to-many relations also are beyond the scope of standard linear models.

To illustrate one-to-many predictions, let us first assume that increasing control costs (or legal restrictions) limit the maximum number of levels to seven, although the need for division of labor necessitates at least three levels. We then have $3 \leqslant \text{LEVELS} \leqslant 7$. Within this range, assume that the minimum and maximum numbers of LEVELS increase logarithmically with SIZE such that $-6 + 3.25\,(\text{LSIZE}) \leqslant \text{LEVELS} \leqslant -2 + 3\,(\text{LSIZE})$. These inequalities define the predicted region shown in Figure 1.5. This one-to-many prediction is consistent with Blau and Schoenherr's data. Although the region appears to give a better description of the bivariate relation than the various functions explored earlier, this one-to-many prediction is much less precise than the one-to-one predictions of those functions. For example, when SIZE = 1000 employees, the proposition shown in Figure 1.5 predicts $3.75 \leqslant \text{LEVELS} \leqslant 7.00$. In developing the prediction analysis of cross classifications in Chapter 3, we take into account the critical role of precision. This is particularly important in the analysis of one-to-many predictions. In contrast, one-to-many predictions lie outside the scope of linear models. Despite this limitation the linear model offers a flexible and precise language for stating predictions about quantitative variables.

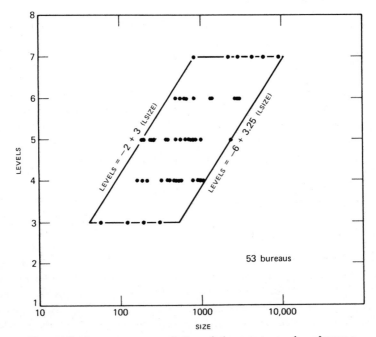

Figure 1.5   A one-to-many prediction relating SIZE to number of LEVELS.

1.6   CRITICAL METHODOLOGICAL ISSUES
IN PREDICTION ANALYSIS

What causes the contrasting weakness of methods for qualitative variables? Are quantitative levels of measurement absolutely necessary for the development of powerful prediction languages and analytic methods? If so, then this is unfortunate for sciences dealing with variables that, in principle or practice, are qualitative. Although qualitative variables are difficult to manipulate, we believe that the "methods gap" is larger than necessary. Current methods for handling qualitative data are severely limited because they cannot handle scientific predictions in a precise and naturally interpreted way. Since one of the goals of science is prediction, the lack of convenient and rigorous methods for dealing with prediction of qualitative events is a serious handicap.

In a sense the word *prediction* sounds somewhat unfamiliar in a qualitative data setting; we are accustomed to thinking of scientific prediction in terms of functional relations among *quantitative* variables. The fact that the relation is not precisely a function (that is, observations having the same values on the independent variables have different values on the dependent variable) is treated in linear prediction by assuming some random variability in the data. In what ways can or should this approach be adapted for prediction of qualitative variables?

There are several problems in formulating an appropriate solution. We sketch some of the basic issues here and propose solutions later in this book.

1.6.1   The Prediction Language

The first problem is the basic prediction language.

Predictive statements in sociology and other social sciences are often of necessity rather crudely worded. Usually this is because we have not reached the interval-scale level of measurement. Thus we might predict that the higher a person's status in the group, the greater his conformity to group norms. ...Notice... that we have said very little about the *form* of this relationship other than that it is positive. Unless we have an interval-scale level of measurement for both variables it becomes very difficult to say much more [Blalock, 1960, p. 274; his emphasis].

Since qualitative variables cannot be manipulated in mathematical functions available for quantitative variables, how does one express a prediction formally? What language is flexible enough to state predictions for qualitative data with precision yet is sufficiently natural that it permits easy statement and manipulation of empirical predictions derived from theory? Even if it were feasible it is not at all clear that the function-with-random-error approach of linear prediction should be adapted to this purpose; as Figure 1.5 suggests even for ratio variables, it may not always be appropriate to restrict the prediction to a single value of the dependent variable.

1.6.2   Accuracy, Scope, Precision, and Differentiation

One crucial issue in evaluating a prediction is the decision as to what constitutes prediction error and how the "extent" of prediction error is to be measured. The venerable and convenient squared error approach of linear models is particularly inappropriate for qualitative variables. The error measure used in linear prediction requires a quantitative measure of distance between values of the dependent variable. No quantitative measure of distance is available for qualitative variables. (For ordinal variables, however, the ordering of states can provide some information about the degree of error. Chapter 4 shows ways in which this information can be used.) What criteria should be adopted?

Answering this question is intimately related to determining what constitutes a successful prediction. If we adopt the simple minimization of incorrect predictions as our criterion of success, we frequently will be trapped into paying too little attention to the amount of "substance" in the prediction. After all, error typically can be minimized by making only "easy" (if trivial) predictions.[13] Thus when two predictions both make an equal number of incorrect predictions, if one applies to a greater range of observations than the other, then it presumably is more successful.

To be specific, event predictions may differ on four dimensions in addition to others discussed in Section 1.2. First is prediction *accuracy*—the extent to which predictions are correct (and errors are minimized). Second is prediction *scope*. The scope of a proposition increases with the proportion of nonvacuous predictions it makes—the proportion for which it is possible for the prediction to be incorrect. Third is the *precision* with which the predicted outcome can be located on the dependent variable. A prediction that specifies a unique state of the dependent variable is more precise than is a proposition predicting that an observation lies in any of several states of the dependent variable. "Rain" is thus more precise than "rain or snow," and the latter is more precise than "rain, snow, sleet, or

---

[13] This has a direct analogue in the construction of confidence intervals in classical statistics; with a fixed sample, high confidence levels can often be achieved only at the cost of specifying very wide bounds. For example, if we predict that the probability of heads in tossing a coin lies in the interval between .01 and .99, we are likely to be right, yet the prediction is nearly vacuous.

hail." Chapter 3 provides quantitative definitions of scope and precision. Fourth is *differentiation* of prediction. This refers to the extent to which *different* predictions are made for the various states of the independent variables. Referring again to Table 1.2, if we predict for every occupation that the person is affiliated with the Conservative Party, then the prediction is of full scope and, moreover, is precise in that it predicts a single state of the dependent variable. Yet this prediction is totally undifferentiated.

Variations in scope, precision, and differentiation of prediction need not be given much attention in the standard linear model. The functions generally have full scope. The predictions are highly differentiated; in a bivariate model with $\beta_1 \neq 0$, each value of X predicts a unique value of Y. The one-to-one point predictions are highly precise. In contrast if one wants to devise general methods for prediction and analysis of qualitative data—or of quantitative data when functional models are inappropriate—it is important that the prediction language and analytic methods address the multidimensional evaluation problem posed by these four dimensions. It is generally true that the greater the prediction's scope, precision, and differentiation, the more difficult it is to achieve prediction accuracy. These trade-offs should be considered in evaluation of prediction success.

These seem to be the fundamental issues. What is the appropriate prediction language? How shall prediction error be measured? In evaluating prediction success how shall we account for scope, precision, and differentiation of a prediction as well as its accuracy? These issues must be confronted in developing an effective research method for prediction analysis of qualitative events. We turn first to specification of a general language for prediction.

1.7 A PREDICTION LOGIC

Perhaps the best way to select a useful prediction language is to listen first to the kinds of predictions scientists make about qualitative variables. For ordinal variables, predictions frequently are stated in the monotone form; for example,

...the higher a man's esteem in a group, the higher his authority is apt to be [Homans, 1961, p. 286].

Later we show that such propositions, restated as degree-2 predictions, can be addressed with our methods. Initially we develop our methods for a second large class of predictions.

In order to identify the characteristics of this second class of predictions, consider the following statements (our emphasis):

Loss in competition *tends* to arouse anger [Homans, 1961, p. 123].

The introduction of universal suffrage led *almost* everywhere (the United States
excepted) to the development of Socialist parties [Duverger, 1954, p. 66].

...a "high" level of education...*comes close* to being a [necessary condition for
democracy] [Lipsit, 1960, p. 57].

...a grand coalition of three players is *likely* to form if and only if the game is
strictly superadditive [Rapoport, 1970, p. 269].

1.7.1 Elementary Formal Logic

If we ignore the italicized "hedging" qualifications for a moment, then
each of the propositions contains an assertion of universal truth stated as
an absolute event prediction. For example, ignoring the "tends," "almost,"
and "comes close" qualifications, the first three propositions in the list are
stated in the form "if x then y"; as such, they can be evaluated as
propositions in elementary formal logic.[14] The statement "if x then y,"
which may be written "$x \rightarrow y$," identifies the truth table shown in Table
1.3A. As the circled entry in this table indicates, the statement "if x then
y" is false if and only if the state "x and not y" (x is true and y is false),
written ($\bar{y}$ & x), occurs. As illustrated in Table 1.3B, a truth table can be
represented as a contingency table in which any falsifying cell, in this case
($\bar{y}, x$), is shaded. Since all the following statements identify the same truth
table shown in Table 1.3, they are regarded as equivalent in formal logic:
"x implies y," "x only if y," "y provided that x," "x is a sufficient
condition for y," and "$\bar{x}$ is a necessary condition for $\bar{y}$." Moreover
additional equivalents can be stated in which the roles of the two variables
are reversed; for example, "$\bar{y}$ implies $\bar{x}$," "$\bar{y}$ is a sufficient condition for $\bar{x}$,"
and "y is a necessary condition for x." Similarly, ignoring the "is likely"
qualification, Rapoport's proposition quoted earlier may be restated in the
form "y if and only if x:" "a grand coalition forms if and only if the game
is strictly superadditive." This biconditional statement and all logical
equivalents such as "$y \leftrightarrow x$," "if x then y, and if $\bar{x}$ then $\bar{y}$," "x is a necessary
and sufficient condition for y," and "y is a necessary and sufficient
condition for x" identify the falsifying events shaded in Table 1.4B. Any

---

[14] Prior knowledge of formal logic is not required to follow the arguments in this book. On the
other hand, it is useful. There are many good texts in formal logic, such as Suppes (1957).
However for the purposes of this book the brief introduction by Kemeny, Snell, and
Thompson (1974) is sufficient. The case for using logic to integrate theory and statistics has
been aptly stated by Ploch (1970).

Table 1.3 Two representations of the truth table for "if *x* then *y*"

(*A*) Conventional Representation

x	→	*y*
true	true	true
true	false	false
false	true	true
false	true	false

(*B*) Contingency Table Representation (falsifying event is shaded)

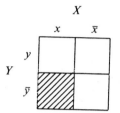

Table 1.4 Two representations of the truth table for "*y* if and only if *x*"

(*A*) Conventional Representation

x	↔	*y*
true	true	true
true	false	false
false	false	true
false	true	false

(*B*) Contingency Table Representation (falsifying events are shaded)

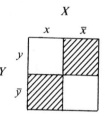

statement in formal logic must be rejected if any falsifying event occurs, even if only one such event is recorded in 1000 observations.

1.7.2 Allowing for Error

Most scientific propositions make some allowance for error, and thus would not be rejected if a single deviant case were observed. Consider the qualifications underlined in the quotations. There is explicit recognition in such qualifications as *almost* and *tends to* that the predictions might commit errors. Yet these qualifications do not convert these (pure strategy) statements into predictions that are overtly probabilistic. That is, these predictions are not similar to the actuarial predictions of, say, Mendelian theories of genetics, which specify the probabilities of various types of offspring given the characteristics of the parents. Rather these statements make event predictions for which some error is allowed. That is, these hypotheses are stated as absolute predictions that, in Coleman's apt phrase (1964, p. 516), are "sometimes true." The form of these predictions appears to conform to John Stuart Mill's belief that it is possible to

...lay down general propositions which will be *true in the main*, and on which, with allowance for the degree of their probable inaccuracy, we may safely ground our expectations and our conduct [Cited by Madge, 1953, p. 67; our emphasis].

The "true in the main" caveat is represented in the quoted propositions by the "almost" and "tends" qualifications. Statements in formal logic assert that exceptions (falsifying events) should *never* occur. In contrast, the "tends to" type of qualification predicts that deviant or error (contrast "falsifying") events *seldom* occur.

1.7.3   A Language for Event Predictions

The language for stating such pure strategy event predictions about qualitative variables, called *prediction logic*, is defined by a formal analogy to elementary logic: just as a proposition in elementary logic may be stated unambiguously by identifying its domain and truth table, a prediction logic proposition may be stated precisely by specifying its domain and the set of error events it identifies. Although the prediction logic is thus defined in terms of this analogy to truth tables, theoretical propositions are conceived in terms of relations among variable states, not error events. Toward this end we use the connective $\rightsquigarrow$ (read "predicts" or "tends to be sufficient for") as the analogue to formal implication ($\rightarrow$), and $\longleftrightarrow\rightsquigarrow$ (read "tends to be necessary and sufficient for") in place of the biconditional connective ($\leftrightarrow$). As in formal logic we use the word *or* (inclusive) as the symbol for disjunction and the symbol & for conjunction. Negation is indicated symbolically by, for example, writing "not x" as $\bar{x}$. Thus all the following prediction logic propositions identify the same error event that is shaded in Table 1.3B, and consequently are equivalent: "x tends to be sufficient for y," "if x predict y," "$x\rightsquigarrow y$," "y tends to be necessary for x," and "$\bar{y}\rightsquigarrow\bar{x}$." Similarly the following equivalent predictions identify the cells shaded in Table 1.4B as error events: "x tends to be necessary and sufficient for y," "$x\longleftrightarrow\rightsquigarrow y$," "predict y if and only x," "$(x\rightsquigarrow y)\&(\bar{x}\rightsquigarrow\bar{y})$," and "$(y\rightsquigarrow x)\&(\bar{y}\rightsquigarrow\bar{x})$."

Referring to Table 1.3B notice that the statement $x\rightsquigarrow y$ makes only the vacuous prediction "y or $\bar{y}$" when $X=\bar{x}$. (As in formal logic such vacuous components of a prediction may be omitted.) We argued in the preceding section that it is important that the prediction language and analytic methods be able to address predictions that differ in scope, precision, and differentiation. It should not be necessary that a nonvacuous prediction be made for every state of the independent variable. Within a given scientific context some states of the independent variable may, in fact, be far more useful than others in predicting the dependent variable, even if the variables are dichotomies. The authors of the following statement emphasized this point:

...the central proposition of the theory is that incompatible authority systems are unstable. The theory does not state that compatible systems are necessarily stable, for there are many other factors which may lead to system instability. For this reason, the theory cannot be advanced as a general explanation of the instability of authority systems; incompatibility is considered to be a sufficient but not necessary condition for instability [Scott et al., 1967, p. 113].

Using the notation shown in Table 1.5, they predicted $\bar{c} \rightsquigarrow \bar{s}$,[15] but made no prediction about the cases in the state c. Inspection of this table suggests that their prediction is supported by the data. The methods developed in Chapter 3 enable us to measure the strength of this support.

Table 1.5   Incompatibility and instability of authority systems ($N = 224$), showing error cell (shaded) for the prediction: $\bar{c} \rightsquigarrow \bar{s}$

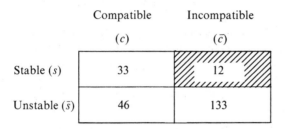

	Compatible (c)	Incompatible ($\bar{c}$)
Stable (s)	33	12
Unstable ($\bar{s}$)	46	133

Source: Laing (1967, p. 155).

1.7.4   The General Bivariate Language

Although the elementary propositions discussed earlier for 2×2 contingency tables provide a convenient introduction, it now is time to define the prediction logic formally for the general bivariate case: the $R \times C$ cross classification. We restrict attention here to pure strategy degree-1 predictions. In Chapters 4 and 5 we extend the bivariate analysis to degree-2 predictions and actuarial propositions, reserving the multivariate prediction logic for Chapter 7.

As in formal logic, define variables as specified sets of mutually exclusive states that exhaust the domain of the prediction under investigation. As before, let Y and X be two independently measured variables specified

---

[15] Seeing the absence of such "hedging" qualifiers as *likely*, *tends*, or *nearly* in the preceding quotation, a reader might question our $\bar{c} \rightsquigarrow \bar{s}$ interpretation. However, these qualifiers are used elsewhere in the article cited, and it is clear that $\bar{c} \rightsquigarrow \bar{s}$ is a valid representation (Laing, 1967, pp. 146–148n). For a more recent statement see Dornbusch and Scott (1975). For a general discussion of propositions consistent with the prediction logic for the special case of 2×2 tables see Francis (1961).

as the finite sets of states $\{y_1,\ldots,y_i,\ldots,y_R\}$ and $\{x_1,\ldots,x_j,\ldots,x_C\}$, where R and $C \geqslant 2$. Suppose that the domain of a prediction logic proposition includes all events in the cross classification $Y \times X$.[16] Treating X as the independent variable, the prediction for observations having $X = x_j$ may be stated in the form

$$\mathcal{P}_{.j} : x_j \rightsquigarrow \mathcal{S}(x_j)$$

where the set of predicted ("success") states $\mathcal{S}(x_j)$ is a set of Y states. The proposition $\mathcal{P}_{.j}$ may be read, "if x_j, predict $\mathcal{S}(x_j)$" or "x_j tends to be sufficient for $\mathcal{S}(x_j)$." The event (y_i, x_j) belongs to the set of *error* events defining $\mathcal{P}_{.j}$,

$$\mathcal{E}_j = \{(y_i, x_j) | y_i \notin \mathcal{S}(x_j)\}$$

if and only if that event would falsify the formal logic statement that x_j implies $\mathcal{S}(x_j)$. Then, considering all states of X, the general prediction logic proposition for $Y \times X$ may be stated in the form

$$\mathcal{P} : \mathcal{P}_{.1} \ \& \ \mathcal{P}_{.2} \ \& \ \cdots \ \& \ \mathcal{P}_{.j} \ \& \ \cdots \ \& \ \mathcal{P}_{.C}$$

We often find it convenient to express this as

$$\mathcal{P} : \mathcal{P}_{.1}, \mathcal{P}_{.2},\ldots, \mathcal{P}_{.j},\ldots \ \& \ \mathcal{P}_{.C}$$

or as

$$\mathcal{P} = \{\mathcal{P}_{.j}\}$$

1.7.5   Error Cells

Equivalently this proposition may be specified solely in terms of the set of error events (*error cells*) it identifies,

$$\mathcal{E}_{\mathcal{P}} = \bigcup_j \mathcal{E}_j$$

---

[16]The domain of prediction, as in elementary logic, can be indicated through the use of universal quantifiers. These define the events for which a statement in the logic applies. To keep matters simple we assume such quantifiers are understood and omit them from prediction statements in this book. However a rigorous statement of any scientific prediction requires precise specification of its domain. For analysis of propositions whose domain excludes certain events in $Y \times X$, see Section 4.5.

The prediction $\mathcal{P}$ is more easily visualized by showing the set of error cells in the appropriate contingency table. Consider the prediction of party affiliation from occupation expressed verbally as "business or professional predicts Conservative, manual worker predicts Labor or other." This takes the form

$$\mathcal{P} : x_1 \rightsquigarrow y_2, x_2 \rightsquigarrow y_2, x_3 \rightsquigarrow (y_1 \text{ or } y_3), \& x_4 \rightsquigarrow (y_1, y_2, \text{ or } y_3)$$

This proposition identifies the error cells that are shaded in Table 1.6.

Table 1.6   Error events for a proposition

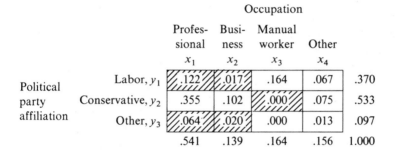

		Professional x_1	Business x_2	Manual worker x_3	Other x_4	
Political party affiliation	Labor, y_1	.122	.017	.164	.067	.370
	Conservative, y_2	.355	.102	.000	.075	.533
	Other, y_3	.064	.020	.000	.013	.097
		.541	.139	.164	.156	1.000

Source: Data adapted from Buck (1963, p. 122).

1.7.6   Admissibility

Several characteristics of propositions we admit for analysis should be stipulated. Note that every Y-state might be included in $\mathcal{S}(x_j)$. Hence, as is the case for x_4 in Table 1.6, admissible propositions need not make (nonvacuous) predictions for all states of the independent variable. The general conditions for admissibility are provided in Chapter 3. The only proposition excluded for all populations is that which has no error cell, although we also exclude propositions for a given population in the degenerate situation in which every error cell has a zero marginal ($P_{i.}=0$ or $P_{.j}=0$). This restriction ensures that empirical evidence is relevant for evaluating predictions admitted for analysis; if no error cell is identified by a statement in the prediction logic, then that statement is a tautology, and thus empirical data are irrelevant.

1.7.7 Equivalence

Note that the error structure shown in Table 1.6 is also identified by the prediction

$$\mathscr{P}' : y_1 \rightsquigarrow (x_3 \text{ or } x_4), y_2 \rightsquigarrow (x_1, x_2, \text{ or } x_4), \& y_3 \rightsquigarrow (x_3 \text{ or } x_4)$$

which reverses the assignment of variables regarded as independent and dependent. Thus there is no implicit assumption about causation in the identification of the error structure. We regard this as fortunate.

By direct analogy to formal logic any two propositions stated in the prediction logic are *equivalent* if and only if they have equivalent error structures. In explaining this definition we need to develop what we mean by *equivalent error structures*. In the simplest case two error structures are equivalent if they identify the same error cells in the same contingency table: $\mathscr{E}_{\mathscr{P}} = \mathscr{E}_{\mathscr{P}'}$. Thus it is obvious that the two propositions $\mathscr{P}$ and $\mathscr{P}'$ identifying the same error structure shown in Table 1.6 are equivalent.

A less obvious form of equivalence is also possible. Suppose that in defining the occupation states, the investigator combined the states "professional" and "business" in Table 1.6 into a single mutually exclusive state, say, "white collar." Suppose further that the same prediction ("Conservative") is made for the new state as was made for the original states. It is clear that this transformation has not changed any prediction success to a prediction error (or conversely). (By symmetry the same argument applies to changes in the Y-states.) In formal logic two statements are logically equivalent if and only if every event that falsifies one falsifies the other. Analogously, *if by transforming variables in one proposition it can be changed into another proposition without affecting the counting of prediction errors (and successes), then the two propositions have equivalent error structures; hence in the prediction logic, they are equivalent propositions*. (One such transformation can be accomplished by permuting the order with which the states of either variable are listed in the prediction and associated contingency table.) The choice among transformations yielding equivalent error structures is arbitrary and has no predictive significance. Consequently this choice should have no effect on the measurement of prediction success. The methods developed in Chapter 3 are designed to ensure that, for the same set of data, equivalent bivariate propositions receive identical evaluations of prediction success.

The prediction logic should be easily understood. It permits flexible and unambiguous specification of predictions of qualitative events. The logic thus makes it possible to specify accurately a wide variety of "shapes" or "forms" of relations in qualitative data—such as monotone, triangular, U- or S-shaped—simply by specifying the error cells in an appropriate contingency table. In the simple 2×2 case there are $2^4 = 16$ logically distinct propositions. Of these, 15 contain one or more error cells and are generally admissible in the prediction logic. The six that appear to have the greatest scientific interest are shown in Table 1.7. Even if we restrict analysis for now to pure strategy, degree-1 predictions defined for the entire domain of $Y \times X$, as the dimensions of the appropriate table increase, the number of admissible propositions grows dramatically. There are 63 admissible predictions in the 2×3 case and 511 in 3×3 contingency tables. There are $2^{RC} - 1$ distinct propositions of this type that are admissible for an $R \times C$ table. The fundamental question is whether we can find an appropriate measure of the prediction success achieved by any proposition in actual scientific applications. The next chapter proposes characteristics such a measure should possess, then evaluates a variety of established measures of statistical association with these criteria.

Table 1.7   Six propositions relating two dichotomous variables

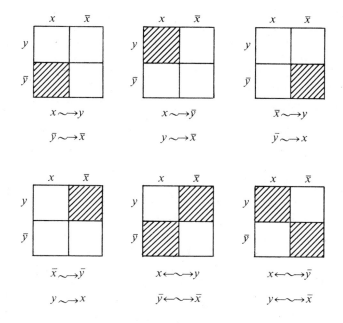

2

PREDICTION AND ASSOCIATION

Conventional methods for analyzing relations among qualitative variables are based on a variety of measures and tests of statistical "association." Each such measure has been designed for a relatively limited purpose. Although in this chapter we review certain of these measures whose features are instructive about prediction analysis of cross classifications, our goal is a general model that will apply to a large class of research problems. Therefore we begin this chapter by proposing criteria that we think should be satisfied by a method if it is to provide an adequate measure of prediction success. Then we review conventional methods and relate them to these criteria. Subsequently Chapters 3 through 6 present a method for bivariate prediction analysis designed to satisfy these criteria, and Chapter 7 extends the method for applications in multivariate analysis. This chapter's discussion of conventional methods is largely limited to bivariate analysis, because most of the fundamental issues must be confronted even when only two variables are involved.

2.1 CRITERIA FOR MEASURING PREDICTION SUCCESS

In empirical research investigators need a method to evaluate the extent to which data support a proposition describing a relation among the variables of interest. In this section we propose seven criteria for evaluating or designing a measure of prediction success. The most important problems in measuring prediction success are encountered in analyzing populations for which all relevant data are available, rather than in making inferences about a population on the basis of a sample. Consequently the first six criteria pertain to descriptive rather than inferential considerations. The undeniable scientific importance of the inferential properties of a measure is recognized in our seventh and last criterion.

Criterion 1. The most important criterion is that the measure should be custom tailored for the specific prediction under investigation, whether that prediction is stated *a priori* or *ex post*. Even in the case of purely *ex post* description of a population, the measure for assessing prediction success should provide a direct evaluation of the specific prediction selected to describe the data. The prediction logic can be used to define $2^{RC} - 1$ alternative predictions for an $R \times C$ table, and there are infinitely many distinct actuarial predictions about the same cross classification. Many of these statements represent potentially meaningful scientific descriptions. (To illustrate, Table 1.7 shows six such statements for the simple 2×2 table.) Each of the many alternative predictions should have its own measure.

The importance of matching the measure to the proposition can be demonstrated simply by considering two hypothetical sets of population results for the cross-classified variables shown in Table 2.1. The proposition "meat$\leadsto$ red" is perfectly supported in (A) since all those who predominantly eat meat drink red wine most frequently. In contrast, the proposition would receive less support were the data as given in (B). There a preference for meat is necessary to relatively heavy consumption of red, but it is hardly sufficient; of every eight who are predominantly meat eaters, five also are predominantly white wine drinkers. So, although "meat$\leadsto$ red" is the perfectly supported statement for (A), "fish$\leadsto$ white" is perfectly supported for (B).

Most conventional techniques cannot register this distinction. For example, the Q, λ, τ, and chi-square based measures reviewed later in this chapter have the same value for both cross classifications. In contrast, we

Table 2.1   Wine and food consumption probabilites in two hypothethical populations

		More frequently eats					More frequently eats	
		Meat	Fish				Meat	Fish
Wine most consumed	Red	.2	.5		Wine most consumed	Red	.3	.0
	White	.0	.3			White	.5	.2
		(A)					(B)	

seek a measure for "meat $\rightsquigarrow$ red" that has different values for (A) and (B) and another measure for "fish $\rightsquigarrow$ white." The value of the latter measure for (B) should equal the value of the former for (A).

The necessary distinctions in the prediction success of "meat $\rightsquigarrow$ red" in the two parts of Table 2.1 are not made by conventional procedures. Note that by reordering (permuting) both the rows and columns, (B) can be made to look like (A). If a measure always remains unchanged whenever rows and columns are reordered, it is said to be *invariant under row and column permutation*. All the conventional measures mentioned earlier, or at least their magnitudes, are invariant in this sense. Proponents for such an invariance property must assume that all that is important in a contingency table is the probability structure it contains, regardless of how the rows and columns are labeled.

The example demonstrates that an invariance requirement would obliterate any connection between probabilities and observable phenomena. *It is clear that state labels are essential information for evaluating a prediction.* Clearly permutation invariance must be dropped.

Any prediction-specific measure must correspond to the prediction statement as well as to the characteristics of the bivariate probability distribution.

Criterion 2. Second, the measure's value should be unchanged by variable transformations that have no predictive significance for the proposition evaluated by the measure. Recall that in prediction logic any two predictions are equivalent if and only if they count the same events as errors (and successes). In particular, one can always transform a variable by permuting the order in which states are listed or by redefining the states. If these variable transformations yield a logically equivalent proposition, then the transformation has no predictive significance; therefore it should have no effect on the measurement of prediction success. That is, *the measure should be identical for all logically equivalent propositions that identify the same independent, intervening, and dependent variables.*

To illustrate with our enological example, consider the category "white" as the combination of two categories "dry white" and "sweet white" found in the original coding of the data. Then the measure for "meat $\rightsquigarrow$ red" computed on the original 3×2 cross classification should be identical to that for the 2×2 table.[1]

---

[1]A word of caution. Were the categories combined before data collection took place, responses to the "most frequently consumed" question can be affected since red can be consumed more frequently than either subcategory but less frequently than the white total. This fundamental change in measurement changes the probability of the error event "meat, white" and thereby destroys equivalence.

Criterion 3. Third, the measure should be capable of evaluating a prediction whatever its scope, precision, and differentiation. Moreover the measure should permit comparison of the success of any two predictions even if they differ in scope, precision, or differentiation.

Prediction accuracy is affected by these dimensions; for example, if the scope of the prediction extends to only one-half of the population, then at most 50% of the observations can be predicted incorrectly. Consequently in order for comparisons to be meaningful the method of evaluation should take these dimensions into account in assessing prediction success. In addition, since theory assessment is a multidimensional problem, separate measures of these dimensions are important complements to the basic measure of prediction success.

Criterion 4. A measure can satisfy the first three criteria without providing an evaluation of a prediction that is readily interpretable by researchers. The fourth criterion prescribes that the measure should have an operational meaning that is directly related to the prediction success of the proposition under investigation.

To meet this criterion we adopt a widely accepted procedure based on comparing error rates for two alternative predictions. The procedure was followed by Goodman and Kruskal, who developed measures based on "probabilistic models of predictive activity" (1954, p. 735). Both this viewpoint and lambda, the measure given most attention in their article, were originally developed by Guttman (1941) in a paper of fundamental importance to the development of cross-classification analysis. Costner (1965, p. 344) clarified and formalized the comparison of error concept by specifying a procedure for developing measures with a *p*roportionate *r*eduction in *e*rror (PRE) interpretation.

PRE measures make a comparison of error rates and permit operational interpretations that are particularly useful when the data exhibit only a moderately strong relation. In the usual convention, a measure equals one when the relation is "strong" or "perfect" and equals zero in the absence of any relation. Interpretive problems arise for intermediate values like .4 or .7. Unless the measure has an operational interpretation, these values will convey little meaning other than that the relation is neither very strong nor very weak. PRE measures are interpreted in terms of the proportion by which prediction errors are reduced.

For the prediction analysis of cross classifications, the PRE approach has a more specific advantage: namely, it permits us to account properly for the scope and precision of a prediction. This accounting is necessary because raw error rates are poor measures of prediction success. An (absolute) error rate of 3 errors per 100 predictions could indicate either that the independent variable is quite useful in predicting the dependent variable or that the prediction addresses a task so trivially easy that very few errors could possibly be made, regardless of the relation between the independent and dependent variables. Conversely a more precise prediction may commit (say) 7 errors per 100 predictions. How do we trade off accuracy and precision in assessing the success of the prediction task undertaken by the proposition? The answer, we think, is to compute the *proportionate* reduction in error achieved by the prediction relative to the error of a reference proposition that has identical scope and precision. Following Costner (1965, p. 344), the PRE procedure is:

1. Specify rule K for making a prediction about each observation's state on the dependent variable *k*nowing its state on the independent variable.
2. Specify rule U, which makes a prediction for each observation when its independent variable state is *u*nknown.
3. Define what constitutes error and how error shall be measured.
4. Define the measure using the form:

$$\text{PRE measure} = \frac{\text{rule } U \text{ error} - \text{rule } K \text{ error}}{\text{rule } U \text{ error}}$$

This procedure is appropriate for the analysis of any prediction, and is followed repeatedly in this book for the analysis of many measures, beginning with λ and τ in Section 2.5. For *a priori* predictions we further constrain Costner's procedure by the following:

5. When evaluating an *a priori* proposition, no knowledge of dependent variable frequencies or probabilities shall be used in formulating either rule K or rule U.

This fifth requirement of our PRE procedure represents a marked departure from conventional modes of analysis. Guttman (1941, pp. 261–262) explicitly indicated the *ex post* use of dependent variable information:

By a measure of the association of variate Y... with variate X... is meant a measure whose absolute value increases with the decrease in amount of errors of prediction of Y from knowledge of the bivariate distribution of Y and X and knowledge of the X values of the individuals, the decrease being from the amount of error of prediction that would exist from knowledge of the univariate distribution of Y alone. [To achieve notational continuity with this book, we have modified the above passage by making Y the dependent and X the independent variable.]

In the case of *a priori* predictions, however, dependent variable probabilities are irrelevant to the *statement* of the theory. Indeed measurement of *Y* (and evaluation of the prediction) may occur long after the prediction is stated. We propose that the prediction to be evaluated should be applied under both rule *K* and rule *U*; the only difference in these applications is whether each observation's independent variable state is known at the time its dependent variable state is predicted. Since the dependent variable probabilities are irrelevant to and, hence, excluded from the theory's representation in rule *K*, it is natural to exclude them also from the comparison prediction of rule *U*. Otherwise the effects on error rates of changing the *X* information available would be confounded by effects associated with use of dependent variable information. For *a priori* predictions, the predictions should be stated under both rules in ignorance of dependent variable probabilities.

Under rule *K* one has knowledge of the *X*-states for each case. Under rule *U* one will frequently use information about the marginal proportions of *X* states. In general, this is the minimum information necessary for rule *U* if the measure is to satisfy criterion 2 with respect to variable transformations. For example, without *X* marginal information one could predict each set of *Y*-states associated with each of the *C* states of *X* with probability $1/C$. Even if the prediction were the same for x_1 and x_2, combining these two categories to yield $C-1$ categories would generally change both rule *U* and the measure.

The distinction between *a priori* and *ex post* prediction made throughout this book is extremely severe. Frequently (see the inflation example in Section 3.2) researchers will state predictions that require knowledge of the marginal distribution of *Y* but not the joint distribution of *X* and *Y*. This information is used to establish cutting points in such predictions as "If *x* is greater than the mean of *X*, then *y* is greater than the mean of *Y*." With such *"ex priori"* predictions one typically cannot specify *a priori* which states of *Y* constitute errors for a given prediction until the cutting points are established from the data. However, once the variable states identified by these cutting points are identified, the prediction can be stated without knowledge of the bivariate distribution. Except for the cutting points, *"ex priori"* predictions are *a priori*.

We do not address *"ex priori"* prediction directly either in terms of population measures or statistical inference from sample data. As discussed in Section 6.3.6 our work on inference has led to the conjecture that inference properties of such predictions come closer to *a priori* prediction than to unconstrained *ex post* analysis. Consequently we treat such predictions as if they were *a priori*.

In summary, the first four, primary criteria concern custom designing to a specific prediction; invariance under variable transformations without predictive significance; responsiveness to scope, precision, and differentiation; and interpretability. The three following criteria prescribe highly desirable if not essential properties.

Criterion 5. The bivariate measure should have multivariate analogues that measure overall prediction success and assess the separate contribution of each subset of independent variables in predicting the dependent variable. Multivariate methods should permit comparison of the prediction success of propositions having the same dependent variable but different independent and intervening variables. Thus, for comparing a trivariate prediction to a bivariate component of that prediction, the multivariate methods must respond to differences in scope, precision, and differentiation when both (*a*) information conditions—represented by the second predictor variable—change and (*b*) the prediction rule changes.

Criterion 6. The sixth criterion pertains to the sensitivity of the measure to minor perturbations of the population distribution. Only slight changes in some or all the cells in a table should lead to a correspondingly minor change in any measure of prediction success.

Criterion 7. The seventh and last criterion concerns statistical inference. Situations where the investigator has data for the entire population or an extremely large sample seldom obtain in practice. (Otherwise, most statisticians would have to seek new work.) A measure that satisfied the preceding criteria but that could not be estimated without enormous amounts of data would be more a methodological curio than a useful research tool. As the seventh criterion, accurate estimation should be feasible with "reasonable" sample sizes.

In summary, these seven criteria stipulate that a measure should

1. Permit custom tailoring to the specific (*a priori* or *ex post*) prediction under investigation.
2. Be invariant under variable transformations that have no predictive significance for that prediction (the logical equivalence property).
3. Account for prediction scope, precision, and differentiation.
4. Permit a proportionate reduction in error interpretation of the values the measure attains.
5. Extend to multivariate analysis.
6. Be reasonably insensitive to minor changes in the probability structure.
7. Permit accurate estimation with modest sample sizes.

In the next sections we evaluate some standard measures of bivariate "association." Although these measures meet some of the preceding criteria and provide a stimulus to further development, none of them was designed to satisfy all the first four, primary criteria. Specifically many of the measures are not prediction specific or PRE; those that are prediction specific do not permit custom designing a measure to an investigator's prediction problem.

2.2 MEASURES OF "ASSOCIATION": AN OVERVIEW

The basic tools for analyzing relations among qualitative variables have consisted of various measures and tests of statistical "association." We have enclosed this term in quotations because the concept it refers to is not well defined. As we survey some of the important measures of "association," it will become clear that the concept is best described by a double negative: *two or more variables are "associated" if they are "not unrelated."* The particular model of nonrelation varies somewhat from one measure to another; usually it is statistical independence or some consequence thereof.

Unfortunately, judging from many data analyses found in published research, the diffuseness of the association concept has not been fully recognized. The dominance of the chi-square test in the social science literature, for example, suggests that rejection of statistical independence is sometimes confused with support for a specific substantive hypothesis. That is, rather than predicting a type of relation and then measuring the success of that prediction, investigators use chi-square to show that the variables are not unrelated, then describe the specific kind of relation exhibited without benefit of statistical clergy.

Often such analysis occurs at the end of exploratory investigations with multiple and perhaps uncertain goals. The investigator may want to rely on a single measure simply to provide manageable summaries of unwieldly data sets (Goodman and Kruskal, 1954, p. 735). Furthermore the use of an omnibus technique reduces intellectual demands; to do custom tailoring, one has to know how to sew.[2] Conveniences notwithstanding, we think that whether the goals of the investigation are specific or not, even if the sole objective is simply compact description, considerably more information should be communicated by a measure than is offered by conventional methods. Knowing that two variables are not unrelated and that they depart from unrelatedness to some specified "degree" says nothing about the *form* of the relation that does obtain. Ordinary chi-square, the first

---

[2]Credit for this metaphor belongs to an anonymous reviewer employed by the publisher.

measure in our review, for example, is used only to determine whether or not there is *any* relation between the variables.

2.3 TESTING FOR NONRELATION AND GOODNESS OF FIT

2.3.1 The Chi-Square Test for Statistical Independence

Perhaps the best-known and most widely used methods are those associated with the chi-square test.[3] For *samples* represented in $R \times C$ contingency tables, chi-square can be defined as

$$\chi^2 = n \sum_{i=1}^{R} \sum_{j=1}^{C} \left[\frac{(f_{ij} - f_{i.}f_{.j})^2}{f_{i.}f_{.j}} \right] \tag{1}$$

where n is the sample size, f_{ij} is the observed proportion of cases in cell (i,j), and $f_{i.}f_{.j}$ is the "expected" proportion in that cell under statistical independence if the population probabilities $P_{i.}$ and $P_{.j}$ are equal to the observed frequencies $f_{i.}$ and $f_{.j}$. (The sample proportions f are directly analogous to the population probabilities P shown in Table 1.1.) Alternatively, letting n's denote numbers of cases,

$$\chi^2 = n \sum_{i} \sum_{j} \left[\frac{(n_{ij} - n_{i.}n_{.j}/n)^2}{n_{i.}n_{.j}} \right] \tag{1'}$$

(From now on, whenever summation is over the entire cross classification, we may omit the summation limits.) Each of the terms in the double summation (1) contains a squared deviation $(f_{ij} - f_{i.}f_{.j})^2$ in the numerator. These squared deviations are then normalized by division by the "expected" proportion. If the variables are indeed statistically independent, the deviations tend to be close to zero and χ^2 tends to be small. Because the deviations are squared, χ^2 is always positive. If statistical independence fails to hold, the size of χ^2 tends to reflect the magnitudes of the squared deviations $(P_{ij} - P_{i.}P_{.j})^2$ of population probabilities.

Once the numerical value of χ^2 is computed from *sample* data one can test if it is significantly large; that is, large enough to have been unlikely to have occurred by chance were the two variables, in fact, statistically independent in the *population*. The test is made by comparing the computed value with tabulated values found in virtually every introductory

---

[3]For details of the test and limitations on its use, see Hays (1973, Chap. 17).

statistics text. The tables, based on the theoretical distribution of χ^2, show, for example, that values of 3.84 or larger would occur only with probability .05 if two dichotomous variables were independent. Therefore we would say that a sample value of 3.84 or greater for a 2×2 table is significant at the .05 level. Consequently if χ^2 is sufficiently large the investigator rejects the null hypothesis of nonrelation (statistical independence).

The χ^2 probability depends not only on the value of the χ^2 statistic but also on the number of *degrees of freedom*. As explained in Chapter 6 and texts such as Hays (1973, Chap. 17), this number for an $R \times C$ table is $(R - 1)(C - 1)$. As a computational example, consider the authority-system data in Table 1.5 as a sample. We have $(2 - 1)(2 - 1) = 1$ degree of freedom. The value of χ^2 is given by (1') as follows:

$$\chi^2 = 224 \left[\frac{(33 - 45 \times 79/224)^2}{45 \times 79} + \frac{(12 - 45 \times 145/224)^2}{45 \times 145} \right.$$

$$\left. + \frac{(46 - 179 \times 79/224)^2}{179 \times 79} + \frac{(133 - 179 \times 145/224)^2}{179 \times 145} \right]$$

$$= 35.74$$

With one degree of freedom the probability of observing this or some greater value under the hypothesis of statistical independence is less than .001. Therefore we can conclude that the two variables—authority system stability and compatibility—are not "unrelated." With a chi-square test that is all we can conclude.

Limitations of Basic Chi-Square Methods

Chi-square meets almost none of our criteria. First, the test focuses not on statistical relation, but rather on nonrelation, statistical independence. The title of the original paper (Pearson, 1900) developing this test describes its purpose well: "On the Criterion That a Given System of Deviations...Is Such That It Can Reasonably Be Supposed to Have Arisen in Random Sampling."

Specifically the value of the chi-square statistic is *not* a measure of the degree of relation. One reason is that the magnitude of this statistic is a function of sample size as well as of the degree of relation. As the sample size grows large, chi-square is approximately

$$n \sum_i \sum_j \frac{(P_{ij} - P_{i.} P_{.j})^2}{P_{i.} P_{.j}}$$

For descriptive purposes this dependence on n is undesirable. One of the standard textbook solutions to this problem is to divide the chi-square statistic by the sample size. Again for large samples this statistic, frequently denoted by ϕ^2 and called the mean-square contingency or phi-square, will closely approximate the population quantity

$$\sum_i \sum_j \frac{(P_{ij} - P_{i.}P_{.j})^2}{P_{i.}P_{.j}}$$

This quantity can only be regarded as a distance measure, indicating "how far" the probability structure is from statistical independence. There is no idea of the direction of this departure or nature of the relation. Like chi-square, ϕ^2 is fixed for the table regardless of what substantive proposition is asserted. Numerous other modifications of chi-square have been suggested (including Pearson's coefficient of contingency C, Tschuprow's T, Cramér's measure), but they all share the same problems: they focus on distance rather than direction, are difficult to interpret, and have no convenient predictive structure.[4]

The basic ideas of the chi-square procedure have been extended in various ways to treat problems involving several variables. Characteristically the focus has been on developing significance tests for various null hypotheses of "no statistical relation" at more complex levels. Essentially these multivariate methods have the same aims and limitations as the bivariate methods.[5]

2.3.2 Using Chi-Square for Goodness of Fit: Custom Designing

Although the classic chi-square test for independence is fixed for a given table, it is also possible to custom design a "goodness-of-fit" statistic that follows the chi-square distribution. Perhaps the simplest illustration comes from considering a theory that specifies a complete set of probabilities P_{ij}^* that are said to determine the bivariate distribution. Then

$$\chi^2 = n \sum_{i=1}^{R} \sum_{j=1}^{C} \frac{(f_{ij} - P_{ij}^*)^2}{P_{ij}^*} \qquad (2)$$

---

[4]For further discussion see Costner (1965) and Goodman and Kruskal (1954, pp. 739–740).
[5]Variants on these procedures have been developed using likelihood ratio concepts (see Mood, Graybill, and Boes, 1974, pp. 452–461) and parallels with information theory (see Kullback, 1968, pp. 155–158). Again the basic emphasis is on hypothesis testing, primarily for independence. For large-sample data, the modified test statistics are approximately equal to chi-square and the same objections apply.

can be used to test the null hypothesis that the true probabilities are the P_{ij}^*. For this goodness-of-fit use of χ^2, the degrees of freedom are $RC-1$. In other respects this statistic resembles ordinary χ^2, except for substitution of the theoretical for the usual "expected" proportions. Despite the formal similarity of (1) and (2), the goodness-of-fit statistic and ordinary χ^2 typically have opposite uses. Large values of ordinary χ^2 indicate deviation from statistical independence. Small values of the goodness-of-fit statistic indicate a good fit between the theoretical probabilities and the actual population probabilities.

Let us return to the example of Table 1.5, but this time assume that, on the basis of some theory, the "expected" probabilities have been assigned as $P_{11}^* = .1, P_{12}^* = .05, P_{21}^* = .2, P_{22}^* = .65$. In this case, using (2),

$$\chi^2 = 224\left[\frac{(33/224-.1)^2}{.1} + \frac{(12/224-.05)^2}{.05} \right.$$

$$\left. + \frac{(46/224-.2)^2}{.2} + \frac{(133/224-.65)^2}{.65} \right]$$

$$= 6.196$$

Since $RC-1=3$, there are three degrees of freedom.

With three degrees of freedom the probability of observing this or some greater value of χ^2 under the hypothesis that the P_{ij}^* hold is less than .20 but greater than .10. Apart from the custom designing to the hypothesized P_{ij}^*, the goodness-of-fit test shows all the deficiencies of ordinary chi-square in evaluating event predictions. As a test it has an undesirable relation with sample size; with modest-sized samples one may well retain the theory even if there are large discrepancies between the theoretical and true probabilities. Moreover the measure is unrelated to assessment of event predictions of a dependent variable through knowledge of an independent variable. After all, the goodness-of-fit test can also be applied to a prediction of the marginal probabilities of just one variable; applying it to the bivariate distribution of $Y \times X$ is indeed identical to studying the univariate probabilities of the composite variable $V = Y \times X$, formed as the Cartesian product of Y and X. The univariate character of the test indicates its inapplicability to bivariate event predictions.

Between completely specified theoretical probabilities and the "nonrelationship" of statistical independence, a variety of partially specified structures may be tested through goodness-of-fit methods, including the quasi-independence methods of Goodman (1968). These are also inappropriate to the analysis of event predictions and lack a PRE interpretation.[6] However goodness-of-fit approaches do have appeal when the basic focus is on predicting probabilities rather than events.[7]

2.3.3 Predicting Probabilities versus Predicting Events

Predicting probabilities, however, has a fundamentally different focus from predicting events. Consider the hypothetical probability structure shown in Table 2.2A and the conditional probabilities of approval given by the following:

$$P(\text{approve}|\text{North-Female}) = \frac{.240}{.25} = .96$$

$$P(\text{approve}|\text{North-Male}) = \frac{.125}{.25} = .50$$

$$P(\text{approve}|\text{South-Female}) = \frac{.125}{.25} = .50$$

$$P(\text{approve}|\text{South-Male}) = \frac{.010}{.25} = .04$$

These probabilities are perfectly fit by the linear model;

$$P(\text{approve}|jk) = .04 + .46j + .46k$$

where $j = 1$ if North, 0 if South, and $k = 1$ if Female, 0 if Male.

Actually, despite the perfect fit, events are predicted poorly for the Northern males and Southern females. Moreover, in the example shown in Table 2.2B, although we still obtain a perfect fit for a second model,

$$P(\text{approve}|jk) = .48 + .02j + .02k$$

accurate event prediction is out of the question. If the population were statistically independent for the given marginals, we would have a cell

---

[6]A detailed technical discussion of the inadequacies of quasi-independence methods for prediction analysis can be found in Hildebrand, Laing, and Rosenthal (1974b).
[7]Other techniques that usefully focus on probabilities include probit analysis (McKelvey and Zavoina, 1975), logit analysis (Theil, 1972), and a variety of log-linear methods (such as Grizzle, Starmer, and Koch, 1969). Basically these models use a linear model type of approach to predicting probabilities. Extensive discussions of this approach are given by Haberman (1974) and Bishop, Feinberg, and Holland (1975).

Table 2.2 Two hypothetical probability structures perfectly fitted by a linear model

(A)

	Northern females	Northern males	Southern females	Southern males	
Approve of the President	.240	.125	.125	.010	.50
Disapprove of the President	.010	.125	.125	.240	.50
	.25	.25	.25	.25	1.00

(B)

	Northern females	Northern males	Southern females	Southern males	
Approve of the President	.130	.125	.125	.120	.50
Disapprove of the President	.120	.125	.125	.130	.50
	.25	.25	.25	.25	1.00

probability of .125 in each cell. Since no actual probability differs from this by more than .005, the two variables are nearly independent and knowledge of region and sex would be of scant help in any prediction of approval.[8] Fitting proportions as in the preceding example may be of interest in some contexts, but those contexts are obviously remote from the prediction of events. The proportion-fitting procedures are simply outside the major focus of this book, event prediction.

2.4 THE CROSS-PRODUCT RATIO AND ITS PROGENY

Another important and closely related class of procedures is related to the cross-product ratio (cpr) for 2×2 tables. Population probabilities are

---

[8]Note that in Table 2.2 the coefficients of j and k are larger for (A) than for (B), a fact related to the greater "predictability" found in the first table. However coefficients of a model predicting proportions do not provide a direct evaluation of the prediction success of an event prediction. Evaluation of probabilistic (actuarial) propositions as event predictions is discussed in Section 4.2.

presented in these tables as follows:

P_{11}	P_{12}
P_{21}	P_{22}

Define the following:

$$\text{cpr} = \frac{P_{11}P_{22}}{P_{12}P_{21}} \qquad \text{for } \textit{populations}$$

$$\text{cpr} = \frac{f_{11}f_{22}}{f_{12}f_{21}} = \frac{n_{11}n_{22}}{n_{12}n_{21}} \qquad \text{for } \textit{samples}$$

(3)

If the variables involved are statistically independent, then the population cross-product ratio is 1. In general, the cpr is $\left\{ \begin{matrix} \geq \\ = \\ < \end{matrix} \right\}$ 1 if the product of the two entries in the diagonal [(1, 1) and (2, 2)] cells is $\left\{ \begin{matrix} \geq \\ = \\ < \end{matrix} \right\}$ the product of the two off-diagonal entries.

The cross-product ratio has been widely used in data analysis, most notably as transformed in Yule's Q:

$$Q = \frac{\text{cpr} - 1}{\text{cpr} + 1} = \frac{[(P_{11}P_{22})/(P_{12}P_{21})] - 1}{[(P_{11}P_{22})/(P_{12}P_{21})] + 1} = \frac{P_{11}P_{22} - P_{12}P_{21}}{P_{11}P_{22} + P_{12}P_{21}}$$

(4)

2.4.1 Multiplication Invariance

The cross-product ratio (hence Q) has an interesting property. Multiplying any row or column by some strictly positive constant k leaves the cross-product ratio unchanged. For example, multiplying the first row by k yields the cpr, $kn_{11}n_{22}/kn_{12}n_{21} = n_{11}n_{22}/n_{12}n_{21}$.

That is, the cross-product ratio is invariant under row and column multiplication, a property that Edwards (1963) and Mosteller (1968) hold to be essential for measures of association. However note that there is only one cpr in a 2×2 table; yet Table 1.7 showed six substantively distinct forms of prediction in 2×2 tables. Since we seek a technique discriminating among these alternative forms, we cannot base our model solely on the cross-product ratio. This can be shown to imply that a suitable prediction logic measure *cannot* be invariant under row and column multiplication.

We are not convinced that this multiplication invariance requirement is desirable scientifically. Implicitly this requirement forces each row state and each column state to have equal weight in the measure. This is plausible when the marginal proportions are arbitrary (perhaps determined by experimental conditions). However if a certain state of the predictor variable occurs very infrequently in nature, it seems unreasonable to give it much weight in assessing the overall prediction effectiveness of a theory. This issue is discussed in more detail in Section 4.3.

2.4.2 Extensions of the cpr Approach

Tables larger than 2×2 can also be analyzed through cpr methods. In larger tables, each 2×2 subtable may have a cross-product ratio of interest. Goodman's *interaction analysis* (1969) is a custom-design method for combining subtable cross-product ratios. However given an appropriate set of $(R-1)(C-1) < 2^{RC} - 1$ cpr's, we can calculate all the subtable cpr's. This implies that just as the single cpr for a 2×2 table cannot discriminate among the $2^4 - 1$ alternative prediction logic statements for the 2×2 table, interaction analysis cannot discriminate among the $2^{RC} - 1$ prediction logic statements for $R \times C$ tables.[9]

Despite the custom-designing and multivariate analysis (Goodman, 1971) features of interaction analysis, the cpr-based measures do not meet our criteria. We can find no PRE interpretation for the cpr measures (criterion 4), and the fact that the measures concern probabilities rather than events means that they cannot be evaluated on the second and third criteria.[10]

2.4.3 Sensitivity to Small Cell Entries

One further problem with cpr-based measures is their sensitivity to small cell entries. Yule's Q appropriately transforms the cross-product ratio in a way that avoids this problem, but the measures for larger tables are based on products of cross-product ratios. The presence of a single zero cell in a table is sufficient either to set the entire product equal to zero (if the cell entry is in the numerator of a ratio) or to render it undefined (if the zero entry is in the denominator). Similarly, small changes in a near-zero cell probability can introduce large fluctuations in a cpr-based measure. The sensitivity problem also arises for goodness-of-fit tests. The sum (2) is undefined whenever observations occur ($f_{ij} > 0$) for events that have a theoretical probability of zero ($P_{ij}^* = 0$). Either multiplying cell probabilities

---

[9]The inappropriateness of interaction analysis for evaluating prediction logic type statements is discussed at length in Laing and Rosenthal (1974).
[10]For the 2×2 case, however, Yule's Q has a PRE interpretation based on a degree-2 prediction. See Section 2.6.

(as in likelihood ratio tests or interaction analysis) or adding ratios of probabilities (as in chi-square) thus violates our sixth criterion. The procedure presented in Chapter 3 avoids this problem by taking a ratio of two sums of probabilities.

Although a variety of measures are related to those examined thus far, this review has indicated the basic problems: simple chi-square methods focus solely on statistical independence, and are incapable of detecting the form of departure from this nonrelation. More elaborate chi-square and cpr methods, although custom designed, lack a predictive structure and are motivated by inferential rather than descriptive criteria. Moreover the various measures are sensitive to apparently minor changes in the probability structure and the values they yield are hard to interpret, since they have no convenient error reduction interpretation that would satisfy the PRE conditions.

We next examine measures of nominal association that were designed explicitly to yield a PRE interpretation.

2.5 PROPORTIONATE REDUCTION IN ERROR MEASURES FOR DEGREE-1 PREDICTIONS

Of established measures for degree-1 predictions, those presented by Guttman (1941) and Goodman and Kruskal (1954) for nominal variables come closest to satisfying the criteria of Section 2.1. The two measures considered in this section share some important features. Based on explicit *ex post* predictions, both measures have a proportionate reduction in error interpretation. Both measures use event predictions. Recall the form of an event prediction: "If these conditions hold, then state so-and-so occurs." One advantage of event predictions is the ease with which error can be specified. If the predicted state does not occur, then there is an error. For any observation an event prediction is either right or wrong. Thus in these two measures *error* is simply defined as an incorrect prediction of the observation's location on the dependent variable.

Assume there are N observations of a particular phenomenon. These N observations may be the entire population; otherwise, in the case of a theoretically infinite population or an indefinitely continuing process, the cases can constitute a very large random sample. A decision maker predicts each observation's location on the dependent variable using rules that depend on the observed probability distributions. Rule U predictions are made in ignorance of the obervation's location on the independent variable, and the predictions depend only on the marginal distribution of the *dependent* variable. In contrast, under rule K the decision maker knows each observation's independent variable state and uses the observed joint probability structure to predict (*ex post*) each observation's state on the dependent variable. Thus in formulating rule K both measures studied in this section use an *ex post* "hypothesis" that the data lie in the distribution as observed, rather than a prediction developed from an *a priori* theory.

Despite their similarities the two measures exhibit one key difference that is instructive for understanding Chapter 3's development of the $\nabla_{\mathcal{P}}$ model: the first measure (λ) uses pure strategy prediction rules, whereas the second measure (τ) uses a mixed strategy prediction.[11]

2.5.1 A Measure of Modal Prediction: λ

Predictions for this measure [due to Louis Guttman (1941) and denoted λ by Goodman and Kruskal (1954, pp. 740–742)] always predict a single state of the dependent variable. If the observation has any state other than the predicted state, an error occurs. We develop the measure by defining the rule K and rule U predictions, calculating their error rates, and forming a PRE measure.

Under rule U (without knowledge of X), predict the modal category of the dependent variable, that is, predict the Y-state having the greatest marginal probability $M \equiv \max_i P_i.$[12] Predicting this *m*odal class for each of the N cases in the population, we are correct for NM cases. Hence

$$\text{Rule } U \text{ errors} = N(1 - M)$$

For the political party example in Table 1.6 "Conservative" is the most likely dependent variable category, so rule U errors $= 597(1.0 - .533) = 279$.

Under rule K (X category known) if $X = x_j$, predict the modal Y category given x_j.[13] The probability of this category is $M_j \equiv \max_i P_{ij}$. Of the NP_j cases with $X = x_j$, we will thus predict NM_j correctly. Conversely we will make $N(P_j - M_j)$ errors.

By summing across the states of X, we find, for the total population,

$$\text{Errors by rule } K = \sum_j N(P_j - M_j) = N\left(1 - \sum_j M_j\right)$$

---

[11]When X is the independent variable, Goodman and Kruskal term the measures λ_a and τ_a. When Y is independent, they are λ_b and τ_b. In general, $\lambda_a \neq \lambda_b$ and $\tau_a \neq \tau_b$.

[12]In case of ties for the modal category, choose one category arbitrarily and proceed as before.

[13]Once more, if there are ties for the modal category, choose one.

Continuing our example the rule K errors are

$$597\left[1-(.355+.102+.164+.075)\right]=597\left[1-.697\right]=181$$

Finally define as required by the PRE model,

$$\lambda=\frac{\text{errors by rule } U-\text{errors by rule } K}{\text{errors by rule } U}=1-\frac{\text{rule } K \text{ errors}}{\text{rule } U \text{ errors}}$$

or

$$\lambda=1-\frac{N\left(1-\sum_{j}M_{j}\right)}{N\left(1-M\right)}=\frac{\sum_{j}M_{j}-M}{1-M} \tag{5}$$

In the example $\lambda=1-181/279=.351$. Thus the value of λ represents the proportionate reduction in errors committed in predicting each case lies in the modal Y-class when one shifts from rule U (without knowledge of the independent variable) to rule K (with knowledge of the independent variable).

Numerical Properties of λ

Let us identify the type of "nonrelation" that λ uses as its zeroing standard. The measure is zero if and only if the same Y-category is modal for all X-states. Statistical independence is sufficient but by no means necessary for this condition.

The other numerical properties of λ are as follows: λ is undefined if there are no errors under rule U. Otherwise $0\leqslant\lambda\leqslant1$. It can equal one only if knowledge of the X category permits perfect prediction of some Y category. [For every j there is one i such that $P(y_i|x_j)=1.0$.]

Note that both rule K and rule U predictions are *ex post*: "Always guess the modal class observed in the data." This *ex post* procedure has been termed optimal by Goodman and Kruskal (1954) since it minimizes the errors of prediction, assuming that only one Y-state is to be predicted for each observation. Yet a rule predicting a unique dependent variable category may not be optimal in a descriptive sense. Other aspects of the probability structure might suggest, for example, the presence of important distinctions among nonmodal categories. To obtain a measure that responded to the entire probability structure and not just the modal probabilities, Goodman and Kruskal proposed a mixed (probability-matching) prediction strategy.

2.5.2   A Measure of Proportional Prediction: τ

A mixed strategy requires that predictions be made (for all cases in a given category) according to an appropriate probability distribution. In developing τ Goodman and Kruskal wished to predict in such a way that the distribution of predicted Y-states would simulate or reproduce the population distribution. (The Goodman and Kruskal measure should not be confused with several other uses of this symbol such as Kendall's τ.) As with λ, we are asked to predict a single state of the dependent variable for each observation.

Under rule K, given an observation with $X = x_j$, predict y_i with probability $P_{ij}/P_{.j}$, the *conditional* probability of y_i given x_j. The probability of being correct is also $P_{ij}/P_{.j}$. Thus the probability given x_j of both predicting y_i and being correct is $(P_{ij}/P_{.j})^2$. By summing this probability over the Y states and multiplying by the total number of $X = x_j$ cases, we find the expected number of correct predictions for $X = x_j$ to be $NP_{.j}\Sigma_i(P_{ij}/P_{.j})^2$. Finally, summing over the X-states, we expect the total number of correct predictions to be $N\Sigma_j P_{.j}\Sigma_i(P_{ij}/P_{.j})^2 = N\Sigma_j\Sigma_i(P_{ij}^2/P_{.j})$ and

$$\text{Expected errors by rule } K = N\left[1 - \sum_i \sum_j \frac{P_{ij}^2}{P_{.j}}\right]$$

The numerical calculation for the Table 1.6 example shows

$$\text{Rule } K \text{ errors} = N\left[1 - \frac{(.122)^2}{.541} + \frac{(.017)^2}{.139} + \cdots + \frac{(.013)^2}{.156}\right]$$

$$= 597(.421) = 251.45$$

Under rule U predict y_i with the *unconditional* probability $P_{i.}$. When y_i is predicted, that prediction is correct with probability $P_{i.}$; thus the probability of both predicting y_i and being correct is $P_{i.}^2$. Summing across the states of Y, the expected number of cases predicted correctly is $N\Sigma_i P_{i.}^2$. Consequently

$$\text{Expected rule } U \text{ errors} = N\left(1 - \sum_i P_{i.}^2\right)$$

Continuing to use Table 1.6 as an example, we find there that expected

$$\text{Rule } U \text{ errors} = N\left[1 - (.370)^2 - (.533)^2 - (.097)^2\right]$$

$$= 597(.570) = 340.17$$

Define the PRE measure

$$\tau = 1 - \frac{\text{expected rule } K \text{ errors}}{\text{expected rule } U \text{ errors}}$$

$$= \frac{\sum_i \sum_j \left(P_{ij}^2 / P_{.j} \right) - \sum_i P_{i.}^2}{1 - \sum_i P_{i.}^2} \tag{6}$$

In the political affiliation example τ equals $1 - 251.45/340.17 = .261$. Thus τ represents the proportionate reduction in expected errors in selecting event predictions probabilistically in accordance with the observed distributions on the dependent variable when one shifts from knowledge of the marginal distribution of the dependent variable to knowledge of each case's location on the independent variable and the joint distribution. Note that τ, like λ, uses *ex post* predictions.

Rule U: Matching Rule K

The rule U error rate of τ may be derived by an alternative procedure that follows the general rule U approach we develop in Chapters 3 and 4. By this approach rule U matches the rule K predictions, but without knowledge of the X category. Select $NP_{.j}$ cases at random from the entire population. As under rule K predict y_i for each of these cases with probability $P_{ij}/P_{.j}$. Since the cases were selected randomly, when y_i is predicted the probability of being correct is the *unconditional* probability $P_{i.}$. Summing across the predicted Y-states, we expect $NP_{.j}\sum_i(P_{ij}/P_{.j})P_{i.}$ correct predictions. Repeating the procedure for each of the component rule K predictions, we expect to be correct for a total of $N\sum_j P_{.j}\sum_i(P_{ij}/P_{.j})P_{i.} = N\sum_i P_{i.}^2$ observations. Therefore whether one probability-matches the Y marginals or, in the alternative procedure, probability-matches the rule K predictions, one expects $N(1 - \sum_i P_{i.}^2)$ rule U errors.

Numerical Properties of τ

Numerically τ is undefined if there exists some y_i such that $P_{i.} = 1$, that is if all N cases fall in the same Y class. Otherwise $0 \leqslant \tau \leqslant 1$. Like λ, it can equal one only if knowledge of the X category permits perfect prediction of some Y category. Finally $\tau = 0$ if and only if the two variables are statistically independent. Thus τ measures the degree of departure from the "nonrelation" of statistical independence.

2.5.3 Limitations of λ and τ

The preceding discussion has shown that both τ and λ are PRE measures for event predictions. In each measure the prediction rule is simple and direct.

Although these measures are important guides to further development, their prediction rules are totally inflexible and *ex post*. *Since the rules depend only on the probability structure, they cannot be custom designed for a particular prediction under investigation.* The researcher has no control whatsoever over the prediction used. Moreover, the prediction rules are constrained to satisfy the highly restrictive condition that exactly one value of the dependent variable must always be predicted for each state of the independent variable.[14] Seeking a measure that will assess predictions that vary across the independent states in the precision with which the dependent variable is predicted, we reject this condition. Suppose, for example, that the probability structure is

	x_1	x_2	
y_1	.47	.03	.50
y_2	.02	.23	.25
y_3	.03	.22	.25
	.52	.48	1.00

Now a good description of this relation can be stated in the prediction logic as $\mathcal{P}$: $(x_1 \rightsquigarrow y_1)$ & $(x_2 \rightsquigarrow y_2$ or $y_3)$. Note that the distinction between y_2 and y_3 has negligible predictive significance. Given this perspective the values of λ and τ seem too small ($\lambda = .400$, $\tau = .424$). If the investigator creates a new dependent variable by combining the second and third rows of this table, then the probability structure becomes

	x_1	x_2	
y_1	.47	.03	.50
y_2'	.05	.45	.50
	.52	.48	1.00

[14]We do not mean to imply that Goodman and Kruskal used this condition thoughtlessly. They state ([1954], pp. 734–735) "if the use to which a measure of association were to be put could be precisely stated, there would be little difficulty in defining an appropriate measure." Our sense of the applications literature, however, is that although the points of interest of a study are often fairly well defined, there has been great difficulty in defining such a measure —it is seldom attempted by researchers.

For this table the values of the measures are substantially increased ($\lambda = .840$, $\tau = .707$). Yet we would describe the second table by $\mathcal{P}'$: $(x_1 \rightsquigarrow y_1)$ & $(x_2 \rightsquigarrow y_2')$, which is logically equivalent to $\mathcal{P}$. Hence by criterion 2 the two tables should receive the same measure. Although aware of the effects of category definitions on λ and τ, Goodman and Kruskal (1954, pp. 737–738) did not regard these effects as undesirable:

> Decisions about the definitions of the classes of polytomy [$R \times C$ contingency table], or changes from a finer to a coarser classification (or vice-versa), can affect all the measures of association of which we know. ...
>
> At first this consideration might seem to vitiate any reasonable discussion of measures of association. We feel, however, that it is in fact desirable that a measure of association reflect the classes *as defined for the data*. Thus one should not speak, for example, of association between income level and level of formal education without specifying particular class definitions. ... That the definition of the classes can affect the degree of association naturally means that careful attention should be given to the class definitions in the light of the expected uses of the final conclusions.

We think that emphasis should be placed on the specification of predictions rather than on the definition of classes. Defining classes *a priori* does allow investigators to constrain the set of prediction rules that can be selected *ex post* for λ and τ. *A priori* constraints, in our view, are better introduced by direct specification of the prediction rules, following the custom-design criterion of Section 2.1.

When *ex post* analysis is attempted, the procedure should typically not be constrained to predicting one and only one dependent variable state for each independent variable state. This constraint can lead to predictions that are inappropriate for the relation found in the data. Even when a virtually errorless prediction can be made for each independent variable state, λ and τ can fail to capture the prediction. The prediction $x_1 \rightsquigarrow (y_1$ or $y_3)$ & $x_2 \rightsquigarrow (y_1$ or $y_2)$ aptly describes the following table:

	x_1	x_2	
y_1	.25	.25	.50
y_2	.01	.24	.25
y_3	.24	.01	.25
	.50	.50	1.00

Yet λ, focusing exclusively on modal classes and ignoring distinctions among nonmodal classes, uses the *ex post* rule $K : x_1 \rightsquigarrow y_1$ & $x_2 \rightsquigarrow y_1$.[15] Accordingly its value is zero. Similarly as the prediction rule underlying τ weights the modal class heavily its value is only .169.

In summary, since they use a PRE model to evaluate event predictions, the λ and τ measures are important guides to further development. Nonetheless they fail to satisfy the criteria of Section 2.1 in important respects. Since they require us to predict only a single Y-state for each X-state, the prediction rules can fail to capture important relations in the data. Moreover rather than permitting custom tailoring to a specific *a priori* proposition, the procedures use the probability structure to establish *ex post* prediction rules.

In part to illustrate the use of *a priori* prediction rules, we next examine measures for pairs of observations.

2.6 MEASURES OF ASSOCIATION FOR DEGREE-2 PREDICTIONS

This section evaluates measures of association for degree-2 predictions with the criteria established in Section 2.1. Most of these measures are for ordinal variables. Recall that degree-2 propositions predict the dependent variable states of each pair of observations.

No ordinal measure known to us can be used for *a priori* degree-1 event predictions.[16] In fact, so far in this chapter we have encountered no *a priori* measures. An *a priori* PRE interpretation can be given to several of the better-known measures of association for ordinal variables, namely, Goodman and Kruskal's γ, Somer's d_{yx}, Kim's d, and Spearman's ρ_s. These are all fixed predictions of degree-2 or 3, and hence fail to meet the custom-design criterion. (For formal definition of these measures and a detailed discussion, see Chapter 5.) Since the fixed prediction is invariably some form of monotonic relation between two variables, such as, "If X increases, then Y tends to increase," these measures are unable to identify the presence of a nonmonotonic relation (Costner, 1965; Lieberson, 1964). That is, these measures indicate at most a weak association when the data distribution is a nonmonotone such as the so-called S, J, or U shapes.

For example, although Goodman and Kruskal's γ is appropriate in

---

[15]In his application of prediction analysis methods to multivariate *ex post* selection problems, Bartlett (1974) indicates that the emphasis on modality is a basic weakness in *ex post* selection techniques based on λ such as THAID (Messenger and Mandell, 1972).

[16]The quadrant measure shown in Table 5.7 comes close to being an exception, but its prediction logic is "*ex priori*" since it requires knowledge of the median state of the dependent variable.

some contexts, γ also

...appears to be useless in the empirical assessment of any proposition more specific in form than "The greater the X, the greater the Y" or "As X increases, Y increases" (and analogous propositions for negative relationships). [Costner, 1965, p. 350].

To illustrate gamma's inability to detect nonmonotones, we consider MacRae's (1967) cross tabulation of two scales of legislative behavior as population data. MacRae used γ to summarize the data shown in Table 2.3. Inspection of the table suggests an inverted U-shaped relation, with relatively few cases in the lower middle section of the table. The near-zero value ($-.015$) of γ for this table indicates that a monotonic relation is absent but fails to detect the nonmonotone.

Table 2.3   Cross tabulation of French Conservative deputies' positions on two legislative roll-call scales

			Mendès-France scale				
			Pro			Con	
			1	2	3	4	
Pro	1		19	17	11	10	57
Europe scale	2		8	2	2	3	15
Con	3		10	1	2	8	21
			37	20	15	21	93

Source: MacRae (1967, p. 192).

Use of the most widely applied ordinal measures is further limited by the fact that their underlying prediction rules are not degree-1.[17] Consequently those rules do *not* enable us to predict an observation's location on the dependent variable given its location on the independent variable. This prevents ordinal measures from distinguishing among a variety of monotone relations. The two examples given in Table 2.4 illustrate this point. By inspection the distribution in (A) is concentrated on and *above* the main diagonal of the table, whereas in (B) the data are concentrated on and *below* the main diagonal. Yet γ and the other standard ordinal measures are identical for the two tables. Both tables do support the

---

[17]The quadrant measure discussed in Kruskal (1958) and shown later in Table 5.7 is an exception.

proposition, "Y tends to increase with X." But they provide differential support for the degree-1 prediction of Y states: high $\leadsto$ (high or medium or low), medium $\leadsto$ (medium or low), & low $\leadsto$ low. In fact, the hypothetical data in (B) support this prediction whereas (A) supports high $\leadsto$ high, medium $\leadsto$ (high or medium), & low $\leadsto$ (high or medium or low).

Table 2.4   Two tables with identical values of γ

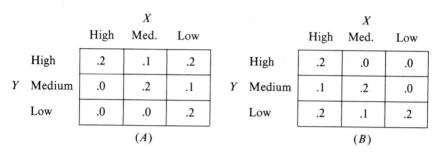

		X					X		
		High	Med.	Low			High	Med.	Low
	High	.2	.1	.2		High	.2	.0	.0
Y	Medium	.0	.2	.1	Y	Medium	.1	.2	.0
	Low	.0	.0	.2		Low	.2	.1	.2
			(A)					(B)	

Although this example illustrates that, like the nominal measures, the various fixed ordinal measures are inadequate for a general model of prediction analysis, the measures do have the important feature of being interpretable as based on *a priori* prediction rules. Analysis of the ordinal measures, however, requires some cumbersome notation that is best deferred until Chapter 5. Yule's Q, discussed in Section 2.4, is a special case of γ for 2×2 tables. Its prediction rules can be developed relatively simply and used to illustrate *a priori* degree-2 prediction.

2.6.1   Yule's Q: A Measure for an *a priori* Degree-2 Prediction

Refer to the probability structure of two variables shown in Table 1.1. Consider drawing an observation at random from the population, recording it, replacing it, and then drawing a second observation. There are N^2 pairs of the N observations. The probability of drawing a given pair of observations is simply the product of the probabilities of the two events. For example, we find the pair $(y_1 x_1), (y_1 x_2)$ with probability $P_{11} P_{12}$. Similarly the probability of $(y_1 x_2), (y_1 x_2)$ is P_{12}^2.

Now we can describe the X-states of the pairs with a new variable X^{*2} with states $x_1 x_1, x_1 x_2, x_2 x_1, x_2 x_2$. The states $y_1 y_1, y_1 y_2, y_2 y_1, y_2 y_2$ define Y^{*2}. We can then represent the probabilites of the events in $Y^{*2} \times X^{*2}$ as the 4×4 cross classification shown in Table 2.5. In this table the probability $P_{11} P_{12}$ represents the probability of the degree-2 event $(y_1 y_1, x_1 x_2)$.

The marginal probabilities of Table 2.5 can be found either by summing the row or column entries or by noting, for example, that the probability the first observation has $X = x_2$ and the second has $X = x_1$ equals $P_{.2}P_{.1}$. As with $Y \times X$, all probabilities sum to one.

Table 2.5   Ordered pairs of two dichotomous variables

$$X^{*2}$$

		$(x_1 x_1)$	$(x_1 x_2)$	$(x_2 x_1)$	$(x_2 x_2)$	
	$(y_1 y_1)$	P_{11}^2	$P_{11}P_{12}$	$P_{12}P_{11}$	P_{12}^2	$P_{1.}^2$
Y^{*2}	$(y_1 y_2)$	$P_{11}P_{21}$	$P_{11}P_{22}$	$P_{12}P_{21}$	$P_{12}P_{22}$	$P_{1.}P_{2.}$
	$(y_2 y_1)$	$P_{21}P_{11}$	$P_{21}P_{12}$	$P_{22}P_{11}$	$P_{22}P_{12}$	$P_{1.}P_{2.}$
	$(y_2 y_2)$	P_{21}^2	$P_{21}P_{22}$	$P_{22}P_{21}$	P_{22}^2	$P_{2.}^2$
		$P_{.1}^2$	$P_{.1}P_{.2}$	$P_{.1}P_{.2}$	$P_{.2}^2$	1.0

Yule's Q predictions can now be analyzed after one further development. We must eliminate all pairs tied on either variable (the first and last row and the first and last column in Table 2.5) from analysis. This leaves us with the 2×2 table outlined in heavy borders within Table 2.5. This 2×2 table contains $N^2(P_{11}P_{22} + P_{12}P_{21} + P_{21}P_{12} + P_{22}P_{11}) = 2N^2(P_{11}P_{22} + P_{12}P_{21})$ pairs. To calculate the probabilities of the events in this new population, we must thus divide the probabilities in the relevant 2×2 table within Table 2.5 by $2(P_{11}P_{22} + P_{12}P_{21})$. For example, we have $N^2 P_{11}P_{22}$ events in the cell $(y_1 y_2, x_1 x_2)$. With ties discarded the probability of this event is $P_{11}P_{22}/[2(P_{11}P_{22} + P_{12}P_{21})]$. The relevant probabilities are shown in Table 2.6 for both the general case and the authority system data of Table 1.5.

For this ties-excluded population, consider the rule K prediction $(x_1 x_2)$ $\longleftrightarrow (y_1 y_2)$. Since $y_2 y_1$ is the error event for $x_1 x_2$ and $y_1 y_2$ is the error event for $x_2 x_1$,

$$\text{Rule } K \text{ error rate} = \left[P(y_2 y_1 | x_1 x_2) P(x_1 x_2) + P(y_1 y_2 | x_2 x_1) P(x_2 x_1) \right]$$

$$= \left[P(y_2 y_1, x_1 x_2) + P(y_1 y_2, x_2 x_1) \right]$$

$$= \frac{P_{12}P_{21}}{P_{11}P_{22} + P_{12}P_{21}}$$

Table 2.6   Ordered pairs of two dichotomous variables with ties eliminated

(*A*)   The General Table

	(x_1,x_2)	(x_2,x_1)	
(y_1,y_2)	$\dfrac{P_{11}P_{22}}{2(P_{11}P_{22}+P_{12}P_{21})}$	$\dfrac{P_{12}P_{21}}{2(P_{11}P_{22}+P_{12}P_{21})}$	.5
(y_2,y_1)	$\dfrac{P_{12}P_{21}}{2(P_{11}P_{22}+P_{12}P_{21})}$	$\dfrac{P_{11}P_{22}}{2(P_{11}P_{22}+P_{12}P_{21})}$	.5
	.5	.5	1.0

(*B*)   Degree-2, Ties-Excluded Representation of Table 1.5

	$(c,\bar{c})$	$(\bar{c},c)$	
$(s,\bar{s})$	$\dfrac{.087}{.197}=.444$	$\dfrac{.011}{.197}=.056$	.5
$(\bar{s},s)$	$\dfrac{.011}{.197}=.056$	$\dfrac{.087}{.197}=.444$	.5
	.5	.5	1.0

Yule's $Q=0.777$

Several alternatives for rule U lead to an expected error rate of $1/2$. These include the following:

1. Predict $y_1 y_2$.
2. Predict $y_1 y_2$ with probability $1/2$; predict $y_2 y_1$ with probability $1/2$.

Paralleling our alternative development of rule U for τ in Section 2.5.2, the latter rule matches the predictions of rule K under the information conditions of rule U.

Finally form the PRE measure

$$Q = 1 - \frac{P_{12}P_{21}/(P_{11}P_{22}+P_{12}P_{21})}{1/2}$$

$$= \frac{P_{11}P_{22}-P_{12}P_{21}}{P_{11}P_{22}+P_{12}P_{21}} \tag{4'}$$

Note that Q can be negative, indicating proportionate *increase* in error. Further note that no characteristic of the dependent variable was used in formulating either rule *K* or rule *U* for *Q*. It is thus an appropriate *a priori* degree-2 measure for dichotomous variables with ties excluded. Despite its widespread use, however, *Q* is inappropriate, even with 2×2 tables, for any *a priori* degree-1 predictions or for *a priori* degree-2 predictions that do not exclude ties.

2.7   CONCLUSION: MEASURES OF ASSOCIATION

In this chapter we proposed seven criteria that should be satisfied by a model for measuring prediction success. In general, the model should

1.   Permit custom tailoring to the specific prediction under investigation, whether that prediction is *a priori* or *ex post*.
2.   Produce equivalent results for logically equivalent propositions once the roles of the variables have been specified.
3.   Apply to predictions regardless of their prediction scope, precison, and differentiation, yet recognize differences in these features as well as prediction accuracy in measuring prediction success.
4.   Permit a proportionate reduction in error interpretation.
5.   Provide measures of multiple and partial prediction success for multivariate analysis.
6.   Focus on the central tendency in the important aspects of the probability structure, being relatively insensitive to minor changes in probabilities.
7.   Permit accurate estimation with modest sample sizes.

This is a rather formidable set of criteria. The measures of association developed for nominal variables fall quite short of the mark. However the PRE feature of λ and τ and the custom-design approaches of goodness of fit and related methods are important aids to further development. Several measures of ordinal association fare somewhat better in that they can be interpreted to be based on *a priori* prediction rules. However they are limited to evaluating the prediction success for a specific degree-2 prediction, and cannot address other degree-2 (or any degree-1) predictions.

In the following chapters we propose a bivariate measure that satisfies all seven criteria. In demonstrating the flexibility of our prediction model we show that many measures of ordinal association are special cases of our measure based on prediction logic propositions that predict degree-2 or 3 relations. Unlike these measures, however, our model can also be used to evaluate the success of the many alternative prediction logic propositions, whether of degree-1 or higher.

PREDICTION ANALYSIS
FOR BIVARIATE
PROPOSITIONS:
POPULATION MEASURES

We now formulate a general model for measuring bivariate prediction success. The model provides a custom-designed, unique measure, $\nabla_{\mathscr{P}}$, for each logically distinct prediction logic statement. This prediction-specific feature of the measure is represented in its symbol: the subscript $\mathscr{P}$ on the basic measure symbol ∇ (read "del") indicates the specific prediction to which the model is applied.

The chapter first presents the measure for the simplest case, the 2×2 table. A second section specifies the domain of admissible predictions for $R \times C$ cross classifications and develops the general model for defining a unique $\nabla_{\mathscr{P}}$ measure for each logically distinct, admissible proposition. The chapter concludes with an evaluation of $\nabla_{\mathscr{P}}$ in terms of the design criteria established in Section 2.1.

Throughout this chapter we focus on measuring bivariate prediction success in a *population*. The population perspective is maintained in Chapter 4, where the basic model is extended to applications concerning the analysis of predictions of events, and in Chapter 5, where we examine predictions about pairs of events. The case for focusing on the population measure before considering statistical inference has been aptly stated by Kruskal (1958, p. 815):

...it seems desirable to decide what we would mean by association if we knew the population of interest completely, before turning to the more complex questions of making inferences about measures of association from a sample. There is little point in estimating a population characteristic if the meaning of that characteristic is unclear.

...[This] approach...differs from the standard textbook approach, which begins with a sample, defines some intuitively reasonable sample measure of association, investigates its distribution, and—sometimes as an afterthought, but more often not at all—finally asks about the underlying population quantity.

Prediction analysis of bivariate sample data is postponed until Chapter 6, where we provide statistical methods for testing hypotheses and establishing confidence intervals.

3.1   MEASURING PREDICTION SUCCESS IN THE 2×2 TABLE

We first consider the joint distribution of two dichotomous variables, $X = \{x_1, x_2\}$ and $Y = \{y_1, y_2\}$. We sometimes use x and $\bar{x}$ (read "not x"), y and $\bar{y}$, as alternative labels for these states. For example, these labels might represent "rain" and "no rain." We begin by developing the $\nabla_{\mathcal{P}}$ measure for predictions identifying a single error cell in a 2×2 table, then address predictions that identify two diagonally opposed error cells.

3.1.1   A Measure for Predictions Identifying a Single Error Cell in the 2×2 Table: $\nabla_{x \rightsquigarrow y}$

The simplest implication in formal logic is $x \rightarrow y$. Of the four possible states of $Y \times X$ only the event $\bar{y}$ & x falsifies this statement. The statement $[x \rightarrow y]$ & $[\bar{x} \rightarrow (y \text{ or } \bar{y})]$ is logically equivalent. It is true if and only if $x \rightarrow y$ is true. The prediction logic analogue of the second statement is $[x \rightsquigarrow y]$ & $[\bar{x} \rightsquigarrow (y \text{ or } \bar{y})]$ or, equivalently, $[x_1 \rightsquigarrow y_1]$ & $[x_2 \rightsquigarrow (y_1 \text{ or } y_2)]$. Only the event y_2 & x_1 is an error for this prediction. As in formal logic it is sometimes convenient in prediction logic to omit those portions of a prediction statement that can commit no errors, in this case $x_2 \rightsquigarrow (y_1 \text{ or } y_2)$. Thus $x_1 \rightsquigarrow y_1$ is the direct analogue of $x \rightarrow y$; it is the simplest prediction we consider.

The statement $x_1 \rightsquigarrow y_1$ is our formalization of "x_1 tends to be a sufficient condition for y_1." It predicts that cases having $X = x_1$ "usually" satisfy $Y = y_1$; equivalently the statement predicts there are "few" cases in the error cell (y_2, x_1). Our task, then, is to evaluate, for a given population, the prediction success of this statement.

The Raw Error Rate

Let us examine some alternative ways of assessing whether $x_1\rightsquigarrow y_1$ commits relatively "few" errors. All the NP_{21} cases with $Y=y_2$ and $X=x_1$ are errors. The simplest error (per case) measure would be $NP_{21}/N=P_{21}$, but this is not even close to an adequate error measure. Consider these two different tables:

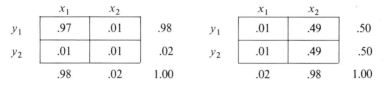

	x_1	x_2			x_1	x_2	
y_1	.97	.01	.98	y_1	.01	.49	.50
y_2	.01	.01	.02	y_2	.01	.49	.50
	.98	.02	1.00		.02	.98	1.00

Although the proposed error measure, P_{21}, equals .01 in both tables, it is only in the left-hand table, where virtually all x_1 cases are also y_1 cases, that knowledge of the X-state is useful in prediction. In the right-hand table, where there are as many (y_2,x_1) cases as (y_1,x_1) cases, the two variables are statistically independent (for example, $P_{21}=.01=.50\times.02=P_2.P._1$). Section 1.3 demonstrates that X has no value for predicting Y if the two variables are statistically independent.

Adjusting for Scope

An obvious difference between the two tables is the very different effective *scopes* of the predictions, reflected in the marginal proportions of x_1 cases. In one table the single state y_1 was predicted for 98% of the cases; in the other, for only 2%. This suggests basing the error measure on just the $NP._1$ cases with $X=x_1$ rather than on all N cases. That is, perhaps we should use the conditional probability $\Pr(Y=y_2|X=x_1)=P_{y_2|x_1}=P_{21}/P._1$. After all, if the only nonvacuous prediction is $x_1\rightsquigarrow y_1$, one might just ignore the x_2 cases. Nonetheless $P_{y_2|x_1}$ still is not adequate, as may be seen from the following pair of tables:

	x_1	x_2			x_1	x_2	
y_1	.48	.25	.73	y_1	.48	.48	.96
y_2	.02	.25	.27	y_2	.02	.02	.04
	.50	.50	1.00		.50	.50	1.00

In both tables $P_{y_2|x_1}=.04$. Although the prediction appears substantially accurate in both tables, apparent support for the prediction in the right-hand table is misleading: knowledge that $X=x_1$ conveys no information about the Y-state, since in the second table X and Y are statistically independent. Thus two tables with identical values for the conditional probability $P_{y_2|x_1}$ differ in the way knowledge of the X-state contributes to prediction success. An even stronger contrast is provided by the table:

	x_1	x_2	
y_1	.48	.49	.97
y_2	.02	.01	.03
	.50	.50	1.00

Here $P_{y_2|x_1} = .02/.50 = .04$ is greater than $P_{y_2} = .03$. In other words, the conditional probability given x_1 that an observation lies in y_2 is greater than the unconditional probability that $Y = y_2$. Consequently predicting y_1 for $X = x_1$ cases would lead to a *greater* proportion of errors than predicting y_1 irrespective of the X-value. If we use the rule of predicting y_1 with knowledge that $X = x_1$, we are worse off than were we to predict y_1 *regardless* of X or even when $X = x_2$. Thus $P_{y_2|x_1}$ seems inappropriate as a measure of prediction success for $x_1 \rightsquigarrow y_1$.

Adjusting for Precision

The problem stems from not taking account of the unconditional probability of y_2. If P_{y_2} is large, the prediction is *precise* in the sense that we are predicting that, given x_1, few observations will occur for a state that is common in the general population. Conversely, if P_{y_2} is small, the prediction is imprecise. We apparently need to take *precision* into account by interpreting $x_1 \rightsquigarrow y_1$ to mean that $P_{y_2|x_1}$ is small *relative to* the marginal probability P_{y_2}. This suggests basing the measure on the ratio of error probabilities $P_{y_2|x_1}/P_{y_2}$. Successful prediction is indicated when this ratio is small, and perfect prediction is indicated when it is zero.

For ease in interpretation, a measure of prediction success should equal 1.0 for a perfect prediction. So we define the $\nabla_{\mathscr{P}}$ measure:

$$\nabla_{x_1 \rightsquigarrow y_1} = 1 - \frac{P_{y_2|x_1}}{P_{y_2}} = 1 - \frac{P_{21}}{P_{2.}P_{.1}} \tag{1}$$

This is the basic $\nabla_{\mathscr{P}}$ measure for a single error cell in a 2×2 table. The measures for the remaining three logically distinct predictions, $x_1 \rightsquigarrow y_2$, $x_2 \rightsquigarrow y_1$, and $x_2 \rightsquigarrow y_2$, are defined by making the appropriate changes of subscripts in (1).

A Numerical Example

Recall from Chapter 1 that Scott, et al. (1967) assert that if an authority system in a formal organization is "incompatible," then it tends to be "unstable." That is, they predict that authority system incompatibility tends to be a sufficient condition for system instability. This *a priori* proposition translates directly into the prediction logic as $\bar{c} \leadsto \bar{s}$. The data have been shown in Table 1.5 for 224 authority systems, treated as a population for illustrative purposes. The prediction is supported:

$$\nabla_{\bar{c} \leadsto \bar{s}} = 1 - \frac{P_{s\bar{c}}}{P_s P_{\bar{c}}}$$

$$= 1 - \frac{12/224}{(45/224)(145/224)}$$

$$= +.588 = \nabla_{s \leadsto c}$$

In fact, the *a priori* prediction $\bar{c} \leadsto \bar{s}$, and its logical equivalent, $s \leadsto c$ ("Stable systems tend to be compatible"), have greater prediction success in these data than any other prediction logic proposition.

Proportionate Reduction in Error Development

The foregoing development of $\nabla_{x_1 \leadsto y_1}$ proceeded intuitively. We now provide a proportionate reduction in error interpretation. For rule K we use the prediction: $x_1 \leadsto y_1$. That is, given that an observation has $X = x_1$, we predict $Y = y_1$. If $X = x_2$ we make the vacuous prediction $Y = (y_1 \text{ or } y_2)$. Note that rule K can be stated *a priori* without knowledge of the probability structure of $Y \times X$, in contrast, for example, to the *ex post* predictions used in Section 2.5 to define τ and λ. Consequently one distinct advantage of this approach is that $\mathscr{P}$ can be a prediction deduced from a theory. With *ex post* prediction rules, in contrast, the probability structure of $Y \times X$ dictates what "prediction" is used in rule K.

Now the prediction logic counts as prediction errors those events that falsify the corresponding statement in elementary formal logic. Assume predictions are made for a population of N cases. The prediction $x_1 \leadsto y_1$ commits errors only for those NP_{21} observations classified as (y_2, x_1). Thus the rule K probability of prediction error equals $NP_{21}/N = P_{21}$. Equivalently rule K makes nonvacuous predictions only for those $NP_{.1}$ cases having $X = x_1$. The probability of prediction error for these cases is $P_{y_2 | x_1}$. Thus under rule K we make $NP_{.1}(P_{y_2 | x_1}) = NP_{21}$ prediction errors.[1] Since in the authority system example the error event for $\bar{c} \leadsto \bar{s}$ is $s\bar{c}$, there are $NP_{s\bar{c}} = 12$ rule K errors.

---

[1] Statements in this section assume a finite population. If the population is infinite, consider drawing a (large) random sample of N cases. Then NP_{21} is the expected number of prediction errors.

We now must choose rule U. Recall that the PRE framework presented in Section 2.1 requires that *in applying* a priori *predictions, rule* U *must not use information about the dependent variable*. Thus, for example, we cannot use the rule underlying λ, "Predict the modal Y class." Instead we seek a rule that does not demand any knowledge of the distribution of Y-states in the population.

We use the predictions of the investigator's theory to define not only rule K, but also the null predictions of rule U. In rule K we predict y_1 for the $NP_{.1}$ observations having $X = x_1$ and we predict (y_1 or y_2) for the rest. In rule U we make those predictions for the same number of observations without knowing the observations' actual X-states; we predict y_1 for $NP_{.1}$ observations selected at random, and (y_1 or y_2) for the remaining observations. *Since both rules K and U make the same number of each type of prediction of Y-states, any difference in their error rates can be attributed solely to the value of information about the state of X when using the specific theory $\mathcal{P}$ to predict Y.*

Note that rule U requires no dependent variable information. The only information required is the marginal probability of the independent variable state x_1. As in rule K, prediction error can occur only for the prediction $Y = y_1$. Since the cases are drawn randomly under rule U, the probability of error for this prediction is the unconditional probability $P_{2.}$. The expected number of rule U errors, therefore, is (the number of times we predict only y_1) $\times$ (probability that $Y = y_2$) $= (NP_{.1})P_{2.} = NP_{2.}P_{.1}$. In the authority system example the corresponding figure is $NP_s P_{\bar{c}} = 224(45/224)(145/224) = 29.129$. Now define the PRE measure,

$$\frac{\text{Rule } U \text{ error} - \text{Rule } K \text{ error}}{\text{Rule } U \text{ error}} = \frac{NP_{2.}P_{.1} - NP_{21}}{NP_{2.}P_{.1}} = 1 - \frac{P_{21}}{P_{2.}P_{.1}} = \nabla_{x_1 \rightsquigarrow y_1}$$

as defined by (1). In our example we have $\nabla_{\bar{c} \rightsquigarrow \bar{s}} = .588$. This value of $\nabla_{x_1 \rightsquigarrow y_1}$ indicates that a 58.8% reduction in error is achieved by applying $x_1 \rightsquigarrow y_1$ given knowledge of X over that expected when the prediction is applied to randomly selected observations whose X-states are unknown. Observe that $\nabla_{x_1 \rightsquigarrow y_1}$ also measures the proportionate decrease in the number expected in the error cell if the two variables were statistically independent.

Statistical Independence and $\nabla_{x \rightsquigarrow y}$

This important relation concerning statistical independence can give rise to misunderstanding. For any of the six 2×2 table propositions shown in Table 1.7, $\nabla_{\mathcal{P}} = 0$ if and only if X and Y are statistically independent. This is a special feature of 2×2 tables *only*. (In the general case of an $R \times C$

table statistical independence implies $\nabla_{\mathcal{P}} = 0$ but the converse does *not* hold.) $\nabla_{\mathcal{P}}$ measures error reduction and not just deviation from statistical independence. Rather than averaging the deviations from statistical independence over all cells, as in measures based on χ^2, $\nabla_{x_1 \leadsto y_1}$ depends on the probabilities pertaining *only to the error cell*. Measures based on ordinary χ^2 only tell us that the data are not unrelated. By focusing on the error cell (y_2, x_1), $\nabla_{x_1 \leadsto y_1}$ assesses the specific prediction $x_1 \leadsto y_1$.

Logical Equivalence and the Roles of Variables

The $\nabla_{\mathcal{P}}$ measure thus incorporates the specific proposition $\mathcal{P}$. The four possible propositions with a single error cell in the 2 × 2 table in general have *different* $\nabla_{\mathcal{P}}$ values. On the other hand, the allocation of the roles *independent* and *dependent* to the two variables does not affect the measure: if $\mathcal{P}$ and $\mathcal{P}'$ are logically equivalent, then $\nabla_{\mathcal{P}} = \nabla_{\mathcal{P}'}$. The prediction $y_2 \leadsto x_2$ is logically equivalent to $x_1 \leadsto y_1$, the prediction we have been examining. They each identify (y_2, x_1) as a single error cell. The rule K error probability for $y_2 \leadsto x_2$ equals P_{21}. Under rule U we predict $X = x_2$ for NP_2 cases and err with probability $P_{.1}$. Thus

$$\nabla_{y_2 \leadsto x_2} = 1 - \frac{P_{21}}{P_{2.}P_{.1}} = \nabla_{x_1 \leadsto y_1} \tag{2}$$

as given in (1). Similarly

$$\nabla_{y_1 \leadsto x_2} = \nabla_{x_1 \leadsto y_2}, \qquad \nabla_{y_2 \leadsto x_1} = \nabla_{x_2 \leadsto y_1}, \qquad \nabla_{y_1 \leadsto x_1} = \nabla_{x_2 \leadsto y_2}$$

This illustrates the general result (19) given later: *If $\mathcal{P}$ and $\mathcal{P}'$ are logically equivalent bivariate predictions, then $\nabla_{\mathcal{P}} = \nabla_{\mathcal{P}'}$.*

This equality should not obscure the fact that $\nabla_{\mathcal{P}}$ is designed to evaluate specific *a priori* predictions in which the variables' roles—independent and dependent—have been assigned. Treating one variable as predictor and the other as predicted appears essential to the concept of error reduction.

Numerical Properties

We now turn to the numerical properties of $\nabla_{x_1 \leadsto y_1}$. Rewriting (1) as

$$\nabla_{x_1 \leadsto y_1} = \frac{P_{2.}P_{.1} - P_{21}}{P_{2.}P_{.1}} \tag{1'}$$

we see that $\nabla_{x_1 \leadsto y_1}$ is defined mathematically unless its denominator is zero. This happens if $P_{.1} = 0$ so that no x_1 cases occur and only tautological predictions are made, or if $P_2 = 0$ so that y_1 always occurs, hence X is useless for predicting Y in the population regardless of the proposition. If the measure is defined, we always have $-\infty < \nabla_{\mathcal{P}} \leq 1$; $\nabla_{\mathcal{P}}$ equals one if there are no rule K errors ($P_{21} = 0$). It is near one if predicting $Y = y_1$ for the $NP_{1.}$ cases having $X = x_1$ eliminates virtually all the errors that would be made by the same prediction for an equal number of cases selected at random. This requires that $P_{21}/(P_2 P_{.1})$, or equivalently $P_{y_2|x_1}/P_{y_2}$, be small. The $\nabla_{\mathcal{P}}$ value of .588 for the $\bar{c} \rightsquigarrow \bar{s}$ example indicates that nearly 59% of the errors made under random application of the prediction rule are eliminated when we make predictions of each system's stability knowing whether or not it is incompatible ($\bar{c}$). On the other hand $\nabla_{\mathcal{P}}$ equals zero if $P_{21} = P_2 P_{.1}$, that is, if the error cell probability satisfies statistical independence.

Interpretation of Negative Values

The $\nabla_{\mathcal{P}}$ measure is negative if more errors for the prediction are observed than would be expected under statistical independence. For the prediction $x_1 \rightsquigarrow y_1$, $\nabla_{\mathcal{P}}$ is negative if $P_{y_2|x_1} > P_{y_2}$; if so, then a $\nabla_{\mathcal{P}}$ value of $-.5$ means that using the information on the X-state with the rule $x_1 \rightsquigarrow y_1$ *increases* the error rate by 50% over random use of the $Y = y_1$ prediction. Obviously if added information about X gives less accurate prediction than random prediction, then the proposition $x_1 \rightsquigarrow y_1$ is a prediction failure.

It is not surprising that $\nabla_{\mathcal{P}}$ can be negative, given that it evaluates a prediction stated *a priori*. We have previously seen in Section 2.6.1 that negative values of Q may be interpreted as proportionate *increases* in error committed by *a priori* predictions.[2] Measures that are always positive, such as R^2 for ordinary linear regressions (discussed in Section 3.2.7) and τ, are based on *ex post* predictions that can never do worse than their corresponding null model. In contrast, an *a priori* prediction can commit greater error than the comparison prediction. Consequently a measure of prediction success for *a priori* propositions should be able to assume negative values. Of course in seeking a prediction to describe data *ex post* the analyst should not select $\mathcal{P}$ if $\nabla_{\mathcal{P}} \leq 0$.

---

[2]A similar comment applies to the measures of association for ordinal variables, γ and d_{yx}. Costner (1965) applies *ex post* prediction rules in interpreting γ. Somers (1968) has taken a similar approach in interpreting his d_{yx}. However, as shown in Section 5.3, both measures can be interpreted on the basis of *a priori* predictions. Negative values of the measures indicate proportionate increases in prediction errors over the benchmark prediction. They also indicate that an "opposite" prediction will be successful.

Historical Context

Having motivated $\nabla_{x \sim \to y}$ and discussed its properties, we close this section by providing historical context. Although we have emphasized differences between our measures and others, $\nabla_{x \sim \to y}$ was actually proposed by Quetelet about 1850 (Goodman and Kruskal, 1959, p. 133). Also, $\nabla_{x \sim \to y}$ is identical to Loevinger's (1947) coefficient of homogeneity when it is applied to the special case of a two-item "scale." Other measures that can be given $\nabla_{x \sim \to y}$ interpretations include Köppen's (1870) measure, the Glass-Rogoff-Carlsson social mobility index (Glass, 1954; Rogoff, 1953; Carlsson, 1958), Schreier's normalized measure of causal strength (1957, pp. 124ff.), Durbin's I (1955), Yasuda's Y_{ii} (1964), and Benini's measure (1928, cited by Goodman and Kruskal, 1959, p. 134).[3] These earlier measures, however, were not generalized beyond the simplest case, $\nabla_{x \sim \to y}$, although, as shown later, other measures of association can be interpreted as special instances of $\nabla_{\mathcal{P}}$ for particular predictions. In the next section we begin the generalization of $\nabla_{\mathcal{P}}$ by considering biconditional predictions for 2×2 tables.

3.1.2 A Measure for Predictions Identifying Two Diagonally Opposed Error Cells in a 2×2 Table: $\nabla_{x \leftarrow \sim \to y}$

We now develop $\nabla_{\mathcal{P}}$ for predictions identifying two diagonally opposed error cells in the 2×2 table. The biconditional statement $x \leftrightarrow y$ in formal logic corresponds to the prediction logic statement $(x \sim \to y)$ & $(\bar{x} \sim \to \bar{y})$, or more simply, $x \leftarrow \sim \to y$. This is the translation into prediction logic of statements such as "y tends to occur if and only if x is present."

Weighted Average Development

The prediction $x \leftarrow \sim \to y$ or, as written in alternative notation, $x_1 \leftarrow \sim \to y_1$, identifies (y_2, x_1) and (y_1, x_2) as error cells. Thus the error set is the union of the error sets for $x_1 \sim \to y_1$ and $x_2 \sim \to y_2$. For both of these propositions (1) gives an appropriate $\nabla_{\mathcal{P}}$ measure. Intuitively one would like to represent $\nabla_{x_1 \leftarrow \sim \to y_1}$ as some weighted combination of the $\nabla_{\mathcal{P}}$ values for the two single-error-cell propositions. One can suggest that the weights should reflect the scope and precision of the two predictions, that is, how much error there is to reduce when one switches from the benchmark or comparison rule U probability structure to the actual probabilities. The weights, then, are the rule U probabilities for the error cells. Weighting in

---

[3]Some of the measures are *ex post* selections of one of the four single-error cell ∇'s for a 2×2 table. Benini's measure is more commonly known as ϕ / ϕ_{max}.

this manner and using (1), we have

$$
\nabla_{x_1 \longleftrightarrow y_1} = \frac{(x_1 \rightsquigarrow y_1 \text{ rule } U \text{ error}) \times \nabla_{x_1 \rightsquigarrow y_1} + (x_2 \rightsquigarrow y_2 \text{ rule } U \text{ error}) \times \nabla_{x_2 \rightsquigarrow y_2}}{(x_1 \rightsquigarrow y_1 \text{ rule } U \text{ error}) + (x_2 \rightsquigarrow y_2 \text{ rule } U \text{ error})}
$$

$$
= \frac{P_{2.}P_{.1}\nabla_{x_1 \rightsquigarrow y_1} + P_{1.}P_{.2}\nabla_{x_2 \rightsquigarrow y_2}}{P_{2.}P_{.1} + P_{1.}P_{.2}}
$$

$$
= 1 - \frac{P_{12} + P_{21}}{P_{1.}P_{.2} + P_{2.}P_{.1}} \tag{3}
$$

As a weighted average $\nabla_{x_1 \longleftrightarrow y_1}$ has the property that its value lies between the values of the ∇ measures for the component predictions $x_1 \rightsquigarrow y_1$ and $x_2 \rightsquigarrow y_2$. This measure is equivalent to Cohen's kappa (Cohen, 1960; Fleiss, 1973) applied to the 2×2 table.

PRE *Development*

Alternatively one can derive $\nabla_{x_1 \longleftrightarrow y_1}$ directly by applying the PRE procedures. This involves finding the error rate for $x_1 \longleftrightarrow y_1$ under rules K and U.

Under rule K, one is given each observation's actual X value. For each of the $NP_{.1}$ cases having $X = x_1$, predict $Y = y_1$. In this situation the argument used earlier for $x_1 \rightsquigarrow y_1$ shows there are NP_{21} prediction errors. Similarly for the $NP_{.2}$ cases having $X = x_2$, predict y_2. Here one makes NP_{12} prediction errors. Thus, there are $N(P_{12} + P_{21})$ total rule K errors.

Under rule U one does not know each observation's location on the independent variable X. Predict y_1 for $NP_{.1}$ randomly selected cases. Earlier we showed that this results in $NP_{2.}P_{.1}$ expected errors. Similarly, predicting y_2 for $NP_{.2}$ randomly selected cases results in $NP_{1.}P_{.2}$ expected errors. The total is thus $N(P_{2.}P_{.1} + P_{1.}P_{.2})$ rule U errors. Then define the PRE measure

$$
\nabla_{x_1 \longleftrightarrow y_1} = 1 - \frac{P_{12} + P_{21}}{P_{1.}P_{.2} + P_{2.}P_{.1}}
$$

Thus $\nabla_{x_1 \longleftrightarrow y_1}$ measures the proportionate reduction in error achieved by the prediction $x_1 \longleftrightarrow y_1$ given knowledge of each observation's location on the independent variable over the expected errors when the prediction is randomly applied according to the marginal distribution on the *independent* variable. Alternatively the measure $\nabla_{x_1 \longleftrightarrow y_1}$ represents the proportionate reduction in the number of cases in the pair of error cells when we shift from statistical independence with the population marginals to the actual probability structure. Note that treating Y rather than X as the independent variable, a logically equivalent proposition is $y_1 \longleftrightarrow x_1$. The measures for these equivalent predictions are equal: $\nabla_{y_1 \longleftrightarrow x_1} = \nabla_{x_1 \longleftrightarrow y_1}$.

Numerical Properties

The basic characteristics of $\nabla_{\mathcal{GP}}$ for two error cells in a 2×2 table parallel those for one error cell. The denominator is zero and thus the measure is undefined only when all the probability is concentrated in one of the nonerror cells. Otherwise $-\infty < \nabla_{x_1 \longleftrightarrow y_1} \leqslant 1$; equality holds only when prediction is perfect ($P_{12} + P_{21} = 0$). $\nabla_{x_1 \longleftrightarrow y_1}$ is zero if the X and Y variables are statistically independent; again, as a special property of 2×2 tables, the converse is true. Once again if the prediction $x_1 \longleftrightarrow y_1$ does worse than the benchmark prediction (that is, if most of the probability is concentrated in the error cells), then $\nabla_{x_1 \longleftrightarrow y_1} < 0$. If this is the case, then the other biconditional prediction $x_1 \longleftrightarrow y_2$ is supported by the probability structure of the population. Obviously the measure for the prediction $x_1 \longleftrightarrow y_2$ in a 2×2 table is

$$\nabla_{x_1 \longleftrightarrow y_2} = 1 - \frac{P_{11} + P_{22}}{P_1.P_{.1} + P_2.P_{.2}} \tag{4}$$

Two Research Examples

We now consider two research examples of $\nabla_{\mathcal{GP}}$ for biconditional predictions. In Chapter 1 we cited Rapoport's (1970) proposition that in *three*-person games the grand coalition (the coalition of all the players) tends to form if and only if the game is strictly superadditive.[4]

Rapoport bases his proposition on data from an experiment by Maschler (1965). In all, Maschler ran 78 three-person games and 45 four-person games. Of these, the grand coalition was "permissible" (and therefore had an official worth greater than zero) in 40 games. Broadening the domain of Rapoport's proposition, we predict for these 40 three- and

---

[4]An *n*-person game is strictly superadditive if and only if the value v of each coalition of players exceeds the sum of the values of disjoint subsets, R and S, of players forming the coalition: $v(R \cup S) > v(R) + v(S)$, for all R and S. The value of a coalition is its security level or the maximum reward it can guarantee itself regardless of how the other players behave.

four-person games that the grand coalition of all players tends to form if and only if the game is strictly superadditive. Maschler's data shown in Table 3.1 strongly support the proposition $g \longleftrightarrow s$: $\nabla_{g \longleftrightarrow s} = 1 - (.125/.475) = .737$.

Table 3.1   The grand coalition tends to form if and only if the game is strictly superadditive

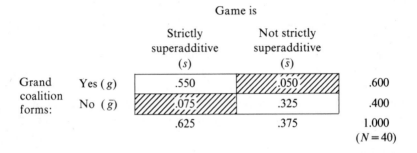

		Game is		
		Strictly superadditive (s)	Not strictly superadditive ($\bar{s}$)	
Grand coalition forms:	Yes (g)	.550	.050	.600
	No ($\bar{g}$)	.075	.325	.400
		.625	.375	1.000
				($N = 40$)

It is instructive to consider a somewhat less strongly supported bicondi-tional propostion. In his study of work organizations in preindustrial societies, Udy (1967, p. 44) asserts:

Tillage, construction, animal husbandry, and manufacturing tend to be carried on by permanent organizations; hunting, fishing, and collection, by temporary organizations.

Taken in context, the words *carried on by* may clearly be replaced by *sufficient for*. Thus this proposition may be stated as $x_1 \longleftrightarrow y_1$ for Table 3.2. The data in the table support Udy's proposition: $\nabla_{x_1 \longleftrightarrow y_1} = +.565$.

The proposition $x_1 \longleftrightarrow y_1$ may be decomposed into the components $x_1 \leadsto y_1$ and $x_2 \leadsto y_2$. These components receive differential support from Udy's data: $\nabla_{x_1 \leadsto y_1} = +.870$, $\nabla_{x_2 \leadsto y_2} = +.418$. Combining these compo-nent measures through the weighted average expression (3) for $\nabla_{x_1 \longleftrightarrow y_1}$ shows:

$$\nabla_{x_1 \longleftrightarrow y_1} = \frac{1}{P_{1.}P_{.2} + P_{2.}P_{.1}} \left(P_{2.}P_{.1}\nabla_{x_1 \leadsto y_1} + P_{1.}P_{.2}\nabla_{x_2 \leadsto y_2} \right)$$

$$= \frac{1}{.478} \left[.155(.870) + .323(.418) \right] = +.565$$

Table 3.2   Nature of work performed by type of organizations in preindustrial societies $(N = 149)^{a,b}$

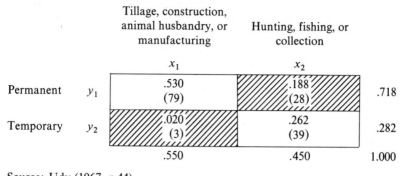

		Tillage, construction, animal husbandry, or manufacturing	Hunting, fishing, or collection	
		x_1	x_2	
Permanent	y_1	.530 (79)	.188 (28)	.718
Temporary	y_2	.020 (3)	.262 (39)	.282
		.550	.450	1.000

Source: Udy (1967, p.44).
[a] Number of cases indicated in parenthesis.
[b] Shading indicates error cells.

Thus although augmenting the scope of the prediction, the second component detracts from Udy's overall prediction success.

Udy reports Yule's $Q = +.95$ for the distribution in Table 3.2. It is easy to show that Yule's Q is a $\nabla_\wp$ measure. One can verify that equation (4) of Chapter 2 is a special case of (4), when (4) is applied to the probability structure indicated in Table 2.5. Consequently *for untied observation pairs*:

$$\nabla_{(x_1,x_2)\longleftrightarrow(y_1,y_2)} = Q \qquad (5)$$

Thus Q is based on a biconditional degree-2 prediction. Since Udy's proposition makes predictions for single organizations, not pairs, we submit that $x_1 \longleftrightarrow y_1$ more faithfully expresses Udy's proposition than $(x_1,x_2) \longleftrightarrow (y_1,y_2)$. Thus $\nabla_{x_1 \longleftrightarrow y_1}$ is a more appropriate measure for Udy's proposition than is Q.

3.1.3   Remaining 2 × 2 cases

In formal logic there are 16 possible truth tables for two dichotomous variables. For each of these tables there is a corresponding statement in prediction logic. In the preceding two sections we developed the measure for the six logically distinct statements that should represent most meaningful scientific theories for two dichotomous variables. For completeness we now give $\nabla_\wp$ for the remaining 10 statements relating two dichotomies.

(a)   Two error cells in the same column

This situation is illustrated by $\mathcal{P}$: $[x_1 \rightsquigarrow \emptyset]$ & $[x_2 \rightsquigarrow (y_1 \text{ or } y_2)]$. (The statement $x_1 \rightsquigarrow \emptyset$ signifies that $\mathbb{S}(x_1)$ is empty; hence, every Y-state is an error for x_1.) Following procedures similar to those used in the preceding sections, we find that if $P_{.1} \neq 0$ then

$$\nabla_{\mathcal{P}} = 1 - \frac{P_{11} + P_{21}}{(P_{1.} + P_{2.})P_{.1}} = 1 - \frac{P_{.1}}{P_{.1}} = 0 \tag{6}$$

Note, once more, that if $\nabla_{\mathcal{P}}$ is not defined, the probability of error under rule U is zero. By analogous developments we can show that $\nabla_{\mathcal{P}} = 0$ or is undefined also for the next two situations:

(b)   Two error cells in the same row

(c)   Four error cells

It is appropriate that $\nabla_{\mathcal{P}}$ is always zero if there are four error cells or two error cells concentrated in either one column or one row. In either case the undifferentiated character of the predictions implies that we can expect no advantage or disadvantage from knowledge of X.

(d)   Three error cells

Four distinct propositions will generate three error cells. We use $[x_1 \rightsquigarrow y_1]$ & $[x_2 \rightsquigarrow \emptyset]$ to illustrate the measure:

$$\nabla_{\mathcal{P}} = 1 - \frac{P_{21} + P_{12} + P_{22}}{P_{2.}P_{.1} + P_{1.}P_{.2} + P_{2.}P_{.2}} \tag{7a}$$

$$= \frac{P_{2.}P_{.1} - P_{21}}{P_{2.}P_{.1} + P_{.2}} \tag{7b}$$

$$= \frac{P_{1.}P_{.2} - P_{12}}{P_{1.}P_{.2} + P_{2.}} \tag{7c}$$

This is defined if either $P_{.2}$ or $P_{2.}$ is nonzero. From (7b) [(7c)] it is seen that the probability $P_{.2}[P_{2.}]$ of the column [row] that contains only error cells does not affect the sign of $\nabla_{\mathcal{P}}$ but only its magnitude. The sign of $\nabla_{\mathcal{P}}$ is determined by whether the cell $(2,1)$ [or alternatively the cell $(1,2)$] has fewer or greater cases than would be expected under statistical independence. If a proposition identifying three error cells in a 2×2 table makes no errors then $\nabla_{\mathcal{P}}$ is undefined since every error cell must have a zero marginal; otherwise it can be shown that $\nabla_{\mathcal{P}} \leqslant \frac{1}{2}$.

The three- and four-error-cell situations take on more interest when weighted errors and actuarial propositions are introduced in Chapter 4.

(e) No error cells

The last of the 16 possibilities for the 2×2 table involves predictions identifying no error cells: $[x_1 \rightsquigarrow (y_1 \text{ or } y_2)]$ & $[x_2 \rightsquigarrow (y_1 \text{ or } y_2)]$. For this "prediction," the measure is undefined since there are no errors:

$$\nabla_{\mathcal{P}} = 1 - \frac{0}{0} \Rightarrow \nabla_{\mathcal{P}} \quad \text{undefined} \tag{8}$$

That $\nabla_{\mathcal{P}}$ is undefined is quite appropriate for such tautologies. This completes the development of $\nabla_{\mathcal{P}}$ for the special case of 2×2 tables. We next develop the general model for evaluating prediction logic propositions about $R \times C$ tables, where $R \geqslant 2$ and $C \geqslant 2$.

3.2 MEASURING PREDICTION SUCCESS IN THE $R \times C$ TABLE

We first present three examples to illustrate the range of problems that can be addressed by the general bivariate population measure. Section 3.2.2 specifies the domain of admissible bivariate predictions. Section 3.2.3 defines the general $\nabla_{\mathcal{P}}$ model. Later subsections discuss its properties. Until Chapter 6 we continue to treat all data in examples as population data.

3.2.1 Three Research Applications of Prediction Analysis

Inflation and the Money Supply in the Netherlands

The basic prediction of the monetarist theory of inflation is that the rate of inflation primarily reflects the rate of increase of the money supply. In his study of inflation in the Dutch economy during 1955–1973, Korteweg (1974) wished to test the monetarist theory and a number of other "impulse" theories in which a single independent variable is given a dominant role in determining inflation and economic growth.[5] We examine here only the relationship between inflation, as measured by changes in consumer prices, and the money supply, the latter narrowly defined as currency and demand deposits lagged one year. Although all his variables were quantitative, Korteweg initially wished to examine the data without relying on specific functional forms needed for regression models. Therefore a uniform procedure was applied to the data to group each variable

---

[5]For a related $\nabla_{\mathcal{P}}$ analysis of Italian data, see Fratianni (1975). For other countries, see the set of studies to be published in Brunner and Meltzer (1978, forthcoming).

into three categories: increase $(+)$, low or no change (0), decrease $(-)$.[6]
Korteweg's prediction was that $+$ predicts $+$, 0 predicts 0, and $-$ predicts
$-$ or, equivalently, that the data of Table 3.3A would tend to concentrate
on the major diagonal.

Table 3.3   Inflation and the money supply.[a,b]

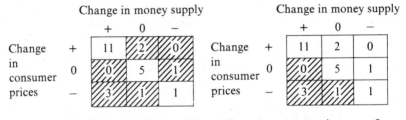

(A) Major Diagonal Prediction (B) Asymmetric Prediction

[a]Error cells (shaded) for two propositions where $+$ represents increase; 0,
small or no change; $-$, decrease.
[b]Numbers are actual numbers of cases for annual data for the Netherlands.
Data provided by Pieter Korteweg.

Like the modal class predictions used in λ (Sec. 2.5.1), the predictions
here predict a unique dependent variable state for each state of the
independent variable. Unlike the λ predictions, however, they are given *a
priori* rather than being determined by the data. Unlike the γ (Sec. 5.3)
predictions, the predictions here are for single events rather than pairs of
events. Moreover the predictions specify a form of relation and not just the
absence of nonrelation as in χ^2.

An alternative prediction to Korteweg's might hold that the effects of
policy variables are not symmetric, since it is easier to have price increases
than decreases. Therefore the asymmetric prediction shown in Table 3.3B
would hold that $+ \leadsto\to +, 0 \leadsto\to (0 \text{ or } +)$, & $(-) \leadsto\to (-, 0 \text{ or } +)$.

In comparison to the major diagonal prediction, this prediction does not
predict a unique dependent variable category for each independent variable category. It is obviously also less precise in that every error event in
Table 3.3B is also an error event in Table 3.3A but not vice versa. The
general $\nabla_{\wp}$ model will account for these differences in precision and permit
comparison of the two models. Since Korteweg's data contained only 24

---

[6]Korteweg identifies the categories' limits *ex post*, and this renders the prediction "*ex priori*"
as discussed in Section 2.5. For illustrative purposes we treat the prediction as *a priori*.

observations, we eventually want to be concerned with inferential questions treated in Chapter 6. In this chapter, we treat these 24 observations as a population.

Political Strategies in Illinois

The next example analyzes game-theoretic propositions predicting the number of candidates nominated in each district by the Democrats and Republicans in the Illinois General Assembly elections for the period 1902–1954. (See Sawyer and MacRae, 1962.)

Each district elects three representatives. The problem facing both parties is to decide whether to nominate one, two, or three candidates. Each player (party) thus has three strategies. Assuming that each party's votes are distributed evenly across its candidates, we can calculate the number of seats won by each party once we know how many candidates each party fields and the division of the vote. For example, if each party nominates two candidates and the Democrats win 46% of the vote in a district, as shown in Table 3.4, then the Democrats win one seat and the Republicans win two.

Table 3.4 Number of legislative seats won by each of two parties in a district, given the numbers of candidates for each party and that the Democrats win 46% of the two-party vote[a]

Number of Republican candidates

		1 (54%)	2 (27%)	3 (18%)
Number of Democratic candidates	1 (46%)	1, 1	(1, 2)	1, 2
	2 (23%)	2, 1	(1, 2)	2, 1
	3 (15 1/3%)	2, 1	1, 2	0, 3

[a]Parentheses enclose the percent of the two-party vote won by each of a party's candidates, assuming that they win equal shares of the party's votes. The candidates with the greatest percentages are elected. For example, the entry (2, 1) in the cell lying in row 2 and column 3 indicates that if the Democrats field 2 candidates in the district and the Republicans field 3, then the Democrats win 2 seats because 23% > 18%. The Republicans then win the remaining seat. The circled cells identify the equilibrium pairs.

In the theory of games, players' strategies are said to be in equilibrium if no player is motivated to change his strategy unilaterally (see Luce and Raiffa, 1957, p. 62). Assume that each party acts as if it forecasts the vote accurately and expects the other to do the same and that each party seeks to maximize the number of seats it wins in the district. For the situation shown in Table 3.4 consider the strategy pair in which each party chooses to present three candidates. As shown in the bottom right-hand cell, this would result in the Republicans winning all three seats. This strategy pair is not in equilibrium since the Democrats could have improved their outcome unilaterally by fielding, instead, two candidates. The two strategy pairs selecting outcomes that are circled in Table 3.4 are equilibrium pairs since neither party has a strict motivation to change its own strategy unilaterally.

This game-theoretic model predicts that the parties tend to choose strategies that are in equilibrium. This model uses the two-party vote split in the district to predict the number of candidates for each party. For the example of Table 3.4, then, the model may be expressed as the proposition "If the Democrats win 46% of the two-party vote, predict that there are one or two Democratic candidates and two Republican candidates." Thus the independent variable is the two-party division of the district vote and the dependent variable is the set of alternative pairs of strategies for the two parties. Suppose we simplify matters by assuming any tie is resolved in favor of the Democrats. Then we may condense the independent variable by using the fact that as the Democrats' share of the two-party vote increases from 0% to 100%, just six distinct patterns of seat allocations result, one for each of the intervals shown as column headings in Table 3.5. (Since the model makes the same predictions for each vote division within any of these intervals, this variable transformation produces logically equivalent results.) Each independent variable category but the last does not include its upper boundary. The dependent variable is the set of nine strategy *pairs* labeling the rows of Table 3.5. The double-hatched cells of this table are the error cells for the game-theoretic model. Note that the situation of Table 3.4 belongs in the third column of Table 3.5.

There is more than one equilibrium strategy pair identified by game theory in each of the six types of district; that is, at least two cells in each column of Table 3.5 are identified as a prediction success for the equilibrium strategy proposition. Sawyer and MacRae introduce additional considerations, such as hedging against an error by the opposing party or against predicting the wrong vote division, to select just one equilibrium strategy pair in each type of district. They predict that the parties tend to select this single equilibrium pair in each type of district. This proposition identifies as error events all cells that are single-*or* double-hatched in Table 3.5.

Table 3.5   Number of candidates nominated by two parties in six types of election districts, Illinois General Assembly elections, 1902–1954[a]

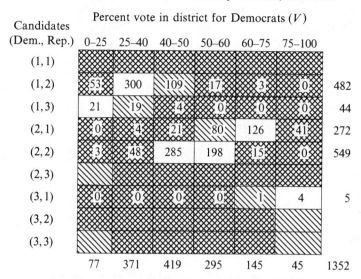

Candidates (Dem., Rep.)	Percent vote in district for Democrats (V)						
	0–25	25–40	40–50	50–60	60–75	75–100	
(1,1)							
(1,2)	53	300	109	17	3	0	482
(1,3)	21	19	4	0	0	0	44
(2,1)	0	4	21	80	126	41	272
(2,2)	3	48	285	198	15	0	549
(2,3)							
(3,1)	0	0	0	0	1	4	5
(3,2)							
(3,3)							
	77	371	419	295	145	45	1352

Source: Data are adapted from Sawyer and MacRae (1962, p. 938). These investigators omitted "…24 elections (many of them involving strong third parties) which display a nomination pattern other than one of …" the five rows containing data above (*ibid.*, p. 940). Due to rounding, we lost one of Sawyer and MacRae's 1353 observations; this has negligible effects on the analysis.

[a]Two error structures are shown in the table. Double-hatching indicates error cells for the prediction that the strategies chosen by the two parties form a pure strategy equilibrium pair. All shaded cells constitute error cells for the Sawyer and MacRae predictions.

Both the Sawyer-MacRae predictions and the game-theoretic predictions can be considered to involve three rather than two variables. These are D, the number of Democratic candidates; R, the number of Republican candidates (R is not the number of rows here); and V, the vote percentage. However no separate predictions are made for D and R. Instead equilibrium *pairs* are predicted. Since the dependent variable is $D \times R$, the problem is bivariate. Moreover, although D and R are both ratio variables, $D \times R$ is a nominal variable.

Still different theories provide alternative treatment of D, R, and V. An alternative to game theory's rational player model is posed by a model that says that one party is naive and gives advance notice of its nomination to the second party. The second party then chooses its number of candidates to maximize its seats, conditional on the vote percentage and the first party's nominations. If the Democrats are the naive party, then R is the dependent variable and $D \times V$ is the independent variable. Returning to the example of Table 3.4, if the Democrats naively and publicly commit themselves in advance to fielding three candidates, then the number of seats won by the Republicans equals the number of Republican candidates. Obviously the Republicans can be predicted in this case to present three candidates for election. The resulting error structure for this model and the relevant data are presented in Table 3.6. On the other hand, if the Republicans are naive, then D is dependent, $R \times V$ independent. The foregoing discussion illustrates that neither the manner of data collection nor the intuitive manner of categorization completely dictates the number of analytical variables. It is the theoretical structure of the analysis that determines whether a problem is bivariate. Similarly, this example shows that which variable is dependent is a function of the theory: the propositions considered above use the observed two-party vote in the election outcome to predict the preelection strategy.

Table 3.6   Illinois candidate nominations, model with Democrats naive, Republicans maximizing[a]

								V											
		0–25			25–40			40–50			50–60			60–75			75–100		
D		1	2	3	1	2	3	1	2	3	1	2	3	1	2	3	1	2	3
	1	0	0	0	0	4	0	0	21	0	0	80	0	0	126	1	0	41	4
R	2	53	3	0	300	48	0	109	285	0	17	198	0	3	15	0	0	0	0
	3	21	0	0	19	0	0	4	0	0	0	0	0	0	0	0	0	0	0

[a] V is Democratic vote percentage, D is number of Democratic candidates, and R is number of Republican candidates.

Coalition Formation in the Triad

Distinct theories also need to be compared when laboratory study data are used to evaluate the prediction success attained by Caplow's (1959) model of coalition formation in three-person situations.[7]

---

[7] We thank Richard J. Morrison for suggesting this example. See R. Morrison (1974a) for a critique of Caplow's model and a formalization that subsumes as special cases both Caplow's model and others' extensions of the latter.

Caplow's independent variable is based on a concept called the *strength* of each player in a three-person coalition game. As shown in the row labels of Table 3.7, each category of this independent variable identifies a distinct *type* of distribution of "strength" across the three players: in type t_2, for example, player A has more "strength" than either of the other (equally "weak") players, B and C (written $S_A > S_B = S_C$), but has less than their combined "strength" ($S_A < S_B + S_C$), and thus A will lose if the coalition (BC) forms. Caplow develops three distinct models that use this independent variable to predict which coalitions form. The domain of each of these models is delimited to one of three types of social situations described by Caplow as "continuous," "episodic," and "terminal." Since Caplow does not operationalize these concepts, it is not clear which model should be applied to a given three-person game. We use the continuous

Table 3.7 Error structure (shaded) for coalition predictions, Caplow's (1959) model for the "continuous situation," where the measurement of the dependent (column) variable depends on the row state

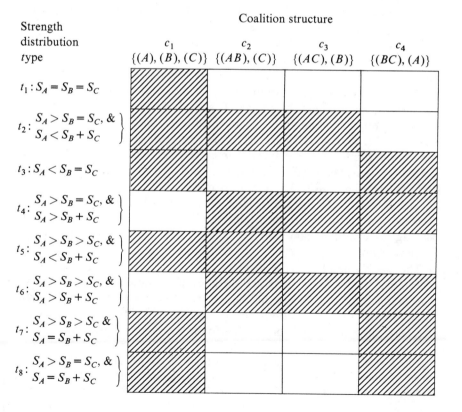

Strength distribution *type*	Coalition structure			
	c_1 $\{(A),(B),(C)\}$	c_2 $\{(AB),(C)\}$	c_3 $\{(AC),(B)\}$	c_4 $\{(BC),(A)\}$
$t_1: S_A = S_B = S_C$	▨			
$t_2: \begin{array}{l} S_A > S_B = S_C, \& \\ S_A < S_B + S_C \end{array}$	▨		▨	
$t_3: S_A < S_B = S_C$	▨			▨
$t_4: \begin{array}{l} S_A > S_B = S_C, \& \\ S_A > S_B + S_C \end{array}$		▨	▨	
$t_5: \begin{array}{l} S_A > S_B > S_C, \& \\ S_A < S_B + S_C \end{array}$	▨			
$t_6: \begin{array}{l} S_A > S_B > S_C, \& \\ S_A > S_B + S_C \end{array}$		▨	▨	
$t_7: \begin{array}{l} S_A > S_B > S_C \& \\ S_A = S_B + S_C \end{array}$	▨			▨
$t_8: \begin{array}{l} S_A > S_B = S_C, \& \\ S_A = S_B + S_C \end{array}$	▨			▨

and episodic predictions, which seem more likely to apply to the game we analyze. We shall treat Caplow's predictions as conditional on either a two-person coalition forming or no coalition forming. Consequently three-person coalitions will be eliminated from both the data and the description of the coalition variable.

Caplow's predictions for the continuous situation identify the error structure shown in Table 3.7. This 8×4 matrix, though, violates an important prerequisite for applying the $\nabla_{\mathscr{P}}$ measure: we cannot name the coalition structure until we know the distribution of "strength." That is, we cannot measure the dependent variable without knowing the observation's location on the independent variable. For example, consider $t_2 \rightsquigarrow c_4$, the prediction that players B and C form a coalition against player A. Since A is so named because he is the strongest of the three, he cannot be identified if we do not know the strength distribution.

The definition of $\nabla_{\mathscr{P}}$ requires independent measurement. We can satisfy this requirement by giving the players names (say, 1, 2, and 3) that do not depend on their strength. The predictions of Caplow's models can also be represented correctly in this alternative naming convention, although we must increase the size of the table to do so. To avoid unnecessary complexity, let us omit strength distribution types 7 and 8, since the data we are examining exclude these types (our omission has no effect on $\nabla_{\mathscr{P}}$ since these types have no predictive significance for the population). Some additional condensation is also possible. For example, the players' names have no predictive significance in types 1, 4, and 6; hence a single category suffices for each of these three types. The resulting error cells for the model of the continuous situation, using independently measured variables, are single- or double-hatched in Table 3.8. The row categories identify applicable permutations of the players' names. For example, the permutation CBA indicates that player 1 is player C in Caplow's naming convention, and players 2 and 3 are B and A, respectively.

The data in Table 3.8 are the coalition outcomes (omitting 32 cases of three-person coalitions) from a laboratory study of 30 male triads, each playing 18 coalition games in sequence (three of each of the first six Caplow types).[8]

The double-hatched cells of Table 3.8 are error events for Caplow's alternative model when the situation is episodic rather than continuous. The predictions of the episodic model are considerably less precise and differentiated than predictions for the continuous situation. As a consequence, many of the distinctions we found necessary to measure predict-

---

[8]We thank W. Edgar Vinacke for supplying his data for our analysis. For a description of the laboratory procedures used, see Vinacke (1962), Cumulative Scores Condition. For illustrative purposes, we treat the 508 games as a population of independent observations.

Table 3.8 Error structures of Caplow's models with independently measured variables, showing coalition outcomes from a laboratory study of 30 male triads

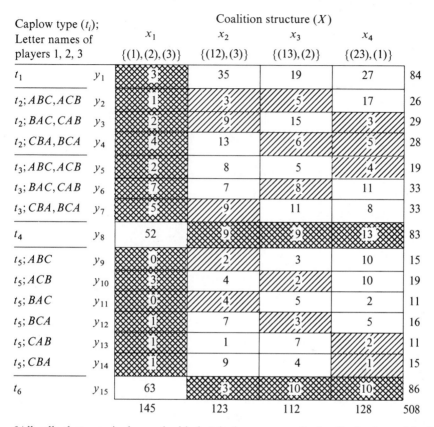

Caplow type (t_i); Letter names of players 1, 2, 3		Coalition structure (X)				
		x_1 {(1),(2),(3)}	x_2 {(12),(3)}	x_3 {(13),(2)}	x_4 {(23),(1)}	
t_1	y_1	3	35	19	27	84
t_2; ABC, ACB	y_2	1	3	5	17	26
t_2; BAC, CAB	y_3	2	9	15	3	29
t_2; CBA, BCA	y_4	4	13	6	5	28
t_3; ABC, ACB	y_5	2	8	5	4	19
t_3; BAC, CAB	y_6	7	7	8	11	33
t_3; CBA, BCA	y_7	5	9	11	8	33
t_4	y_8	52	9	9	13	83
t_5; ABC	y_9	0	2	3	10	15
t_5; ACB	y_{10}	3	4	2	10	19
t_5; BAC	y_{11}	0	4	5	2	11
t_5; BCA	y_{12}	1	7	3	5	16
t_5; CAB	y_{13}	1	1	7	2	11
t_5; CBA	y_{14}	1	9	4	1	15
t_6	y_{15}	63	3	10	10	86
		145	123	112	128	508

[a]All cells that are single- or double-hatched are error cells for Caplow's model of the continuous situation. All double-hatched cells are errors for the episodic model.

ion success under the continuous situation are not needed for the episodic. Specifically the players' names have no predictive significance. Also, the only necessary distinction as to types of strength distribution for our data is between two sets of types, $\{t_1, t_2, t_3, t_5\}$ and $\{t_4, t_6\}$. Thus we can combine variable states to create the 2×2 error structure shown in Table 3.9 without affecting the value of $\nabla_{\mathcal{P}}$.

For these data the predictions of Caplow's episodic situation model may be restated simply in prediction logic as $y_2^* \longleftrightarrow x_1^*$.

We will reanalyze these data and those from the Korteweg and Sawyer-MacRae studies after presenting a general model for the measurement of

Table 3.9   Error structure of Caplow's model for the "episodic situation," showing coalition outcomes from a laboratory study of 30 male triads

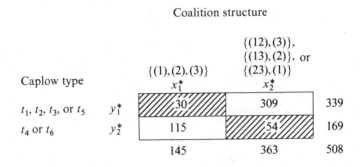

Coalition structure

Caplow type		$\{(1),(2),(3)\}$ x_1^*	$\{(12),(3)\},$ $\{(13),(2)\}$, or $\{(23),(1)\}$ x_2^*	
t_1, t_2, t_3, or t_5	y_1^*	30	309	339
t_4 or t_6	y_2^*	115	54	169
		145	363	508

prediction success in $R \times C$ tables. We begin by specifying the domain of prediction logic propositions.

3.2.2   The Domain of Bivariate Propositions

We wish to develop a general $\nabla_{\mathcal{P}}$ measure for any admissible prediction logic proposition. Section 1.7.4 indicated that, when X is the independent variable, these propositions are stated in the form

$$\mathcal{P} = \{\mathcal{P}_j\} = \{x_j \rightsquigarrow \mathcal{S}(x_j)\} \qquad (9a)$$

where $j \in \{1,\ldots,C\}$ and $\mathcal{S}(x_j) \subseteq \{y_1, y_2, \ldots, y_i, \ldots, y_R\}$. That is, the proposition predicts that if an observation has $X = x_j$, then its Y-state belongs to $\mathcal{S}(x_j)$, a subset of the Y-states. The proposition commits a prediction error for each observation having $X = x_j$ but having a Y-state that does not lie in $\mathcal{S}(x_j)$. For example, in the Sawyer-MacRae game-theoretic model shown in Table 3.5, we have

$$\mathcal{P} : 0 \leqslant V < 25 \rightsquigarrow \{(1,3),(2,3),(3,3)\}$$

$$25 \leqslant V < 40 \rightsquigarrow \{(1,2),(1,3)\}$$

$$40 \leqslant V < 50 \rightsquigarrow \{(1,2),(2,2)\}$$

$$50 \leqslant V < 60 \rightsquigarrow \{(2,1),(2,2)\}$$

$$60 \leqslant V < 75 \rightsquigarrow \{(2,1),(3,1)\}$$

$$75 \leqslant V \leqslant 100 \rightsquigarrow \{(3,1),(3,2),(3,3)\}$$

Note that if $\mathcal{S}(x_j)$ includes all of Y, then no observation having $X = x_j$ can be a prediction error. On the other hand, (9a) also permits $\mathcal{S}(x_j)$ to be empty; in this event all cases having $X = x_j$ are prediction errors.

The proposition stated in (9a) uses X as the independent variable to identify the set of error cells $\mathcal{E}_{\mathcal{P}} = \cup_j \mathcal{E}_{\cdot j}$, where $\mathcal{E}_{\cdot j}$ contains every cell (y_i, x_j) for which y_i is not in $\mathcal{S}(x_j)$. Alternatively, using Y as the independent variable we have predictions stated in the form

$$\mathcal{P}' = \{\mathcal{P}'_{i\cdot}\} = \{y_i \rightsquigarrow \mathcal{S}'(y_i)\} \qquad (9b)$$

where $i \in \{1, \ldots, R\}$ and $\mathcal{S}'(y_i) \subseteq \{x_i, x_2, \ldots, x_j, \ldots, x_C\}$. The proposition $\mathcal{P}'$ identifies the set of error cells $\mathcal{E}_{\mathcal{P}'} = \cup_i \mathcal{E}_{i\cdot}$, where $\mathcal{E}_{i\cdot}$ contains every cell (y_i, x_j) for which x_j does not belong to $\mathcal{S}'(y_i)$. In the Caplow example of Table 3.8, $y_{12} \rightsquigarrow \mathcal{S}'(y_{12}) = \{x_2, x_4\}$ and $\mathcal{E}_{12\cdot} = \{(y_{12}, x_1), (y_{12}, x_3)\}$.

Each logically distinct proposition in the prediction logic identifies a unique set of error cells in $Y \times X$. Thus, we can specify a logically distinct prediction simply by specifying the set of error cells it identifies, $\mathcal{E}_{\mathcal{P}}$. *Any two propositions $\mathcal{P}$ and $\mathcal{P}'$ are logically equivalent if $\mathcal{E}_{\mathcal{P}} = \mathcal{E}_{\mathcal{P}'}$.*

Admissibility

In general, a proposition is admissible for prediction analysis of a population if and only if the number of expected errors (or the probability of error) under rule U is greater than zero. Obviously a proposition is never admissible if it has no error cells: $\mathcal{E}_{\mathcal{P}} = \varnothing$. Such propositions cannot be rejected by scientific observation and do not merit consideration. As to the remaining $2^{RC} - 1$ propositions, their admissibility depends on the population marginal distributions, as we have seen for the 2×2 case.

> Given the definition of $\nabla_{\mathcal{P}}$ for the general $R \times C$ case in the following section, a proposition is not admissible for prediction analysis of a population if and only if for every cell belonging to $\mathcal{E}_{\mathcal{P}}$, either $P_{i\cdot}$ or $P_{\cdot j} = 0$. $\qquad (10)$

3.2.3   A General Model for Measuring Bivariate Prediction Success: $\nabla_{\mathcal{P}}$

This section provides a general model defining a unique measure of prediction success for each logically distinct, admissible proposition. Measures presented in Section 3.1 for the 2×2 table represent particular applications of this general model.

Presentation of the model is facilitated by defining ω_{ij} as the error cell indicator: $\omega_{ij} = 1$ if $(y_i, x_j) \in \mathcal{E}_{\mathcal{P}}$, and, otherwise, $\omega_{ij} = 0$. For example, in the major diagonal prediction of inflation rates, $\omega_{ii} = 0$ and $\omega_{ij} = 1$ for all $i \neq j$, where i, $j \in \{1,2,3\}$. For the alternative prediction with the triangular pattern of error cells shown in Table 3.3B, $\omega_{21} = \omega_{31} = \omega_{32} = 1$, and all other $\omega_{ij} = 0$.

Following the design criteria of Section 2.1, we use the proportionate reduction in error model to define $\nabla_{\mathcal{P}}$.

Under rule K, knowing each observation's X-state we use $\mathcal{P}_{.j}$ for the $NP_{.j}$ cases having $X = x_j$ and err with the *conditional* probability $\sum_i \omega_{ij}(P_{ij}/P_{.j})$. Summing across the states of the independent variable, X, we commit a total of

$$NK_{\mathcal{P}} = N \sum_j P_{.j} \sum_i \omega_{ij}(P_{ij}/P_{.j}) = N \sum_i \sum_j \omega_{ij} P_{ij}$$

rule K errors. In the asymetric inflation proposition shown in Table 3.3B, there are $24[0/24 + 3/24 + 1/24] = 4$ rule K errors.

Under the benchmark, rule U, we do not know the X-state of each observation. Rather than predicting $S(x_j)$ for the $NP_{.j}$ cases with $X = x_j$, we predict $S(x_j)$ for $NP_{.j}$ *randomly drawn observations* and err with the unconditional probability $\sum_i \omega_{ij} P_{i.}$. Summing across the various $S(x_j)$ predictions, we expect to commit

$$NU_{\mathcal{P}} = N \sum_i \sum_j \omega_{ij} P_{i.} P_{.j}$$

errors under rule U. For Table 3.3B, these amount to $24[(14/24)(6/24 + 5/24) + (8/24)(5/24)] = 8.083$ expected errors.

Of the two sets of marginal probabilities that enter the rule U error rate, *only the $P_{.j}$ or independent variable probabilities need be known in formulating the prediction rule.* The dependent variable probabilities enter only the *ex post* computation of error. Thus rule U *does not require dependent variable information.*

Now define the proportionate-reduction-in-error measure

$$\nabla_{\mathcal{P}} = 1 - \frac{\text{Errors under rule } K}{\text{Errors under rule } U} = 1 - \frac{\displaystyle\sum_i \sum_j \omega_{ij} P_{ij}}{\displaystyle\sum_i \sum_j \omega_{ij} P_{i.} P_{.j}} \tag{11}$$

Equation (11) represents our basic population measure of prediction success for any admissible bivariate proposition, $\mathcal{P}$. $\nabla_{\mathcal{P}}$ measures the PRE attained by the prediction $\mathcal{P}$ given knowledge of each observation's location on the independent variable over that expected when the prediction is randomly applied according to the marginal distribution on the independent variable. Alternatively, $\nabla_{\mathcal{P}}$ measures the proportionate reduction in the number of cases observed in the set of error cells for $\mathcal{P}$ over the number expected under statistical independence, given the marginals of the actual probability structure. For Table 3.3B, $\nabla_{\mathcal{P}} = .505$.

Note that the measures developed for the 2×2 table are particular applications of (11). For example, let $\mathcal{P}$ be the prediction $x_1 \longleftrightarrow y_1$. This proposition identifies the set of error cells, $\mathcal{E}_{x_1 \longleftrightarrow y_1} = \{(y_1, x_2), (y_2, x_1)\}$. Thus $\omega_{12} = \omega_{21} = 1, \omega_{11} = \omega_{22} = 0$. Then for the 2×2 table, (11) yields

$$\nabla_{x_1 \longleftrightarrow y_1} = 1 - \frac{P_{12} + P_{21}}{P_{1.}P_{.2} + P_{2.}P_{.1}}$$

as given in (3).[9]

Evaluating Strictly Inadmissible, Trivial, or Weak Propositions

In general, $-\infty < \nabla_{\mathcal{P}} \leqslant 1$ if the measure is defined. For certain types of propositions $\nabla_{\mathcal{P}}$ must be less than one or be undefined *for any* population. First, a proposition $\mathcal{P}$ is said to be *strictly inadmissible* if it identifies no error event: $\mathcal{E}_{\mathcal{P}} = \varnothing$. If a proposition is strictly inadmissible, then $\nabla_{\mathcal{P}}$ is undefined.

Second, a proposition is said to be *trivial* if it is not strictly inadmissible and if all its error events lie in one or more rows or columns that consist entirely of error events, *and* all its success events lie in one or more rows or columns consisting entirely of success events. It can be shown that every trivial proposition is logically equivalent to a trivial proposition for a 2×2 table. In the 2×2 table the trivial propositions identify error structures consisting of four error cells or just two error cells that comprise an entire column or row. If the proposition $\mathcal{P}$ is trivial, then $\nabla_{\mathcal{P}} = 0$ or is undefined for all populations.

A proposition is said to be *weak* if it is neither strictly inadmissible nor trivial, but all its error cells comprise one or more entire rows or columns of the cross classification. In the case of 2×2 tables, for example, weak propositions identify exactly three error cells. It can be shown that every weak proposition is logically equivalent to a weak proposition for the 2×2 table. If the proposition $\mathcal{P}$ is weak, then $\nabla_{\mathcal{P}}$ must be (strictly) less than one or be undefined for any population.

A proposition is *strong* if it is neither strictly inadmissible, trivial, nor weak. If the proposition $\mathcal{P}$ is strong, then it is possible that $\nabla_{\mathcal{P}}$ can equal one. In this book we focus primarily on strong propositions for which $-\infty < \nabla_{\mathcal{P}} \leqslant 1$.

[9]Cohen's kappa equals $\nabla_{\mathcal{P}}$ for any square table when the proposition $\mathcal{P}$ predicts the main diagonal states: $\{x_i \leadsto y_i\}$ (Cohen, 1960, 1968; Fleiss, 1973).

Accounting for Prediction Scope and Precision

In well-developed fields of scientific inquiry, one may well be less interested in comparing a proposition $\mathcal{P}$ from a new theory to some neutral null model than in comparing the predictive capacity of $\mathcal{P}$ to that of a proposition $\mathcal{P}'$ generated from an alternative theory.

A possible solution would simply be to compare the number of rule K errors ($N\sum_i\sum_j\omega_{ij}P_{ij}$ and $N\sum_i\sum_j\omega'_{ij}P_{ij}$). This is clearly inappropriate, if only because the scope of the two theories may differ. In the inflation example the asymmetric model is guaranteed to be error-free when the money supply is decreased. In contrast, the main diagonal proposition may err for such cases. Therefore the rule K error rates at least ought to be normalized to reflect the total number of predictions being made that conceivably could be in error: that is, the scope of the prediction.

Let us measure the scope of the component prediction $\mathcal{P}_{.j}$ as follows:

$$\text{scope of } \mathcal{P}_{.j} = P_{.j} \text{ if } \mathcal{E}_{.j} \neq \varnothing$$
$$= 0 \text{ otherwise} \tag{12}$$

Then the overall scope of the proposition equals the sum of the scope measures for the component predictions:

$$\text{scope of } \mathcal{P} = \sum_j (\text{scope of } \mathcal{P}_{.j}) \tag{13}$$

Therefore, the scope of a proposition equals the proportion of the population for which non-tautological predictions are made.

Normalizing for the number of nontautological predictions is not enough. Even if the two theories both identify an error event for a given x_j, the precision of the predictions may differ. One theory may predict that any of a large number of Y-states occurs while the other may predict a single Y-state. Again referring to the inflation example, if the money supply remains unchanged, the asymmetric model errs only if prices fall whereas the main diagonal model is also in error when they rise. Let us measure the precision of a component prediction $\mathcal{P}_{.j}$ by the unconditional (marginal) probability that it is in error:

$$\text{precision of } \mathcal{P}_{.j} = \sum_i \omega_{ij} P_{i.} \tag{14}$$

The *overall precision* of the proposition then equals the weighted average of the precisions of the component predictions when each component is weighted by its scope:

$$\text{precision of } \mathcal{P} = \sum_j \left(\text{scope of } \mathcal{P}_{.j}\right)\left(\text{precision of } \mathcal{P}_{.j}\right)$$

$$= \sum_i \sum_j \omega_{ij} P_{i.} P_{.j} \tag{15}$$

$$= U_{\mathcal{P}}$$

Thus we may interpret the rule U error rate as a measure of the prediction's precision. By this interpretation, in $\nabla_{\mathcal{P}} = 1 - K/U$, the rule K error rate for $\mathcal{P}$ is normalized by the measure of $\mathcal{P}$'s precision. This interpretation provides one justification for using the PRE model to define ∇.

Of the many factors that might be considered in comparing predictions, we will emphasize $\nabla_{\mathcal{P}}$ and the measure of precision $U_{\mathcal{P}}$ when summarizing the performance of a proposition. Additional information, such as the PRE achieved in each error cell, also is useful for evaluating a prediction.

The Three Research Examples: Trade Offs in Prediction Evaluation

The ideas of this section may be illustrated with the three research examples of Section 3.2.1. If we compare the asymmetric prediction with the main diagonal prediction in the inflation example or the game-theoretic prediction with the Sawyer-MacRae prediction in the election example or the episodic prediction with the continuous prediction in the coalition example, in each comparison we see that the error set of the first prediction is a subset of the error set of the second prediction. Consequently we are guaranteed to make at *least* as many rule K errors in the second situation. In fact, the Korteweg data of Table 3.3 show four errors for the first proposition, seven for the second. The 418 errors of the Sawyer-MacRae prediction exactly double the 209 errors of the strictly game-theoretic model. Caplow's continuous model has 154 errors as against only 86 for the episodic model. Clearly we cannot compare these error rates directly. Normalizing by rule U errors, as shown in Table 3.10, reduces the disparity in the rule K error rates for the pairs of theories. In all three pairs, however, $\nabla_{\mathcal{P}}$ is greater for the theory with fewer error cells.

For example, one would attribute greater prediction success ($\nabla_{\mathcal{P}} = .61$) to Caplow's episodic model than to the continuous model ($\nabla_{\mathcal{P}} = .50$). It is interesting that, on the basis of piecemeal χ^2 tests, Vinacke (1962, p. 9)

Table 3.10   Comparing pairs of theories

Measure	Inflation		Illinois		Coalitions	
	Upper triangle	Main diagonal	Game theory	Sawyer-MacRae	Episodic	Continuous
rule K error	.167	.291	.155	.309	.169	.302
rule U error	.337	.583	.500	.664	.438	.599
$\nabla_{\mathscr{P}}$	.505	.500	.691	.534	.614	.496
N	24		1352		508	

reached the opposite conclusion that the data better supported the continuous model.[10] Simple χ^2 tests designed to test only whether the variables are unrelated are irrelevant to any analysis for *a priori* prediction logic propositions. The use of inappropriate methods is especially likely to produce conflicting results in data structures as complex as those of Table 3.8. For such tables it appears extremely difficult to compare models on an *ad hoc* basis. Applying the $\nabla_{\mathscr{P}}$ model has the advantage of directly custom designing the evaluation to the predictions of the models.

Sometimes one model has greater precision *and* success: $\mathscr{P}$ clearly dominates $\mathscr{P}'$ in these respects if both $\nabla_{\mathscr{P}} > \nabla_{\mathscr{P}'}$ and $U_{\mathscr{P}} > U_{\mathscr{P}'}$. Often, however, the investigator must make a trade off. For the comparisons in Table 3.10 the theories with the higher $\nabla_{\mathscr{P}}$ values also have less precision. Data such as Vinacke's therefore ought not to be assessed solely on the basis of a $\nabla_{\mathscr{P}}$ value. The continuous model has lower prediction success but greater precision. The sources of success and precision can be identified through a decomposition of prediction success.

3.2.4   Decomposition of Prediction Success

The success and precision of a proposition may be decomposed into elementary components. Let $\mathscr{P}_{ij}$ identify the single error cell (y_i, x_j), and let

$$\nabla_{ij} = 1 - \frac{P_{ij}}{P_{i.}P_{.j}}$$

denote the prediction success of the proposition $\mathscr{P}_{ij}$. Then $\nabla_{\mathscr{P}}$ is a weighted

[10]This is not a criticism of Vinacke, who, of course, had to use methods available circa 1960.

average of the prediction success it achieves in its various error cells:

$$\nabla_{\mathcal{P}} = \sum_i \sum_j \left(\frac{\omega_{ij} P_{i.} P_{.j}}{U_{\mathcal{P}}} \right) \nabla_{ij} \tag{16}$$

where, as before, $\omega_{ij} = 1$ if $(y_i, x_j) \in \mathcal{E}_{\mathcal{P}}$, and, otherwise, $\omega_{ij} = 0$. If $P_{i.} P_{.j} = 0$ then ∇_{ij} is undefined; in this case we can define the product $\omega_{ij} P_{i.} P_{.j} \nabla_{ij} = 0$. Similarly,

$$\nabla_{\mathcal{P}} = \sum_j \left[\frac{\sum_i \omega_{ij} P_{i.} P_{.j}}{U_{\mathcal{P}}} \right] \nabla_{.j} \tag{17}$$

where $\nabla_{.j} = 1 - \sum_i \omega_{ij} P_{ij} / \sum_i \omega_{ij} P_{i.} P_{.j}$. That is, the overall prediction success for proposition $\mathcal{P}$ is a weighted average of the $\nabla_{.j}$ measures for its component propositions, $\mathcal{P}_j : x_j \rightsquigarrow \mathcal{S}(x_j)$. Each weight in this average is the ratio of rule U error for proposition $\mathcal{P}_j$ to $U_{\mathcal{P}}$, the rule U error for the overall proposition $\mathcal{P}$. By (17) the greater the precision of $\mathcal{P}_j$ relative to the precision of the other component predictions, the greater the contribution of $\nabla_{.j}$ to the overall measure of prediction success $\nabla_{\mathcal{P}}$.

To illustrate the importance of the weighted average feature, consider the two probability structures shown in Table 3.11. In both tables assume the proposition is the asymmetric prediction of inflation, which can be represented as $\mathcal{P} : x_1 \rightsquigarrow y_1, x_2 \rightsquigarrow (y_1 \text{ or } y_2)$. In Table 3.11$A$ the measures for the component predictions are $\nabla_{\mathcal{P}_1} = 1 - (.00 + .00)/(.0625 + .0625) = 1.0$ and $\nabla_{\mathcal{P}_2} = 1 - .10/.125 = .20$. The component measures have the same respective values for the probability structure of Table 3.11B. Despite these equalities, there is an important difference between (A) and (B). In (A) the expected errors under each of the two component predictions are exactly one-half of the total rule U errors for the composite prediction $\mathcal{P}$. Hence for (A) $\nabla_{\mathcal{P}} = \frac{1}{2}\nabla_1 + \frac{1}{2}\nabla_2 = +.625$. In contrast, in (B) expected rule U errors for the component prediction $\mathcal{P}_1$ contribute $104/105$ of the total expected rule U errors for the composite prediction $\mathcal{P}$. Hence for (B) $\nabla_{\mathcal{P}} = (104/105)\nabla_1 + (1/105)\nabla_2 = +.992$. Although the component measures are equal in the two parts of the table, the value of the overall measure for (B) is higher since the more successful component prediction is more heavily weighted, in accordance with (17). For a second illustration of the import of (17), refer to Section 3.1.2, where we decomposed the prediction success of $x_1 \leftarrow\!\!\!\rightarrow y_1$ into its components $x_1 \rightsquigarrow y_1$ and $x_2 \rightsquigarrow y_2$ to analyze Udy's population of work organizations in preindustrial societies.

Table 3.11   Sensitivity of $\nabla_{\mathcal{P}}$ to relative precisions of two component predictions

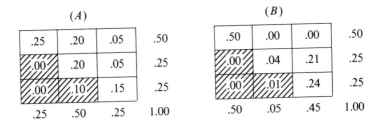

	(A)		
.25	.20	.05	.50
.00	.20	.05	.25
.00	.10	.15	.25
.25	.50	.25	1.00

	(B)		
.50	.00	.00	.50
.00	.04	.21	.25
.00	.01	.24	.25
.50	.05	.45	1.00

Decomposition of Caplow's Continuous Model

Decomposition for larger tables can be illustrated with Vinacke's data for Caplow's continuous model $\mathcal{C}$. The investigator might be interested in how well the model predicts for each of the six types of strength distribution (t_i) shown in Table 3.8. Using (16) to develop the analogous expression to (17) when the rows identify states of the independent variables, we obtain for the data of Table 3.8

$$\nabla_{\mathcal{C}} = \sum_{i=1}^{6} \left(\frac{\text{rule } U \text{ error for type } t_i}{\text{rule } U \text{ error for model } \mathcal{C}} \right) \nabla_{t_i}$$

$$= .080\left(\nabla_{t_1} = .875\right) + .213\left(\nabla_{t_2} = .400\right) + .149\left(\nabla_{t_3} = .210\right)$$

$$+ .199\left(\nabla_{t_4} = .477\right) + .153\left(\nabla_{t_5} = .559\right) + .206\left(\nabla_{t_6} = .626\right)$$

$$= .496$$

Clearly the success of the model varies widely with the type. The individual ∇_{t_i} values may therefore be suggestive as to where the model needs further elaboration. Note that although the predictions for t_1 are most successful, $(\nabla_{t_1} = .875)$, t_1 also contributes least (.080) to overall precision.

Comparing Two Theories with Decomposition Techniques

In comparing two distinct theories it is useful to formulate another corollary of expression (16). Let two propositions $\mathcal{P}$ and $\mathcal{P}'$ have error sets $\mathcal{E}_{\mathcal{P}}$ and $\mathcal{E}_{\mathcal{P}'}$. Let $\mathcal{E}_+$ denote the set of all cells in $\mathcal{E}_{\mathcal{P}'}$ but not in $\mathcal{E}_{\mathcal{P}}$. Similarly let $\mathcal{E}_-$ designate the set of all cells in $\mathcal{E}_{\mathcal{P}}$ but not in $\mathcal{E}_{\mathcal{P}'}$. Let U_+ and U_- be the rule U error rates associated with $\mathcal{E}_+$ and $\mathcal{E}_-$. Then it follows from (16) that

$$\nabla_{\mathcal{P}'} = \frac{1}{U'}\left(U\nabla_{\mathcal{P}} + U_+\nabla_+ - U_-\nabla_- \right) \tag{18}$$

If any U is zero, then we define the product $U\nabla$ to be zero also. When, as in the three examples of Section 3.1.1, all the error cells of $\mathcal{P}$ are also errors for $\mathcal{P}'$, we have

$$\nabla_{\mathcal{P}'} = \frac{1}{U'}(U\nabla_{\mathcal{P}} + U_+\nabla_+)$$

In the Sawyer-MacRae case let $\mathcal{P} = \mathcal{G}$ denote the game theory predictions and $\mathcal{P}' = \mathcal{SM}$ represent the Sawyer-MacRae predictions. We then have

$$\nabla_{\mathcal{SM}} = \frac{1}{.664}\left[.500\nabla_{\mathcal{G}} \quad + 1.64\nabla_+\right]$$

$$= \frac{1}{.664}\left[.500(.691) + .164(.055)\right]$$

$$= .534$$

Since in this case ∇_+ is nearly zero, we can state an unambiguous preference for the purely game-theoretic model. The Korteweg inflation results are quite different. With $\mathcal{A}$ denoting the asymmetric model and $\mathcal{D}$ the diagonal model, we have

$$\nabla_{\mathcal{D}} = \frac{1}{.583}\left[.337\nabla_{\mathcal{A}} \quad + .256\nabla_+\right]$$

$$= \frac{1}{.583}\left[.337(.505) + .256(.474)\right]$$

$$= .500$$

The Vinacke data represent an intermediate case:

$$\nabla_{\mathcal{C}} = \frac{1}{.599}\left[.438\nabla_{\mathcal{E}} \quad + .161\nabla_+\right]$$

$$= \frac{1}{.599}\left[.438(.614) + .161(.177)\right]$$

$$= .416$$

where $\mathcal{C}$ denotes the continuous and $\mathcal{E}$ the episodic version of Caplow's model. The value (.177) of ∇_+ is considerably less than $\nabla_{\mathcal{E}}$ for the episodic model. Consequently $\nabla_{\mathcal{C}} < \nabla_{\mathcal{E}}$; but the additional error cells, $\mathcal{E}_+$, make a substantial contribution to the precision of the continuous model's predictions.

3.2.5 Differentiation

Our illustrations of (18) relate to the problem of adjusting for prediction precision. This is accomplished in the $\nabla_{\mathscr{P}}$ measure through the term $U_{\mathscr{P}}$. In Section 1.6 we defined the differentiation of a proposition as the extent to which it makes different predictions for different states of the independent variable. Differentiation has a more subtle relationship to the measure than does overall precision since rule U errors may or may not increase with differentiation. In Table 3.12, for example, the predictions for (A) and (C) are undifferentiated whereas (B) is differentiated. Yet the rule U error rate for (B) is greater than that for (C) but less than that for (A).

Table 3.12   Relationship of rule U errors to differentiation

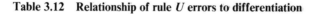

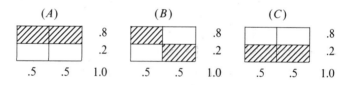

(A)			(B)			(C)		
		.8			.8			.8
		.2			.2			.2
.5	.5	1.0	.5	.5	1.0	.5	.5	1.0

The effect of differentiation on the ∇ measure is apparent when a prediction is totally undifferentiated. In such cases the prediction is trivial, and $\nabla_{\mathscr{P}}$ equals zero if it is defined. Now consider an error structure that has some undifferentiated rows containing only error cells and some differentiated rows that contain both error cells and success cells. We may define $\omega_{i.} = \sum_j \omega_{ij}$ to express $\nabla_{\mathscr{P}}$ as

$$\nabla_{\mathscr{P}} = 1 - \frac{\displaystyle\sum_{i|\omega_{i.}<C}\sum_{j} \omega_{ij} P_{ij} + \mu}{\displaystyle\sum_{i|\omega_{i.}<C}\sum_{j} \omega_{ij} P_{i.} P_{.j} + \mu} \qquad \text{where } \mu = \sum_{i|\omega_{i.}=C} P_{i.}$$

Note that the greater is the probability in the rows consisting entirely of error cells μ, the more difficult it is for $\nabla_{\mathscr{P}}$ to be large. Thus for large μ the proposition is penalized for its lack of differentiation and $\nabla_{\mathscr{P}}$ will not assess the prediction as successful.

3.2.6 Numerical Properties of $\nabla_{\mathscr{P}}$

For every admissible proposition $-\infty < \nabla_{\mathscr{P}} \leq 1$. The measure equals one only if no prediction errors are committed when applying the prediction $\mathscr{P}$ knowing each observation's state on the independent variable (rule K). If

$0 \leqslant \nabla_{\mathcal{P}} \leqslant 1$, then the value of $\nabla_{\mathcal{P}}$ is the proportion of rule U errors that are eliminated by applying $\mathcal{P}$ with knowledge of the independent variable.

If two variables are statistically independent, then $\nabla_{\mathcal{P}} = 0$. However, in general, $\nabla_{\mathcal{P}} = 0$ does *not* imply statistical independence. (This implication does apply in the special case of a 2×2 table, but that is an exception to the general rule.) To demonstrate why this is so, consider Korteweg's major diagonal inflation proposition specifying the set of error cells shaded in Table 3.3A. Now assume the hypothetical population shown in Table 3.13. For the marginals shown in this table, if the variables were statistically independent, we would have $P_{ij} = P_{i.}P_{.j} = 1/9$ for all i,j. Although this condition is not met in Table 3.13, the *total* proportion of cases falling within the *set* of six error cells does equal the *total* proportion expected in this set under statistical independence. Thus $K_{\mathcal{P}} = U_{\mathcal{P}}$; hence $\nabla_{\mathcal{P}} = 0$. This example demonstrates that $\nabla_{\mathcal{P}} = 0$ does *not* imply statistical independence. Nor does $\nabla_{\mathcal{P}} = 0$ imply that $P_{ij} = P_{i.}P_{.j}$ in every error cell. The measure $\nabla_{\mathcal{P}}$ equals zero if the *total* proportion of cases in the *set* of all error cells equals the *total* proportion expected in this set of cells under statistical independence.

Table 3.13   Error structure yielding $\nabla_{\mathcal{P}} = 0$ **when** X **and** Y **are** *not* **statistically independent**

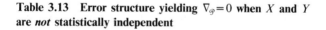

Change in money supply

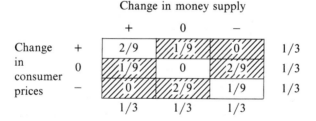

3.2.7   Prediction Analysis and the Linear Model

In Sections 1.1.1 and 1.5 we indicated that linear functions have been used to great advantage in the analysis of quantitative variables. We now digress to indicate how the prediction analysis of qualitative data both parallels and differs from the standard bivariate linear model. This section is mainly intended for those readers already acquainted with the linear model.

Our comments apply only to the analysis of *population* data. Consider the linear prediction

$$y = \beta_0 + \beta_1 x$$

when the parameters β_0 and β_1 have been selected *ex post* to minimize the sum of squared errors. In this case the development of the coefficient of determination R^2 as a measure of prediction success is well known. Rule K errors are simply

$$\text{rule } K \text{ errors} = \sum_{t=1}^{N} (y_t - \beta_0 - \beta_1 x_t)^2 = N\sigma_{Y|X}^2$$

where $t = 1, \ldots, N$ indexes the observations in the population and $\sigma_{Y|X}^2$ is the average squared error given X or, alternatively, the variance of Y given X.

The rule U prediction is "Always predict the population mean of the dependent variable"

$$\mu_Y = \left(\frac{1}{N}\right) \sum_t y_t$$

This *ex post* prediction minimizes squared errors in the absence of knowledge of X. We find rule U errors $= \sum_{t=1}^{N}(y - \mu_Y)^2 = N\sigma_Y^2$, where σ_Y^2 is the variance of Y. (The square root of this term, σ_Y, is called the standard deviation or standard error of Y; it measures the amount of variation of the dependent variable.) In sum, σ_Y^2 is the minimum squared error that can be achieved predicting Y without information about independent variable values. Then

$$R_{YX}^2 = 1 - \frac{\sigma_{Y|X}^2}{\sigma_Y^2} = 1 - \frac{\text{rule } K \text{ error}}{\text{rule } U \text{ error}} \qquad (19)$$

(The subscripts on R^2 are usually dropped when the meaning is clear.) R^2, the *coefficient of determination, measures the proportionate reduction in squared prediction errors when we shift from the null prediction $y = \mu_Y$ to the fitted linear prediction $y = \beta_0 + \beta_1 x$.* Like $\nabla_{\mathcal{P}}, R^2$ is thus a PRE measure. Also parallel to $\nabla_{\mathcal{P}}$, R^2 is undefined if the null model commits no prediction error (y always has the same value) since this would cause the denominator in (19) to be zero. Otherwise in *ex post* linear prediction we are guaranteed that $0 \leqslant R^2 \leqslant 1$. Furthermore it can be shown for the bivariate case that $R_{YX}^2 = R_{XY}^2$, indicating that, as with $\nabla_{\mathcal{P}}$, the measure of prediction success does not depend on which of the two variables is assigned the role of independent variable in the model.

Overall Error Reduction as a Weighted Average

The R^2 measure also has a weighted average interpretation similar to equation (16) for $\nabla_{\mathcal{P}}$. To simplify the discussion in a way that loses none of the interpretation's substance, suppose the independent variable X has a finite set of C values $\{x_1, x_2, \ldots, x_j, \ldots, x_C\}$. For example, this set could be $\{3, 4, 5, 6, 7\}$, the distinct values for LEVELS observed in the Blau and Schoenherr data in Figure 1.4. The average squared error for the linear model can be computed just for those $N_{.j} > 0$ observations having the same value, $X = x_j$, on the independent variable:

$$\sigma_{Y|x_j}^2 = \frac{1}{N_{.j}} \sum_{\substack{\text{each observation} \\ t \text{ with } X = x_j}} (y_t - \beta_0 - \beta_1 x_t)^2$$

Thus the total error for cases having $X = x_j$ equals

$$N_{.j}\, \sigma_{Y|x_j}^2$$

Also, if we use the null model to predict μ_Y for $N_{.j}$ cases selected at random and without knowledge of X, the expected amount of squared error will be $N_{.j}$ times the amount expected for a single case, or $N_{.j}\sigma_Y^2$. To compare the expected errors in the two situations, define

$$R_{Yx_j}^2 = 1 - \frac{N_{.j}\, \sigma_{Y|x_j}^2}{N_{.j}\, \sigma_Y^2} = 1 - \frac{\sigma_{Y|x_j}^2}{\sigma_Y^2}$$

Noting that

$$\sigma_{Y|X}^2 = \frac{1}{N}\left(N_{.1}\sigma_{Y|x_1}^2 + N_{.2}\sigma_{Y|x_2}^2 + \ldots + N_{.j}\sigma_{Y|x_j}^2 + \ldots + N_{.C}\sigma_{Y|x_C}^2\right)$$

simple algebra shows that

$$R_{YX}^2 = \sum_{j=1}^{C} \frac{N_{.j}}{N} R_{Yx_j}^2$$

Thus analogous to (17) for $\nabla_{\mathcal{P}}$, R_{YX}^2 is a weighted average of the proportionate reduction in squared error achieved for each X-value by the linear model. Error is said to be *homoscedastic* if the average squared prediction error is the same for all X-states:

$$\sigma_{Y|x_j}^2 = \sigma_{Y|X}^2 \qquad \text{for all } x_j$$

Otherwise the error is called *heteroscedastic*. The heteroscedastic case is analogous to $\nabla_{\mathcal{P}}$, since the $\nabla_{.j}$ values typically differ for each x_j.

Our comments have all applied to *ex post* linear predictions. In contrast, $\nabla_{\mathcal{P}}$ has been designed to permit the evaluation of *a priori* predictions. When the parameters of the linear function are specified *a priori*, and these parameter values are used to define rule K and rule U error, the R^2 statistic defined by (19) loses its conventional interpretation. Like $\nabla_{\mathcal{P}}$ for *a priori* predictions, it can be negative.

3.3 $\nabla_{\mathcal{P}}$ as a Solution to the Design Objectives

In Section 2.1 we specified a demanding set of criteria for a measure of prediction success. Later, in Chapters 6 and 7, we demonstrate that the $\nabla_{\mathcal{P}}$ model satisfies criteria 5 and 7 in Section 2.1 which, prescribe effective applications in the prediction analysis of multivariate propositions and convenient statistical properties. The other criteria pertain directly to bivariate measurement. These criteria require that the model yields a unique, operationally interpretable measure for each logically distinct bivariate proposition regardless of the proposition's scope, precision, and differentiation. This section discusses how the $\nabla_{\mathcal{P}}$ measure meets these design objectives.

Criterion 1 requires that the model yield a measure that is *custom designed* for the particular proposition under investigation. Recall that each logically distinct proposition can be stated unambiguously by specifying a set of error cells, $\mathcal{E}_{\mathcal{P}}$. By definition, each distinct set of error cells has a unique set of error cell indicators, $\{\omega_{ij}\}$. Since $\nabla_{\mathcal{P}}$ is defined solely in terms of $\{\omega_{ij}\}$ and the probability structure of the $R \times C$ cross classification, $\nabla_{\mathcal{P}}$ yields a unique measure for each admissible, logically distinct proposition. Thus $\nabla_{\mathcal{P}}$ is custom designed, as required by Criterion 1.

Criterion 3 requires that the measure be able to evaluate a prediction *regardless of its scope, precision, and differentiation*. Propositions vary on these dimensions. (Scope increases with the number of independent variable states for which nonvacuous predictions are made, precision increases the more narrowly the predicted outcomes can be located on the dependent variable, and differentiation increases the more different are the predictions made for the various states of the independent variable.) Since the prediction logic permits propositions that differ in scope, prediction, and differentiation, $\nabla_{\mathcal{P}}$ satisfies criterion 3 as stated previously. Criterion 3 further requires that the *scope, precision*, and *differentiation* of a proposition be considered in measuring its prediction success. Obviously as the precision of a prediction is increased, more error cells are included in the error set and, consequently, there is an increase in the rule U expected errors.[11] Sections 3.2.3 and 3.2.4 illustrated how this normalizing factor is used to

---

[11]This statement of course does not apply to degenerate situations with zero marginals.

allow $\nabla_{\mathcal{P}}$ to reflect variations in scope and precision. Similarly, Section 3.2.5 demonstrated that $\nabla_{\mathcal{P}} = 0$ for any proposition that is both admissible and totally undifferentiated. Moreover, if some variable state is always an error event, regardless of X, then the magnitude of $\nabla_{\mathcal{P}}$ is dampened.

In addition to the first criterion's requirement that the model provide different measures for different propositions, *criterion 2* stipulates that ∇ be equal for logically equivalent propositions, once the roles of the variables have been specified. Specifically we must verify that ∇ is invariant when, by transforming variables in a proposition, the counting of prediction errors (and successes) is unaffected. For example, we have seen that the predictions of Caplow's episodic model could be represented in either the 15×4 table of Table 3.8 or the 2×2 table of Table 3.9. Clearly the two representations are logically equivalent, since any event that is an error in one must be an error in the other and vice versa.

In general, let two propositions $\mathcal{P}$ and $\mathcal{P}'$ identify the sets of error cell indicators $\{\omega_{ij}\}$ and $\{\omega'_{i'j'}\}$ for domains $Y \times X$ and $Y' \times X'$, respectively, where the variables of one proposition can be created by transforming the variables of the other; then if $\mathcal{P}$ and $\mathcal{P}'$ are logically equivalent, admissible propositions,

$$\sum_i \sum_j \omega_{ij} P_{ij} = \sum_{i'} \sum_{j'} \omega'_{i'j'} P_{i'j'} \quad \text{and} \quad \sum_i \sum_j \omega_{ij} P_{i.} P_{.j} = \sum_{i'} \sum_{j'} \omega'_{i'j'} P'_{i'.} P'_{.j'}.$$

Therefore, by (11), $\nabla_{\mathcal{P}} = \nabla_{\mathcal{P}'}$. Thus we state the general result:

If $\mathcal{P}$ and $\mathcal{P}'$ are logically equivalent, admissible propositions, then $\nabla_{\mathcal{P}} = \nabla_{\mathcal{P}'}$. $\qquad\qquad$ (20)

$\nabla_{\mathcal{P}}$ satisfies criterion 2 by indicating the same prediction success in a population for logically equivalent propositions, whether the propositions identify the same set of error cells in a particular table, or identify error structures that can be shown to be equivalent by appropriate transformations of the variables.

This feature has a number of important ramifications. Result (20) ensures that the perhaps arbitrary identification and ordering of the classes for each variable have no effect on the measurement of prediction success. This means, for example, that if one simply permutes the order in which the variable states are listed in the cross classification, the value of $\nabla_{\mathcal{P}}$ is unaffected if the corresponding changes also are made in the statement of the proposition, $\mathcal{P}$. The result also implies that, in the bivariate case, $\nabla_{\mathcal{P}}$ is equal for two logically equivalent propositions even if the variable that is independent in one proposition is dependent in the other. Since the number of rule K errors is simply a count of error events, it obviously will not be affected by the roles of the variables. The rule U situation is a little more complex. We have seen that the number of rule U errors for a proposition is simply the sum of the rule U errors for those $\mathcal{P}_{ij}$ such that $\omega_{ij} = 1$. Now in each $\mathcal{P}_{ij}$, we have a symmetry. The scope of $\mathcal{P}_{ij}$ when X is the dependent variable is its precision when Y is dependent, and vice versa. Since rule U errors for $\mathcal{P}_{ij}$ are the product of its scope and precision, these errors are not affected by the roles of the variables. Hence any sum of $\mathcal{P}_{ij}$ rule U errors over various $\mathcal{P}_{ij}$'s is also invariant. A similar symmetry fails to exist in the multivariate case; in general the benchmark will be affected by which variable is predicted, but not by variable transformations that yield logically equivalent predictions.

Criterion 4 requires that the measure of prediction success be *operationally interpretable within the context* of the research problem to which it is applied. This requirement is satisfied by using the proposition $\mathcal{P}$ under two conditions of knowledge about the independent variable in developing a proportionate reduction in error measure. Not only does $\nabla_{\mathcal{P}}$ share the advantages possessed by all PRE measures, but also it has the additional advantage of being based on the specific prediction that the investigator chooses to study, rather than being based on a statistician's choice of some "generally useful" *a priori* prediction, or worse, on the *ex post* "prediction" that the data are distributed as given in the population's probability structure.

Criterion 6 specifies that the measure should be *relatively insensitive to minor perturbations of the probability structure*. The $\nabla_{\mathcal{P}}$ model combines the probabilities for the error cells by adding rather than multiplying. This is an important source of stability; in contrast, multiplicative models are highly sensitive to minor perturbations of the probability structure.

In summary, $\nabla_{\mathcal{P}}$ satisfies those design criteria that pertain to the measurement of prediction success attained in a population by a bivariate proposition. This chapter has considered only *a priori* degree-1 predictions with binary specifications of error and success events. In the next chapter we extend the basic model to encompass a wider variety of research problems.

4

BIVARIATE PREDICTION ANALYSIS: APPLICATIONS

Extending the basic $\nabla_{\mathscr{P}}$ model for degree-1 event predictions, this chapter explores research applications. We continue to focus on data describing an entire population or representing a large enough sample that sampling questions need not be of concern.

In Section 4.1 we discuss the assignment of different weights to various error events in order to account for differences in their importance or severity. Section 4.2 considers the analysis of actuarial, mixed strategy predictions discussed earlier in Section 1.2.4; the relation between pure and mixed strategy predictions and the comparative analysis of the two types of prediction; the distinction between the prediction success of an actuarial proposition and the proposition's goodness of fit to population probabilities. Section 4.3 is concerned with the relation between prediction success and marginal probabilities for populations with identical conditional probabilities. Section 4.4 considers the *ex post facto* selection of a degree-1 prediction for a population. Section 4.5 discusses the analysis of cross classifications where certain events are either logically impossible or irrelevant to the analysis.

4.1 DIFFERENTIAL WEIGHTING OF PREDICTION ERRORS

4.1.1 Weighted Errors

Some error events might be regarded as more important or extreme than others. The investigator may want to recognize this by assigning a set of weights to a proposition's error events or by evaluating the results under various alternative sets of weights. The bivariate $\nabla_{\mathcal{P}}$ model can be extended to permit differential weighting or costing of various error events in assessing prediction success.

The extension is quite direct. For any $R \times C$ probability distribution for a population of size N, let $\mathcal{W} = \{\omega_{ij}\}$ be a set of error weights, for $i = 1, \ldots, R$ and $j = 1, \ldots, C$, where $\omega_{ij} = 0$ if (y_i, x_j) is *not* an error event and otherwise $0 < \omega_{ij} < \infty$. Applying $\mathcal{P}$ with knowledge of each observation's location on the independent variable, we commit

$$N \sum_i \sum_j \omega_{ij} P_{ij}$$

weighted rule K errors. Similarly under rule U we expect

$$N \sum_i \sum_j \omega_{ij} P_{i.} P_{.j}$$

weighted prediction errors. Then the weighted error reduction measure is defined as

$$\nabla_{\mathcal{P}, \mathcal{W}} = 1 - \frac{\displaystyle\sum_i \sum_j \omega_{ij} P_{ij}}{\displaystyle\sum_i \sum_j \omega_{ij} P_{i.} P_{.j}} \tag{1}$$

The model (1) measures the proportionate reduction in weighted prediction errors achieved in a population by a proposition $\mathcal{P}$ for error weights $\mathcal{W}$. This variation of the basic $\nabla_{\mathcal{P}}$ model also satisfies the design criteria we specified earlier. For example, if $\mathcal{P}$ and $\mathcal{P}'$ are logically equivalent propositions with equivalent error weights, then $\nabla_{\mathcal{P}, \mathcal{W}} = \nabla_{\mathcal{P}', \mathcal{W}'}$.[1]

Multiplicative Transformations, Units of Measurement, and Precision

Several additional points should be made about the use of the weighted measure: The value of $\nabla_{\mathcal{P}, \mathcal{W}}$ is not changed by multiplicative transformations of $\mathcal{W}$. That is, $\nabla_{\mathcal{P}, \mathcal{W}} = \nabla_{\mathcal{P}, k\mathcal{W}}$, where $0 < |k| < \infty$. However, adding a nonzero constant to all ω_{ij} *does* modify the value of $\nabla_{\mathcal{P}}$. In contrast, rule U errors (and rule K errors as well) are affected by multiplicative transformations. Thus we can arbitrarily affect the measure of prediction precision. The problem is similar to that which arises in comparing the variance of a

---

[1] If $\mathcal{P}$ and $\mathcal{P}'$ are logically equivalent propositions, then an observation is a rule K error for $\mathcal{P}$ if and only if it is also a rule K error for $\mathcal{P}'$. The corresponding matrices of error weights, $\mathcal{W}$ and $\mathcal{W}'$, are *equivalent* if and only if they assign the same weight to each observation.

quantitative variable such as height. If height is measured for one population in inches and for another in centimeters, we generally are unwilling to compare the variances until we have expressed both measurements in equivalent standard units. However if we want to compare a theory predicting height in inches to another predicting weight in pounds, there is no way to compare the variances unless we adopt a common unit of measurement or *numeraire*. Similarly we must specify a numeraire in comparing the precision of two predictions. For predictions with unweighted errors the observation is the natural unit; an error in one theory has the same value as an error in the other. In contrast, in the case of weighted errors investigators may wish to specify some multiplicative transformation that reflects an externally specified numeraire. It is often useful to select a numeraire such that the largest error weight assigned equals one. This guarantees that the measure of precision U satisfies $0 \leqslant U \leqslant 1.0$. We emphasize, however, that the basic ∇ measure, like R^2, is a ratio variable that is unaffected by the choice of numeraire.

Negative Weights

The measure might be extended to permit $\omega_{ij} < 0$, thus weighting the prediction *successes* differentially. In general, if the set $\mathcal{W}$ contains both positive and negative numbers, then $-\infty < \nabla_{\mathcal{P}} < +\infty$. In contrast, if $\mathcal{W}$ contains only positive numbers then $-\infty < \nabla_{\mathcal{P}} \leqslant 1.0$. Moreover mixing positive and negative weights can make $\nabla_{\mathcal{P}}$ very sensitive to small changes in the probability structure, thus violating criterion 6 of Section 2.1. It is easy to construct a 2×2 example with both positive and negative weights where an infinitesimal change in the P_{ij} would change the (limit) value of $\nabla_{\mathcal{P}}$ from $-\infty$ to $+\infty$! Consequently we prefer to restrict $\mathcal{W}$ to positive numbers.

"Trivial" Propositions and Weighted Errors

The remarks in Section 3.2.3 on $\nabla_{\mathcal{P}}$ for trivial propositions must be modified for weighted errors. Recall that the error events of a trivial proposition comprise entire rows or columns of the cross classification, and the same is true for the success events if there are any. If only binary error weights are used, then $\nabla_{\mathcal{P}}$ equals zero or is undefined for a trivial proposition. On the other hand, if error events of a trivial proposition are differentially weighted, then (as for weak propositions with weighted or unweighted errors) $\nabla_{\mathcal{P}}$ is less than one or is undefined. If error weights are used, then "trivial" propositions—including those that identify all events as errors—can be of substantial scientific interest. This becomes more apparent in Section 4.2 where pure strategy error weights are related to methods for prediction analysis of actuarial propositions.

4.1.2   Ordinal Error Weights and $\nabla_{\mathscr{P}}$: An Example from Survey Research

Although precise error weights can arise naturally in certain research contexts, more typically the differential weighting of error events is at least partially arbitrary. In his study of potential for violence in political behavior based on survey data (treated here as an entire population), Muller (1972) scaled questionnaire responses to obtain two ordinal variables: approval of political violence (APV) and intention to engage in political violence (IEPV). *A priori*, Muller stated (p. 396):

> I shall hypothesize that (1) APV and IEPV will show a strong degree of fit to a monotonic function, and (2) cases which deviate from direct monotonicity will show higher rank-order on APV than on IEPV.

The Triangular Hypothesis

Muller thus predicted a tendency for the data shown in Table 4.1 to concentrate on the main diagonal and for those cases not on the diagonal

Table 4.1   Intention to engage in political violence and approval of political violence[a]

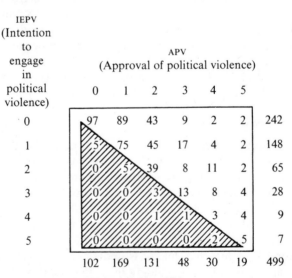

IEPV (Intention to engage in political violence)	APV (Approval of political violence)						
	0	1	2	3	4	5	
0	97	89	43	9	2	2	242
1	5	75	45	17	4	2	148
2	0	5	39	8	11	2	65
3	0	0	3	13	8	4	28
4	0	0	1	1	3	4	9
5	0	0	0	0	2	5	7
	102	169	131	48	30	19	499

Source: Muller (1972, Table 2, p. 936).
[a] The shaded triangle indicates the error set for Muller's *a priori* hypothesis, $\mathscr{P}$.

to lie above it ($\mathcal{P}$). Thus all cells lying in the shaded triangle of Table 4.1 are error events for Muller's hypothesis. By inspection, there are relatively few observed errors for this prediction. The value for (unweighted) $\nabla_{\mathcal{P}}=+.835$ indicates strong support for the prediction. Muller used two measures. The first, a symmetric version of Somers' d, is for degree-2 analysis. The second, R^2, was computed by treating the variables as interval variables with "values" equal to the category indices ($y_i=i, x_j=j$). We discussed R^2 in Section 3.2.7. In the next chapter we analyze d. Neither measure is designed to evaluate the success of predicting that the data tend to lie on or above the main diagonal; hence the values of these measures are considerably less informative than the value of $\nabla_{\mathcal{P}}$ for this investigation.

Weighting the Errors

Muller's remarks, as well as his use of d and R^2, indicate that he regards errors as more severe the "further" they are below the main diagonal. For example, by using R^2, Muller adopted a squared distance measure of prediction error. Suppose one specifies error weights that give greater weight to events "far" from the set of predicted cells. For example, let us regard cells in the diagonal band of cells immediately below the main diagonal as being one unit of "distance" ($v=1$) from the predicted set, cells in the next immediately lower diagonal band of cells as being two units of distance ($v=2$), and so on, such that the lowest left-hand cell is regarded as five units of distance ($v=5$) from the main diagonal. In the spirit of Muller's tacit use of squared distance as a measure of prediction error (through his reporting of R^2), one might set the error weight of any cell lying v units of distance below the main diagonal as $\omega_v=v^2/25$, where the normalizing factor of 25 is the maximum value of v^2 for the cross classification. The maximum error weight assigned is thus 1.0, with the entire matrix of error weights being as follows:

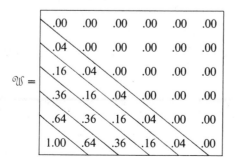

$$\mathcal{W} = \begin{array}{|cccccc|} \hline .00 & .00 & .00 & .00 & .00 & .00 \\ .04 & .00 & .00 & .00 & .00 & .00 \\ .16 & .04 & .00 & .00 & .00 & .00 \\ .36 & .16 & .04 & .00 & .00 & .00 \\ .64 & .36 & .16 & .04 & .00 & .00 \\ 1.00 & .64 & .36 & .16 & .04 & .00 \\ \hline \end{array}$$

Then $\nabla_{\mathscr{P}, \mathscr{U}} = .945$ indicates a greater proportionate reduction in weighted error than in unweighted error ($\nabla_{\mathscr{P}} = .835$) for Muller's proposition. However under the weighting scheme his proposition is considerably less precise (weighted rule U error, $U = .029$) than when errors are unweighted ($U = .206$).

Bounding ∇ When Weights Are Specified Ordinally

Of course assignment of specific numerical values for error weights typically is arbitrary. However, investigators unwilling to specify numerical weights might be willing to order the various error events according to their severity. Consider the following ordinal specification of error. Set $\omega_{ij} = \omega_v$ for any event representing the degree of error v, and order the weights such that $\omega_v > \omega_{v'}$ if and only if $v > v'$; as before, the error weight for a predicted event is $\omega_{ij} = \omega_0 = 0$. For example, referring again to Muller's proposition and using the unit of distance v described earlier to indicate degree of error, this procedure specifies the order $\omega_5 > \omega_4 > \omega_3 > \omega_2 > \omega_1 > \omega_0 = 0$ for the proposition. How sensitive is the value of $\nabla_{\mathscr{P}, \mathscr{U}}$ to variations in the weights consistent with this particular ordering? Expression (16) given in Section 3.2.4 for decomposing prediction success implies that

$$\nabla_{\mathscr{P}, \mathscr{U}} = \sum_{v=1}^{5} \frac{\omega_v U_v}{U_{\mathscr{P}, \mathscr{U}}} \nabla_v \tag{2}$$

In this expression, letting $\mathscr{E}_v$ denote the set of all error cells assigned the weight ω_v, ∇_v is the value of ∇ for the proposition whose set of error cells is $\mathscr{E}_v$,

$$\nabla_v = 1 - \frac{\sum\limits_{\mathscr{E}_v} P_{ij}}{\sum\limits_{\mathscr{E}_v} P_{i.} P_{.j}}$$

and U_v is the unweighted rule U error rate for the error cells $\mathscr{E}_v$. Using (2) it is easy to prove that

> If $\nabla_v \geqslant \nabla_{v'}$ for all v, v' such that $v > v'$, then for any set of error weights $\mathscr{U}$ consistent with the ordinal specification of error, $\nabla_{\mathscr{P}, \mathscr{U}}$ must be greater than or equal to $\nabla_{\mathscr{P}}$, the measure for unweighted errors. (3)

Obviously the more heavily weighted are certain error events, the more important it is that the proposition make few such errors. For Muller's data $\nabla_5 = \nabla_4 = \nabla_3 = 1.00 > \nabla_2 = .96 > \nabla_1 = .73$; therefore (3) guarantees that weighted $\nabla_{\mathcal{P},\mathcal{U}}$ for any set of weights ordered as specified earlier will be bounded below by .835, the value of unweighted $\nabla_{\mathcal{P}}$. Sensitivity analysis of the Muller data therefore shows that his proposition attains a high level of prediction success for all weights consistent with the ordinal specification of error. Furthermore since his data satisfy the $\nabla_v > \nabla_{v'}$ condition needed to apply (3), additional, microlevel support is discovered for his proposition: the component propositions identifying the most heavily weighted error events have the greatest prediction success.

The Weighted Main Diagonal Hypothesis

The proposition $\mathcal{P}$, however, does not quite capture all the content of Muller's statement. The two phrases "strong degree of fit to a monotonic function" and "cases that deviate from direct monotonicity" suggest that we also consider a proposition $\mathcal{P}'$ with $\mathcal{E}_{\mathcal{P}'} = \mathcal{E}_{\mathcal{P}} \cup \mathcal{E}_+$ where $\mathcal{E}_+$ contains all cells above the diagonal. For $\mathcal{P}'$, cells above the main diagonal are also error cells but have less weight than those below it. Consider, for example,

$$\omega_{ij} = 0 \qquad \text{if } j = i$$

$$\omega_{ij} = 1 \qquad \text{if } i > j$$

$$\omega_{ij} = \omega', \quad 0 \leqslant \omega' \leqslant 1 \qquad \text{if } i < j$$

Now since $\nabla_{\mathcal{P}} = .835 > \nabla_{\mathcal{E}_+} = .094$, theorem (3) guarantees that weighted $\nabla_{\mathcal{P}'}$ is at least .295, the value of unweighted $\nabla_{\mathcal{P}'}$. Thus the $\mathcal{P}'$ version of Muller's proposition is supported. Further $\nabla_{\mathcal{P}'}$ is at most $.835 = \nabla_{\mathcal{P}}$, its value when $\omega' = 0$. In general, increasing the value of ω' decreases prediction success but increases the precision of the predictions.

This section has discussed analysis of propositions with differentially weighted error events. We have indicated that the weighted average feature of ∇ can in some cases permit a relatively sharp assessment of prediction success even when the weights can only be ordered. To this point we have focused on *pure* prediction strategies: if x_j then *always* predict the set of Y-states $\mathcal{S}(x_j)$. The next section extends the methods to evaluate actuarial propositions, which predict the various dependent variable states probabilistically. We show that differentially weighted errors for pure strategy predictions can be interpreted in terms of mixed strategy propositions that are ∇-equivalent.

4.2 PREDICTION ANALYSIS OF ACTUARIAL PROPOSITIONS

4.2.1 Strategies of Actuarial Prediction

An actuarial proposition specifies, for each state of the independent variable(s), probabilities that the various dependent variable states will occur: rain with probability $1/3$, no rain with probability $2/3$. Typically, actuarial propositions can be written as a set of statements of the form "If $X = x_j$, then $Y = y_i$ with conditional probability q_{ij}." In order to discuss these propositions we begin by reviewing the three alternative ways to apply actuarial propositions that were identified in Section 1.2.4.

Predicting Probabilities

First, an actuarial proposition predicts the RC conditional probabilities (or proportions) for the cross classification, that is, predicts $P(y_i|x_j) = q_{ij}$. Thus applied, an actuarial proposition can be evaluated by the goodness of fit of the theoretical probabilities $\{q_{ij}\}$ to the population probabilities $\{P(y_i|x_j)\}$. For example, the prediction "rain with probability $1/3$" perfectly fits data in which rain is recorded for exactly one-third of the observations. In this application an actuarial proposition predicts aggregate proportions rather than making an event prediction for each observation. A variety of methods, most notably maximum likelihood techniques (Hoel, Port, and Stone, 1971, Sec. 3.5), and the χ^2 goodness-of-fit test we discussed in Section 2.3.2 can be used to evaluate the fit achieved by actuarial propositions. (One difficulty in applying these methods results from the fact that they are based on functions that are undefined if any event predicted to have zero probability has even a single occurrence in the data.)

Pure Strategy Prediction

A second approach converts the actuarial proposition into an absolute event (pure strategy) proposition predicting that dependent variable state the actuarial proposition asserts to be most likely: "If $X = x_j$, predict $Y = y_i^*$," where y_i^* is that Y-state for which q_{ij} is the greatest.[2] By the second approach the actuarial proposition is translated into an implied pure strategy proposition that can be evaluated using the methods described earlier. (Variations to this approach can be used. For example, the actuarial proposition also can be used to predict that set of states deemed "not least likely.") An important disadvantage of this approach is that it

---

[2] If more than one Y-state maximizes q_{ij} when $X = x_j$, there is a choice between either arbitrarily predicting one such state or predicting all such states. Naturally the ∇ values will differ for these two choices.

treats identically all actuarial propositions that agree as to which state is most likely. Thus "no rain" is the prediction for any actuarial proposition that assigns the greatest probability to this event. Consequently this approach cannot discriminate among alternative actuarial propositions that it converts into equivalent pure strategy predictions.

Mixed Strategy Prediction

We emphasize a third approach that applies an actuarial proposition by using a mixed prediction strategy to make an event prediction for each observation. More formally we interpret an actuarial proposition as the mixed strategy prediction "If $X = x_j$, then, with probability q_{ij}, predict $Y = y_i$." The q_{ij} satisfy

$$\sum_i q_{ij} = 1$$

Returning to the example, the investigator would select for any observation the event prediction "rain" with probability $1/3$ and otherwise make the alternative event prediction "no rain". Unlike the first approach, which predicts aggregate proportions, the mixed strategy approach yields an event prediction for each observation. Unlike the pure strategy approach, the mixed strategy application maintains the distinctions among alternative actuarial propositions, thus facilitating comparative analysis of their success in predicting the dependent variable state of each observation.

4.2.2 An Example: Chertkoff's Actuarial Model of Coalitions in the Triad

Chertkoff (1967) presents an actuarial revision of Caplow's (1959) continuous model of coalition formation discussed in Section 3.2.1. The asymptotic probabilities of the four coalition outcomes according to Chertkoff's actuarial model are shown for each type of strength distribution in Table 4.2. Note that the dependent variable defines the columns of this table. The entry q_{ij} in the cell (t_i, c_j) is the asymptotic probability under Chertkoff's model that coalition structure c_j forms given that the strength distribution is of type t_i. Thus for type t_5 the model states that coalition (BC) forms with probability $2/3$, coalition (AC) forms with probability $1/3$, and all other outcomes occur with probability zero.

Now consider the three alternative ways of applying an actuarial model. It is clear that Chertkoff (1967, p.175) interprets his model to predict population proportions. By this approach, for example, the asymptotic model predicts for type t_5 that coalition (BC) will form in two-thirds of the cases and that (AC) will form in the other one-third. By the second

Table 4.2   An actuarial prediction of coalitions in the triad[a]

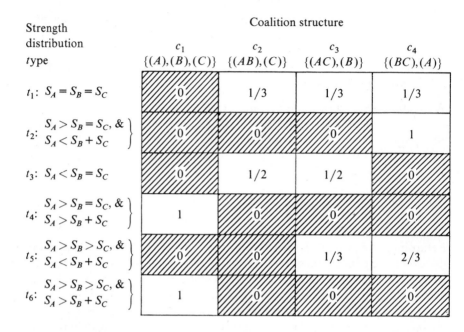

Strength distribution type	Coalition structure			
	c_1 $\{(A),(B),(C)\}$	c_2 $\{(AB),(C)\}$	c_3 $\{(AC),(B)\}$	c_4 $\{(BC),(A)\}$
t_1: $S_A = S_B = S_C$	0	1/3	1/3	1/3
t_2: $S_A > S_B = S_C$, & $S_A < S_B + S_C$	0	0	0	1
t_3: $S_A < S_B = S_C$	0	1/2	1/2	0
t_4: $S_A > S_B = S_C$, & $S_A > S_B + S_C$	1	0	0	0
t_5: $S_A > S_B > S_C$, & $S_A < S_B + S_C$	0	0	1/3	2/3
t_6: $S_A > S_B > S_C$, & $S_A > S_B + S_C$	1	0	0	0

[a] The number in cell (t_i, c_j) is q_{ij}, the asymptotic probability according to Chertkoff's (1967) model that coalition structure c_j forms, given t_i, the distribution of strength among players. Since the coalition structure is the dependent variable, $\sum_j q_{ij} = 1.0$ for these data; the probabilities are conditional within each row. Cell (t_i, c_j) is shaded if and only if it is an error event for Caplow's (1959) continuous model.

approach the actuarial model is transformed into a pure strategy prediction by including within the predicted set of coalitions in each row only those coalitions that are assigned the greatest probability in that row by the actuarial model. Note that in all but type t_5 this transformation of Chertkoff's model yields pure strategy predictions identical to those of Caplow's (1959) model for the continuous situation. In type t_5, however, this pure strategy application of Chertkoff's model predicts only coalition (BC). [For all types this transformation of Chertkoff's model yields predictions that are identical to the pure strategy predictions of Gamson's (1961) model.] Finally, the third interpretation uses mixed strategy predictions: applying Chertkoff's model under type t_5, for example, select the prediction "coalition (BC)" with probability 2/3 and otherwise predict "(AC)". In the next section we develop a model for measuring the prediction success of such mixed strategy propositions.

4.2.3 Measuring Prediction Success of Mixed Strategy Propositions

Let $\mathcal{2}_{YX}$ be a mixed strategy proposition predicting Y from X. Thus $\mathcal{2}_{YX}$ is a set of statements relating each of the states of Y and X by the proposition "If $X = x_j$, then, with probability q_{ij}, predict $Y = y_i$." We now apply the general PRE model described in Section 2.1 to define a measure of mixed strategy prediction success.

Rule K Errors

The PRE model first requires a rule K for predicting each observation's dependent variable state knowing its state on the independent variable. Rule K follows directly from the statement of the mixed strategy proposition: "If $X = x_j$, then, with probability q_{ij}, predict $Y = y_i$." Given that both y_i is predicted and $X = x_j$, the expected proportion of successes is just $P(y_i|x_j) = P_{ij}/P_{.j}$, the conditional probability of y_i; therefore the error rate, being one minus the success rate, is $1 - P_{ij}/P_{.j}$. Summing over the predictions made for $X = x_j$ and weighting by the probability with which each y_i is predicted, we find

$$
\begin{aligned}
\text{Expected rule } K \text{ error rate using mixed strategy predictions associated with } x_j &= \sum_i q_{ij}\left(1 - \frac{P_{ij}}{P_{.j}}\right) \\
&= 1 - \sum_i \frac{q_{ij}P_{ij}}{P_{.j}} \\
&= \sum_i \frac{(1 - q_{ij})P_{ij}}{P_{.j}}
\end{aligned}
$$

Then considering all X states,

$$
\begin{aligned}
\text{Expected rule } K \text{ error rate for } \mathcal{2}_{YX} &= \sum_j P_{.j}\sum_i \frac{(1 - q_{ij})P_{ij}}{P_{.j}} \\
&= \sum_i \sum_j (1 - q_{ij})P_{ij}
\end{aligned}
$$

Rule U Errors

The PRE model next requires that a rule U be specified for predicting each observation's Y-state when its X-state is unknown. Following procedures analogous to those of Section 3.2.3, with probability $P_{.j}$ we use the prediction associated with x_j. When we use the prediction associated with x_j, y_i is predicted, as in rule K, with probability q_{ij}. For example, with probability $P_{.2}$ we select $\{q_{i2}\}$, and given this selection, choose with probability q_{32} to predict y_3. If y_i is predicted, the probability of error is the unconditional probability $1 - P_{i.}$. Therefore

$$\begin{aligned}\text{Expected rule } U \text{ error rate using mixed} \atop \text{strategy predictions associated with } x_j &= \sum_i q_{ij}(1 - P_{i.}) \\ &= 1 - \sum_i q_{ij} P_{i.} \\ &= \sum_i (1 - q_{ij}) P_{i.}\end{aligned}$$

And now considering all states of X, the overall

$$\begin{aligned}\text{Expected rule } U \text{ error rate for } \mathcal{Q}_{YX} &= \sum_j P_{.j} \sum_i (1 - q_{ij}) P_{i.} \\ &= \sum_i \sum_j (1 - q_{ij}) P_{i.} P_{.j}\end{aligned}$$

The ∇ Measure for Mixed Strategy Propositions

Finally, define the PRE measure for the mixed strategy prediction $\mathcal{Q}_{YX} = \{q_{ij}\}$ as

$$\nabla_{\mathcal{Q}} = 1 - \frac{\text{expected } \mathcal{Q}_{YX} \text{ error in predicting } Y, \text{ given } X}{\text{expected } \mathcal{Q}_{YX} \text{ error in predicting } Y, \text{ given random } X}$$

$$= 1 - \frac{\displaystyle\sum_i \sum_j (1 - q_{ij}) P_{ij}}{\displaystyle\sum_i \sum_j (1 - q_{ij}) P_{i.} P_{.j}} \tag{4}$$

Expression (4) defines $\nabla_{\mathcal{Q}}$ for any actuarial proposition having q_{ij} which sum to one across states of the dependent variable for each state of the independent variable. If X is the independent variable, then $\mathcal{Q}_{YX} = \{q_{ij}\}$ where $\sum_i q_{ij} = 1$. On the other hand, if Y is the independent variable, then $\mathcal{Q}_{XY} = \{q_{ij}\}$ where $\sum_j q_{ij} = 1$. In either case $\nabla_{\mathcal{Q}}$ may be computed by inserting the appropriate q_{ij} in expression (4).

∇-Equivalent Pure Strategies

By comparing expression (4) with the definition of $\nabla_{\mathcal{P}, \mathcal{U}}$ for pure strategy predictions given in expression (1), it is clear that for every actuarial proposition $\mathcal{2}_{YX} = \{q_{ij}\}$ there is a ∇-equivalent pure strategy proposition with weighted errors.

$$\text{If } \{\omega_{ij}\} = \{(1 - q_{ij})\}, \text{ then } \nabla_{\mathcal{P}, \mathcal{U}} = \nabla_{\mathcal{2}} \tag{5}$$

Thus if no $q_{ij} = 0$, the largest implicit error weight is less than one. However, the use of unweighted errors in the component predictions represents a numeraire that investigators would regard as identical to the numeraire used in pure strategy predictions with unweighted errors for the same data. Therefore normally no standardization of the error weights implied by a mixed strategy prediction will be made in comparing the precision of alternative predictions.

Because of (5) and (6), given later, our comments in Section 4.1 about weighted ∇ generally apply to $\nabla_{\mathcal{2}}$. For example, the procedures for prediction analysis using ordinally specified error weights can also be used for ordinally specified actuarial predictions such as "Coalition (BC) or (AC), but (BC) more than (AC)." Our subsequent comments about $\nabla_{\mathcal{2}}$ can be applied to $\nabla_{\mathcal{P}, \mathcal{U}}$ as well. However the type of mixed strategy prediction we have considered only represents a special type of prediction logic proposition. If $\Sigma_i q_{ij} = 1.0$, the ∇-equivalent pure strategy proposition has error weights satisfying

$$\sum_i \omega_{ij} = \sum_i (1 - q_{ij}) = R - 1$$

Many pure strategy propositions, such as the Caplow model of Section 3.2, do not satisfy this condition.

Mixed Strategies Generalized to Set Predictions

This subsection is technically detailed and may be skipped without loss of continuity.

In the most general form a mixed strategy prediction $\mathcal{2}_{YX}$ represents a probability mixture of T set predictions of the form $\{$"Predict $S^{(t)}(x_j)$ with probability $q_{*j}^{(t)}$"$\}$ where $\Sigma_t q_{*j}^{(t)} = 1$ for $j = 1, \dots, C$. If desired, error weights $\{\omega_{ij}^{(t)}\}$ may be associated with each prediction. (We use q_{*j} simply to indicate, for example, that q_{*2} refers to the prediction probabilities for the second column and not for the second row.) Then applying the PRE model yields

$$\nabla_{\mathcal{2}} = 1 - \frac{\displaystyle\sum_{j=1}^{C} \sum_{i=1}^{R} \sum_{t=1}^{T} q_{*j}^{(t)} \omega_{ij}^{(t)} P_{ij}}{\displaystyle\sum_{j=1}^{C} \sum_{i=1}^{R} \sum_{t=1}^{T} q_{*j}^{(t)} \omega_{ij}^{(t)} P_{i.} P_{.j}} \tag{4'}$$

In this general situation (1) and (4′) imply

$$\text{If } \{\omega_{ij}\} = \left\{ \sum_{t=1}^{T} q_{*j}^{(t)} \, \omega_{ij}^{(t)} \right\}, \text{ then } \nabla_{\mathcal{P}, \mathcal{U}} = \nabla_{\mathcal{Q}} \tag{5′}$$

Expression (5′) establishes part a of theorem (6):

> $a.$ Every mixed strategy proposition is ∇-equivalent to a pure strategy proposition with weighted errors.
> $b.$ Every pure strategy proposition is ∇-equivalent to a mixed strategy proposition with unweighted errors.   (6)
> $c.$ Every mixed strategy prediction where X is treated as the independent variable has a ∇-equivalent mixed strategy prediction where Y is the independent variable.

Parts b and c are established in Appendix 4.1, which also shows how to obtain a mixed strategy proposition that is ∇-equivalent to a given pure strategy proposition.

4.2.4   Mixed Strategy Analysis of the Coalition Data

To illustrate applications of $\nabla_{\mathcal{Q}}$, consider again Chertkoff's (1967) actuarial model of coalition formation in certain three-person games. As presented in Table 4.2, Chertkoff's model relates the type of strength distribution t_i to coalition structure c_j. By the same technique discussed in Section 3.2.1, Table 4.3 reformulates the predictions of Table 4.2 for two independently measured variables.

Table 4.3 represents Chertkoff's model $\mathcal{Q}_{XY}$ by indicating $\omega_{ij} = (1 - q_{ij})$ for each cell. Earlier we presented data in Table 3.8 for 508 coalition outcomes observed in Vinacke's (1962) laboratory study of 30 male triads. On the basis of its rule U error rate (.746), Chertkoff's actuarial model is highly precise for these data and achieves modest reduction in this error: $\nabla_{\mathcal{Q}} = .320$. Using expression (5) and equation (17) of Chapter 3, we can decompose $\nabla_{\mathcal{Q}}$ for Chertkoff's model into additive components associated with each of the six types of strength distribution t_i:

$$\nabla_{\mathcal{Q}} = .169 \nabla_{\mathcal{Q}_1} + .167 \nabla_{\mathcal{Q}_2} + .171 \nabla_{\mathcal{Q}_3} + .157 \nabla_{\mathcal{Q}_4} + .175 \nabla_{\mathcal{Q}_5} + .162 \nabla_{\mathcal{Q}_6}$$

$$= .169(.109) + .167(.400) + .171(.072) + .157(.477) + .175(.264) + .162(.626)$$

Table 4.3 Error structures for alternative models predicting coalition formation[a]

Caplow type (t_i); Letter names of players 1,2,3		Coalition structure (X)			
		x_1 {(1),(2),(3)}	x_2 {(12),(3)}	x_3 {(13),(2)}	x_4 {(23),(1)}
t_1	y_1	1	2/3	2/3	2/3
t_2; ABC, ACB	y_2	1	1	1	0
t_2; BAC, CAB	y_3	1	1	0	1
t_2; CBA, BCA	y_4	1	0	1	1
t_3; ABC, ACB	y_5	1	1/2	1/2	1
t_3; BAC, CAB	y_6	1	1/2	1	1/2
t_3; CBA, BCA	y_7	1	1	1/2	1/2
t_4	y_8	0	1	1	1
t_5; ABC	y_9	1	1	2/3	1/3
t_5; ACB	y_{10}	1	2/3	1	1/3
t_5; BAC	y_{11}	1	1	1/3	2/3
t_5; BCA	y_{12}	1	1/3	1	2/3
t_5; CAB	y_{13}	1	2/3	1/3	1
t_5; CBA	y_{14}	1	1/3	2/3	1
t_6	y_{15}	0	1	1	1

[a] The number in cell (y_i, x_j) is $\omega_{ij} = 1 - q_{ij}$, where q_{ij} is the asymptotic (conditional) probability that, given $Y = y_i$, Chertkoff's (1967) model identifies coalition structure x_j as a success event. All shading indicates (unweighted) error cells for Caplow's (1959) model of the continuous situation [double hatching indicates error cells for Caplow's (1959) model of the episodic situation].

Inspection of this decomposition indicates that the rule U error rate for Chertkoff's model is distributed very evenly across the six types of situations; this is not surprising since, regardless of the situation type, the mixed strategy proposition predicts a single coalition outcome for each observation. The component $\nabla_{\mathcal{Q}_i}$ terms for each of the types t_i indicate that Chertkoff's prediction success is greater for the even-numbered types. A second decomposition summarizes this pattern:

$$\nabla_2 = .486 \nabla_{2,\, t \text{ even}} + .514 \nabla_{2,\, t \text{ odd}}$$

$$= .486(.500) + .514(.150)$$

It is interesting that Chertkoff's actuarial model achieves greater success in precisely those types (t_2, t_4, and t_6) for which it is in fact a pure strategy model with predictions identical to those of Caplow's continuous model. The next subsection discusses comparative analyses of these and related alternative models in greater detail.

4.2.5 Comparative Analyses of Actuarial or Weighted Error Propositions

In Chapter 3 we introduced methods for comparative analysis of the prediction success achieved in the same data by alternative pure strategy propositions identifying unweighted error events. This section extends those methods for comparative evaluation of pure strategy predictions for which differential weighting of error events may (but need not) be specified. Given the implicit error weights established by (5′), our remarks also apply to the mixed strategy analysis of actuarial predictions. In developing these extensions we continue the previous section's comparative analysis of the asymptotic version of Chertkoff's model of coalitions.

Let $\{\omega_{ij}\}$ and $\{\omega'_{ij}\}$ represent error weights for two propositions $\mathcal{P}$ and $\mathcal{P}'$ about the same $Y \times X$ cross classification. [Of course it is possible, paralleling our analysis of the Muller data in Section 4.1, that one is interested in comparing alternative weighting schemes for the same proposition. In this case $\mathcal{P} = \mathcal{P}'$ but $\{\omega_{ij}\} \neq \{\omega'_{ij}\}$.]

Use the symbol $\mathcal{E}_+$ to denote the set of all cells for which the error weight of $\mathcal{P}'$ is greater than for $\mathcal{P}$:

$$\mathcal{E}_+ = \left\{ (y_i, x_j) \,\middle|\, \omega'_{ij} > \omega_{ij} \right\}$$

Analogously, define

$$\mathcal{E}_- = \left\{ (y_i, x_j) \,\middle|\, \omega'_{ij} < \omega_{ij} \right\}$$

If (y_i, x_j) lies in $\mathcal{E}_+$, let $\omega_{ij}^+ = \omega'_{ij} - \omega_{ij}$; otherwise, $\omega_{ij}^+ = 0$. Similarly if (y_i, x_j) belongs to $\mathcal{E}_-$, then let $\omega_{ij}^- = \omega_{ij} - \omega'_{ij}$, and otherwise $\omega_{ij}^- = 0$. Thus the matrices $\{\omega_{ij}^+\}$ and $\{\omega_{ij}^-\}$, respectively, indicate the error weight added to and subtracted from the ω_{ij} to form the alternative proposition $\mathcal{P}'$. Now for the respective sets of error weights $\{\omega_{ij}\}$, $\{\omega'_{ij}\}$, $\{\omega_{ij}^+\}$, and $\{\omega_{ij}^-\}$, denote the corresponding rule U error rates by U, U', U_+, and U_- and the ∇

measure by $\nabla_{\mathcal{P}}$, $\nabla_{\mathcal{P}'}$, $\nabla_{\mathcal{E}_+}$, and $\nabla_{\mathcal{E}_-}$. Then,

$$\nabla_{\mathcal{P}'} = \frac{U}{U'}\nabla_{\mathcal{P}} + \frac{U_+}{U'}\nabla_{\mathcal{E}_+} - \frac{U_-}{U'}\nabla_{\mathcal{E}_-} \tag{7}$$

By expression (7) the value of ∇ for the alternative proposition $\mathcal{P}'$ equals the weighted sum of the PRE achieved by the original model $\mathcal{P}$ plus that added in those cells $\mathcal{E}_+$ in which $\mathcal{P}'$ increases the error weights, minus the PRE "lost" by decreasing the error weights in $\mathcal{E}_-$. Applying expressions (5) and (17) in order to compare Chertkoff's actuarial model $\mathcal{Q}$ with Caplow's model $\mathcal{C}$, the result for the experimental data analyzed earlier is

$$\nabla_{\mathcal{Q}} = \frac{.586}{.746}\nabla_{\mathcal{C}} + \frac{.160}{.746}\nabla_{\mathcal{E}_+}$$

$$= \frac{.586}{.746}(.496) + \frac{.160}{.746}(-.328)$$

$$= .320$$

Since in this case $\mathcal{E}_-$ is empty, U_- equals zero and $\nabla_{\mathcal{E}_-}$ is undefined; consequently the term $(U_-/U')\nabla_{\mathcal{E}_-}$ can be omitted. This decomposition reveals that by adding .160 weighted rule U error in the set of cells $\mathcal{E}_+$, Chertkoff's model increases prediction precision over that of Caplow's model from .586 to .746, but is penalized for this by a decrease in prediction success.

Using a Mixed Strategy to Predict a Single Outcome

In order to develop additional insight into the relation of these two models, recall that they agree in predicting a single outcome in the even-numbered types of strength distribution, t_2, t_4, and t_6. In fact, by using a mixed prediction strategy, Chertkoff's model predicts a single outcome for each observation in every type t_i, although that predicted outcome is selected probabilistically. On the other hand, Caplow's pure strategy model includes three outcomes in the predicted set for type t_1 and two outcomes for types t_3 and t_5 (see Table 4.3). Suppose we transform Caplow's model $\mathcal{C}$ into an actuarial version "$\mathcal{QC}$" that uses a uniform probability distribution across the set predicted by model $\mathcal{C}$ to predict a single coalition outcome for each observation. For example, given any observation in strength type t_1, model $\mathcal{C}$ predicts the set $\{x_2, x_3, x_4\}$ whereas model $\mathcal{QC}$ selects just one of these outcomes, each with probability $1/3$. Thus, like Chertkoff's (and unlike Caplow's) model, the actuarial proposition $\mathcal{QC}$ predicts a single outcome for each observation.

Model $\mathcal{QC}$ disagrees with Chertkoff's model $\mathcal{Q}$ only in strength distribution type t_5: Chertkoff predicts coalition (BC) with probability 2/3 and coalition (AC) with probability 1/3, whereas model $\mathcal{QC}$ predicts each with probability 1/2. Comparing model $\mathcal{Q}$ with $\mathcal{QC}$ using the decomposition procedures,

$$\nabla_{\mathcal{Q}} = \frac{.746}{.746}\,\nabla_{\mathcal{QC}} + \left[\frac{.007}{.746}\,\nabla_{\mathcal{E}_+} - \frac{.007}{.746}\,\nabla_{\mathcal{E}_-} \right]$$

$$= \frac{.746}{.746}\,(.307) + \left[\frac{.007}{.746}\,(.081) - \frac{.007}{.746}\,(-1.294) \right]$$

$$= .320$$

Note first that models $\mathcal{Q}$ and $\mathcal{QC}$ are, on the basis of their rule U error rates ($=.746$, rounded to three places), equally precise. The terms within brackets refer to the changes in error weight of model $\mathcal{Q}$ over that of model $\mathcal{QC}$ in the only situation, type t_5, for which they disagree. These bracketed terms indicate that model $\mathcal{Q}$ achieves slightly greater overall prediction success by predicting in type t_5 coalition (AC) with lower probability (hence greater error weight in $\mathcal{E}_+$) than model $\mathcal{QC}$, and coalition (BC) with greater probability (hence lower weight in $\mathcal{E}_-$). However, since the two models mostly agree, their ∇ values are nearly equal.

Dominance

Let us now use this example to return to the theme that more than one dimension should be considered in evaluating scientific propositions. With regard only to the two dimensions emphasized in the foregoing analyses, the model $\mathcal{P}$ is said to *dominate* the model $\mathcal{P}'$ in a population if $\nabla_{\mathcal{P}} > \nabla_{\mathcal{P}'}$ and $U_{\mathcal{P}} \geqslant U_{\mathcal{P}'}$ or if $\nabla_{\mathcal{P}} \geqslant \nabla_{\mathcal{P}'}$ and $U_{\mathcal{P}} > U_{\mathcal{P}'}$. In the data analyzed earlier, the set predictions of Caplow's model $(\mathcal{C})$ for the continuous situation are more successful but less precise than the single event, mixed strategy predictions of Chertkoff's model $(\mathcal{Q})$:

$$\nabla_{\mathcal{C}} = .496 > \nabla_{\mathcal{Q}} = .320, \text{ but } U_{\mathcal{C}} = .586 < U_{\mathcal{Q}} = .746$$

Consequently, considering only PRE and precision, neither model can be said to dominate the other in these data. On the other hand, transforming Caplow's model into the mixed strategy, single outcome predictions of model $\mathcal{QC}$, we found that $\nabla_{\mathcal{Q}} = .320 > \nabla_{\mathcal{QC}} = .307$ and $U_{\mathcal{Q}} = U_{\mathcal{QC}}$. Thus Chertkoff's actuarial model dominates $\mathcal{QC}$ in these data, although neither achieves substantial prediction success.

4.2.6 Actuarial Prediction Success *versus* Goodness of Fit

The goodness of fit achieved by an actuarial model is evaluated by comparing the model's postulated conditional probabilities to the corresponding conditional probabilities in the population. We now contrast goodness of fit with the success an actuarial model achieves in making event predictions for each observation.

There typically is little relation between an actuarial model's prediction success and its goodness of fit. Treating Y as the dependent variable, the best-fitting (*ex post*) actuarial model for a bivariate population is $\mathcal{Q}_{YX}^* = \{q_{ij}^*\}$, where for every y_i and x_j, q_{ij}^* equals $P(y_i|x_j)$, the conditional probability of y_i given x_j. (The rule K prediction developed in Section 2.5.2 for Goodman and Kruskal's τ is identical to $\mathcal{Q}_{YX}^*$. Moreover $\nabla_{\mathcal{Q}^*} = \tau$.) For example, if two variables are statistically independent in the population, then $\mathcal{Q}_{YX}^* = \{q_{ij}^*\} = \{P_{i.}\}$. Although this $\mathcal{Q}^*$ perfectly fits the data, it is unsuccessful in making mixed strategy event predictions for each observation: $\nabla_{\mathcal{Q}^*} = 0$. (Of course if the two variables are statistically independent, then $\nabla_{\mathcal{Q}}$ must be zero or undefined for any proposition $\mathcal{Q}$.)

Returning to the coalition example we have been considering and continuing to treat the experimental data shown in Table 3.8 as a population, the best-fitting actuarial model attains $\nabla_{\mathcal{Q}_{XY}^*} = .198$, and $U_{XY}^* = .748$. Recall that for Chertkoff's (1967) *a priori*, actuarial model $\mathcal{Q}$, $\nabla_{\mathcal{Q}} = .320$ and $U_{\mathcal{Q}} = .746$. Thus, although Chertkoff's model fits the data less well, his *a priori* mixed strategy predictions are nearly as precise and substantially more successful than are the event predictions of the perfectly fitting actuarial model.

More generally the solution to (13), developed later in this chapter, implies that *for a given level of precision* (rule U error), *we can always find a pure strategy prediction with unweighted errors that has a ∇ value at least as great as any actuarial prediction.* (There is one unimportant qualification to this statement: to achieve a target level of precision exactly, we may need to include one error cell with a weight less than one.) The statement indicates that designing a model for the prediction of individual events can be quite different from designing a model for predicting aggregate proportions. Moreover $\nabla_{\mathcal{Q}}$ for an *a priori* actuarial model typically is not maximized when the population proportions perfectly match the theory. Consequently a high value of $\nabla_{\mathcal{Q}}$ cannot be interpreted as indicating a good fit of $\mathcal{Q}$ to the proportions.

Our discussion of the implications of the solution to (13) is not intended to convey a sense that absolute theories are always preferable. In such areas as the quantum theory of atomic particle decay an actuarial statement is clearly an accurate description of the phenomena. In fact, the quantum theory contains an uncertainty principle which explicitly states one cannot predict submicroscopic events. However, in the social sciences many phenomena now treated in an actuarial fashion may eventually be subject to absolute prediction—or at least some probabilities will be driven in the direction of one or zero. From this viewpoint much social science theory is simply incomplete, and the probabilities represent a form of hedging against incomplete theory or information.

How does such hedging influence the evaluation of prediction success? Consider the (2×2) proposition $\mathcal{Q}$: "If x occurs, y is predicted with probability .8 and, if $\bar{x}$ occurs, $\bar{y}$ is predicted with probability .8." This can be regarded as a hedging of the absolute prediction $x \longleftrightarrow y$. Now if the data support $x \longleftrightarrow y$, then $0 < \nabla_{\mathcal{Q}} < \nabla_{x \longleftrightarrow y}$; in fact, $\nabla_{\mathcal{Q}}$ cannot be greater than .6 in this example, even if $\nabla_{x \longleftrightarrow y} = 1.0$. So one always has less positive success with the actuarial theory. Similarly if the data fail to support the absolute prediction $(\nabla_{x \longleftrightarrow y} \leqslant 0)$, then $\nabla_{\mathcal{Q}}$ again lies between zero and $\nabla_{x \longleftrightarrow y}$, indicating that $\mathcal{Q}$ has less negative prediction success. Thus although actuarial theories *a priori* limit the possible level of prediction success, they also protect the investigator against really disastrous results.

In conclusion, the methods presented in this section for the analysis of mixed strategy propositions are distinct from more traditional proportion-fitting methods. The two types of methods serve different purposes. Prediction analysis methods are not designed to evaluate the fitting or predicting of probabilities. Similarly an exclusive reliance on goodness-of-fit methods may lead researchers to ignore relevant questions of predictive content. For an extreme example, an actuarial model that predicts statistical independence or something close to it may fit the data, but it has little value for predicting events.

The remainder of this book deals primarily with pure strategy predictions. However since theorem (6) establishes an equivalence between pure and mixed strategies, our subsequent comments also pertain to the mixed strategy analysis of actuarial propositions.

4.3 PREDICTION ANALYSIS OF UNCONDITIONAL AND CONDITIONAL PROBABILITY STRUCTURES

We now turn to a classic question in the literature on measuring *association*.[3] Should the analysis be based on the unconditional probabilities

---

[3] The seminal papers on this topic are Yule (1912) and Greenwood and Yule (1915). The problem is the subject of continuing attention; see Mosteller (1968).

$\{P_{ij}\} = \{P(y_i, x_j)\}$ or should it focus on a conditional probability structure such as $\{P_{ij}/P_{\cdot j}\} = \{P(y_i|x_j)\}$? A primary thesis of this volume is that the answers to such questions should not be imposed by an arbitrary methodological fiat but rather reflect the investigator's own decisions within a particular research context. This section discusses alternative treatments of the probability structure in applying the $\nabla_\mathcal{P}$ model.

Alternative treatments may be viewed in terms of evaluating the proposition $\mathcal{P}$ in a "derived" population with probabilities P'_{ij} rather than in the actual population characterized by our now familiar P_{ij}. The P'_{ij} bear some substantive relation to the original P_{ij}, a relation that can be expressed by a mathematical function. One situation where one does not want to deal with the original P_{ij} arises when the investigator deems certain events irrelevant to the analysis. That situation is the topic of Section 4.5. Another situation occurs when the investigator wishes to maintain the conditional probabilities such that $P'(y_i|x_j) = P(y_i|x_j)$ while the marginal probabilities are modified to $P'_{\cdot j}$. This modification has special appeal in the comparative analysis of two or more populations, and has been a focus of discussion in the previous literature.

We enter that discussion by presenting two tables found in Goodman and Kruskal (1954), showing the 2×2 cross classification of treatment and survival for Hospitals I and II, treated here as populations. Table 4.4 presents their population probabilities and probabilities for a third hypothetical population, Hospital III. In interpreting measures of success for these tables, we make two assumptions that appear consistent with the earlier literature. First, differences in survival rates are attributable only to hospital procedures and not to other interpopulation differences, such as

Table 4.4 Hypothetical data for three hospitals

Survival		Treated t	Not treated $\bar{t}$			Treated t	Not treated $\bar{t}$			Treated t	Not treated $\bar{t}$	
Lived	s	.84	.03	.87	s	.42	.14	.56	s	.42	.02	.44
Died	$\bar{s}$	.04	.09	.13	$\bar{s}$	.02	.42	.44	$\bar{s}$	.14	.42	.56
		.88	.12			.44	.56			.56	.44	
		Hospital I				Hospital II				Hospital III		

Proposition	∇	U		∇	U		∇	U
$\bar{t} \rightsquigarrow \bar{s}$:	.713	.104		.554	.314		.897	.194
$t \rightsquigarrow s$:	.650	.114		.897	.194		.554	.314
$t \longleftrightarrow s$:	.680	.219		.685	.507		.685	.507

the initial severity of the disease or genetic differences. Second, the *conditional* probabilities of life and death within each hospital are independent of the proportion of patients treated. This implies that if no patients were treated, $100\% \times (.42/.44) = 95.5\%$ would die in the third hospital whereas only 75% would die in each of the first two hospitals. A possible scenario is that the third hospital serves less nutritious food than the first two.

4.3.1 Prediction Analysis of the Original Probabilities

Let us first examine the hospitals with the basic method given in Chapter 3. An optimistic researcher might postulate one of the three propositions: treatment tends to be *necessary* to survival, $\bar{t} \rightsquigarrow \bar{s}$; *sufficient*, $t \rightsquigarrow s$; *necessary* and *sufficient*, $t \longleftrightarrow s$.

For the first proposition we use the relation $P_{ij} = P(y_i|x_j)P_{.j}$ to find:

$$\nabla_{\bar{t} \rightsquigarrow \bar{s}} = 1 - \frac{P_{\bar{t}}P_{s|\bar{t}}}{P_{\bar{t}}P_s} = \frac{P_t(P_{s|t} - P_{s|\bar{t}})}{P_s} \tag{8}$$

Now our second assumption implies that the quantity $P_{s|t} - P_{s|\bar{t}}$ equals the proportion of treated patients whose survival can be attributed to the treatment. Consequently $\nabla_{\bar{t} \rightsquigarrow \bar{s}}$ represents the proportion of surviving patients whose survival can be attributed to the treatment.

The conditional probabilities of survival are identical in Hospitals I and II: $P_{s|t} = .955$ and $P_{s|\bar{t}} = .250$. A little algebraic manipulation of (8) shows that whenever conditional probabilities are identical across populations, $\nabla_{\bar{t} \rightsquigarrow \bar{s}}$ is a simple function of the proportion of patients treated. The function is increasing if treatment improves chances for survival. Here $\nabla_{\bar{t} \rightsquigarrow \bar{s}} = .705P_t/(.250 + .705P_t)$. Thus $\nabla_{\bar{t} \rightsquigarrow \bar{s}}$ is greater for Hospital I (.713) than for Hospital II (.554), which less frequently employs the treatment.

If the efficacy of treatment is narrowly defined as $P(s|t) - P(s|\bar{t})$, the proportion of treated patients whose survival can be attributed to the treatment, then the three hospitals are equally effective (efficacy = .705). Yet in Hospital III the general conditions of survival are so poor that for the relatively small proportion of patients who are not treated there, not being treated is nearly sufficient for death to occur (or treatment is necessary to living). Consequently the greatest value (.897) of $\nabla_{\bar{t} \rightsquigarrow \bar{s}}$ occurs there.

A similar analysis for the *sufficiency* prediction allows us to interpret $\nabla_{t \rightsquigarrow s}$ as the proportion of deceased patients whose death can be attributed to the absence of treatment:

$$\nabla_{t \leadsto s} = \frac{P_{\bar{t}} \left(P_{\bar{s}|\bar{t}} - P_{\bar{s}|t} \right)}{P_{\bar{s}}}$$

Given identical conditional probabilities and efficacious treatment, $\nabla_{t \leadsto s}$ decreases with the proportion of patients who are treated. Consequently $\nabla_{t \leadsto s}$ is lower in Hospital I (.650) than in Hospital II (.897). It is lowest in Hospital III (.554). Finally the prediction success of the necessary and sufficient proposition $t \longleftrightarrow\!\!\!\!\leadsto s$ is a weighted average of the success of $t \leadsto s$ and $\bar{t} \leadsto \bar{s}$.

The three hospitals are obviously quite different, at least from the viewpoint of prospective patients if not from that of the traditional measurement of association. For many purposes of evaluation it therefore is useful to have a numerical measure that discriminates among the hospitals. This role cannot be played by measures that are functions solely of the cross-product ratio described in Section 2.4. All three hospitals have a cross-product ratio of 63.0, reflecting the fact that they have equal differences in conditional probabilities. (Similarly Yule's Q equals .969 in all three hospitals.) Only measures that, like ∇, depend on the marginals can make these discriminations.

4.3.2 Prediction Analysis and Conditional Probabilities

In some circumstances, however, investigators may wish to have measures that are invariant across populations whenever conditional probabilities are invariant. Referring to the first two hospitals, Goodman and Kruskal (1954, p. 746) observe,

...the *conditional* probabilities of life, given treatment (non-treatment), are exactly the same for both hospitals, namely .955 (.250). The reason that the conditional probabilities are the same while the λ_b [λ] values are different is, of course, that the two hospitals treated very different proportions of their patients. And the proportions treated were probably determined by factors having nothing to do with 'inherent' association between treatment and cure.

Apparently what Goodman and Kruskal had in mind by "'inherent' association" is that, in the goodness-of-fit sense described in the previous section , both hospitals are perfectly described by the actuarial proposition $\mathcal{Q}^* = \{ q_{11} = .955, q_{21} = .045, q_{12} = .250, q_{22} = .750 \}$. However, as pointed out in that section, a good fit to proportions is a concept quite distinct from prediction success. That section also tells us that $\nabla_{\underline{\mathbb{2}}\bullet}$ equals Goodman and Kruskal's τ (.463 in Hospital I and .496 in Hospital II). Again we stress that such prediction differences appear relevant. If another hospital had the same conditional probabilities as I and II but treated only 1% of its patients, marginal differences would be of more interest than a common "inherent" association.

Normalizing on the Independent Variable

Nonetheless if the investigator desires a prediction measure that eliminates the effects of the independent variable marginals, he can transform population probabilities to have desired marginals $P'_{\cdot j}$ before computing ∇. The transformation is done by computing $P'_{ij} = (P_{ij}/P_{\cdot j})P'_{\cdot j}$. Note that if two populations have identical conditional probabilities $P_{ij}/P_{\cdot j}$ and are further both transformed to some common $P'_{\cdot j}$, they will also have identical cell probabilities P'_{ij} and dependent variable marginals $P'_{i\cdot}$. Consequently ∇ values will be equal for the two transformed populations.

Uniform Normalization

One particularly prominent choice for normalization, advocated recently by Goodman and Kruskal (1974, pp. 191–192), is to make all the independent variable marginals equal; that is, set $P'_{\cdot j} = 1/C$ for all j. Denote this transformation by $Y \times X_u$, indicating *u*niform marginals on X. By this method the transformed probabilities for the first two hospitals are as shown in Table 4.5:

Table 4.5 Transformed probabilities for the first two hospitals

	t_u	$\bar{t}_u$	
s	$.5P_{s\mid t}=$ $.5(.955)$	$.5P_{s\mid\bar{t}}=$ $.5(.250)$	$.5(P_{s\mid t}+P_{s\mid\bar{t}})=.603$
$\bar{s}$	$.5P_{\bar{s}\mid t}=$ $.5(.045)$	$.5P_{\bar{s}\mid\bar{t}}=$ $.5(.750)$	$.5(P_{\bar{s}\mid t}+P_{\bar{s}\mid\bar{t}})=.398$
	$.5$	$.5$	

We can apply ∇ to this probability distribution for various propositions, and the results will be invariant with the unconditional probability of treatment. For example, to measure the extent to which the conditional probability structure supports the prediction that treatment tends to be a necessary condition for survival, define

$$\nabla_{\bar{t}_u \leadsto \bar{s}} = 1 - \frac{.5 P_{s|\bar{t}}}{(.5)^2 (P_{s|t} + P_{s|\bar{t}})} = \frac{P_{s|t} - P_{s|\bar{t}}}{P_{s|t} + P_{s|\bar{t}}} \tag{9}$$

Comparison of (8) and (9) shows how normalization removes the effect of P_t, the proportion of patients treated. Consequently $\nabla_{\bar{t}_u \leadsto \bar{s}} = .585$ in Hospitals I and II. In contrast, $\nabla_{\bar{t}_u \leadsto \bar{s}} = .886$ for Hospital III, reflecting the lower conditional probabilities of survival in that hospital.

Similarly we can measure the extent to which, independent of the proportion of patients treated in the two hospitals, treatment tends to be a necessary *and* sufficient condition for survival:

$$\nabla_{t_u \longleftrightarrow s} = 1 - \frac{.5 (P_{s|\bar{t}} + P_{\bar{s}|t})}{(.5)^2 (P_{s|t} + P_{s|\bar{t}} + P_{\bar{s}|t} + P_{\bar{s}|\bar{t}})}$$

$$= P_{s|\bar{t}} - P_{\bar{s}|t} = P_{s|t} - P_{s|\bar{t}} \tag{10}$$

For all three hospitals $\nabla_{t_u \longleftrightarrow s} = .705$. Note that $\nabla_{t_u \longleftrightarrow s}$ equals the reduction in the conditional probability of death (and the increase in the conditional probability of life) associated with treatment. This measure has considerable intuitive appeal. In fact, Greenwood and Yule (1915, p. 185) proposed this index to measure the "advantage" of a treatment. However, unlike $\nabla_{\bar{t}_u \leadsto \bar{s}}$ or $\nabla_{t_u \leadsto s}$ it does not permit discrimination between populations with equal *differences* in conditional probabilities that have unequal *magnitudes* of these probabilities (Hospitals I and II versus Hospital III).

Dependent Variable Normalization

In other contexts the investigator may wish to normalize *dependent* variable marginals. For example, suppose the investigator wished to define an index of the effectiveness of a diagnostic test that does not depend on the proportion diseased in the population. Let $x =$ diagnosed diseased, $\bar{x} =$ diagnosed not diseased, $y =$ diseased, $\bar{y} =$ not diseased. Transforming so that the Y marginals both equal .5, we obtain, by symmetry with (10),

$$\nabla_{x \longleftrightarrow y_u} = P_{x|y} - P_{x|\bar{y}}$$

$$= 1 - (P_{\bar{x}|y} + P_{x|\bar{y}}) = J$$

Thus $\nabla_{x \leftarrow \rightarrow y_u} = J$, the index proposed by Youden (1950) to measure the effectiveness of a diagnostic test.

The transformation so that one marginal is uniform in a 2×2 table is appealing since these transformations yield measures that are simple functions of the conditional probabilities. Greater complexity occurs with larger tables. Moreover, even in a 2×2 table this particular transformation may be inappropriate within the research context. For example, in evaluating a diagnostic test for a rare disease, to base the verdict on a transformed distribution in which the unconditional probability of the disease is artificially set to $1/2$ is to evaluate predictions in a context bearing little resemblance to any natural population.

Effect of Alternative Category Definitions

In the general $R \times C$ case, category definitions can have undue influence on analyses based on such transformations of the original probabilities. Table 4.6A shows an error set and probabilities for a hypothetical population. Combining x_1 and x_2, the two categories that have no distinctions of predictive significance, results in Table 4.6B.

In both tables $\nabla_{\mathscr{P}} = .375$. If, as shown in Table 4.7, the original two tables are transformed to have uniform X marginals, then $\nabla = .333$ when just two

Table 4.6   Two logically equivalent error structures in $Y \times X$

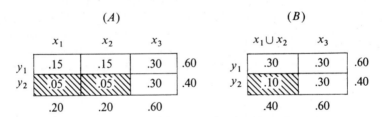

Table 4.7 $Y \times X_u$ transformations of Table 4.6

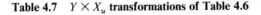

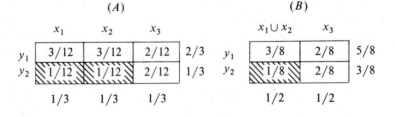

states of X are identified, but $\nabla = .250$ when X has three states. Not only do the ∇'s for the transformed data differ from ∇ for the original table, but they also differ between themselves. Since the results based on the transformations are thus affected by distinctions that do not affect whether an event is considered an error or success, the analysis of transformed probability distributions can violate criterion 2 of Section 2.1.

Pooled Normalization

In attempting to cope with these difficulties that arise from an artificial transformation, suppose we adopt the following perspective. Assume that an investigator holds *a priori* not only a proposition $\mathcal{P}$ but also that probabilities conditional on X are equal in the populations of interest. Then he expects not only a positive ∇ in each population but also that the populations ∇'s will be equal under *any* transformation to identical marginals on X. If so, then the untransformed ∇ for the pooled data combining the two populations can serve as a useful benchmark. If the separate populations have unequal marginals, then each of the marginals can be transformed to equal the pooled marginals. Differences between population-transformed ∇ values, untransformed ∇ values, and the pooled ∇ value provide an indication of the extent to which the differences in conditional probabilities (as against differences in marginal probabilities) account for the different levels of prediction success of $\mathcal{P}$ in the populations. This type of analysis seems more appropriate to evaluating prediction success than does the use of artificial marginals that differ from the marginals of the natural populations. However, there seem to be no pat answers to these questions—each type of analysis can be useful in particular research contexts.

Normalizing Both Marginals

Techniques also have been developed to transform $Y \times X$ to obtain desired marginals for both variables (Deming and Stephan, 1940a, 1940b; Mosteller, 1968). Transformations are made so that the cross-product ratios discussed in Section 2.4 are kept constant in each 2×2 subtable within the cross classification. The problems discussed earlier for transformations of one variable's marginals can be even more troublesome with double transformations. Moreover, although single transformations may be interpreted readily in terms of original conditional probabilities such as $P(y_i|x_j)$, interpretation of doubly transformed tables in these terms is less direct.

4.4 *EX POST FACTO* SELECTION OF AN OPTIMAL PROPOSITION

In this section we consider the application of the $\nabla_\mathcal{P}$ model in choosing, *ex post*, an admissible proposition $\mathcal{P}^*$ that, in a well-defined sense, is an "optimal" description of a bivariate population. We emphasize that our present interest is in population description and not in sample estimation; the latter is postponed until Chapter 6. This qualification is important because the analysis in this section depends critically on the separate cell proportions, rather than on sums that, because of central tendency achieved through aggregation, are less sensitive to sampling variations.

The discussion in this section continues to rely on the weighted average interpretation of $\nabla_\mathcal{P}$. Recall that the overall measure of prediction success for a proposition $\mathcal{P}$ is a weighted average of the prediction success achieved by the elementary predictions, as shown in equation (16) of Chapter 3:

$$\nabla_\mathcal{P} = \sum_i \sum_j \frac{\omega_{ij} P_{i.} P_{.j}}{U_\mathcal{P}} \nabla_{ij} \tag{11}$$

As in Chapter 3, if $\omega_{ij} P_{i.} P_{.j}$ is zero, then we define $\omega_{ij} P_{i.} P_{.j} \nabla_{ij}$ to be zero. All further statements assume the admissibility of all RC cells in a cross classification.[4]

4.4.1 A Naive Concept of Optimality

We next consider the criteria by which we can define an admissible proposition as being "optimal." There is a large variety of alternative conceptions of optimality. By one conception, a proposition $\mathcal{P}$ is an optimal description of an $R \times C$ probability distribution if and only if $\nabla_\mathcal{P} \geqslant \nabla_{\mathcal{P}'}$ for every other admissible proposition $\mathcal{P}'$. By this approach we would identify as optimal any proposition $\mathcal{P}^*$ that maximizes $\nabla_\mathcal{P}$ for the population of interest. That is, $\mathcal{P}^*$ would be optimal if it were a solution to the maximization problem

$$\underset{\mathcal{P}}{\text{Max}}\ \nabla_\mathcal{P} \tag{12}$$

This maximization problem is easy to solve. Just pick the biggest ∇_{ij}. Since, by (11), $\nabla_\mathcal{P}$ is a weighted average of the prediction success of the component predictions, $\nabla_\mathcal{P}$ can not exceed the value of the largest compo-

---

[4]If, *ex post*, some population marginals are found to be zero, the corresponding rows or columns may be deleted and the analysis conducted on the resulting subtable.

nent ∇_{ij}. Let $\mathcal{P}^*_{ij}$ be any elementary proposition such that $\nabla^* = \max_{i,j} \nabla_{ij}$. Then $\mathcal{P}^*_{ij}$ is optimal in the sense of (12).

However, the conception of optimality given in (12) has an obvious weakness. Note that unless there is more than one optimal elementary proposition, the optimal proposition $\mathcal{P}^*$ identifies only a single error cell. This solution is increasingly unsatisfactory as the number of cells in the table increases. For example, let us refer once more to Sawyer and MacRae's (1962) data shown in Table 3.5. For this population, consider the elementary proposition: "If the Democratic vote exceeds 75%, then predict the two-party strategy pair will not be one Democratic candidate and three Republican candidates." This proposition identifies only one error cell and is highly imprecise for these data: rule U error equals .001. Nonetheless, since $\nabla_{ij} = 1$ for the population, this elementary proposition is optimal in the sense of (12).

Do we wish to call this proposition optimal? We must if the concept of optimality is defined by (12). Yet an optimality concept that leads almost inexorably to propositions of such limited precision is unsatisfactory if investigators seek propositions of greater substance.

What if an alternative proposition of greater scope and precision achieved only slightly less prediction success as measured by $\nabla_{\mathcal{P}}$? Might it not be preferable to an elementary prediction? We think so. Presumably the investigator is willing to relinquish some prediction accuracy if that leads to a proposition containing more substance. If so, then the optimality criterion considered above should be replaced by one that recognizes this willingness to make trade offs.

4.4.2 *U*-Optimality

We now examine a second concept of optimality that adds rule U error constraints to the simple criterion given by (12). In Section 3.2.3 we interpreted the probability of rule U error, $U_{\mathcal{P}} = \sum_i \sum_j \omega_{ij} P_{i.} P_{.j}$, as a summary measure of the precision of the prediction task undertaken by a proposition in a population. The greater the U-value, the less vacuous is the proposition for the population of interest. In searching *ex post* for an optimal proposition we propose to maximize $\nabla_{\mathcal{P}}$ subject to obtaining a target level of precision U^*, where $0 < U^* < 1$; that is,

$$\text{Max}_{\mathcal{P}} \nabla_{\mathcal{P}}$$

subject to (13)

$$\sum_i \sum_j \omega_{ij} P_{i.} P_{.j} = U^*$$

The following procedure for solving (13) follows directly from the weighted average equation (11) as do the other results and solutions in this section. Rank the RC cells in order of *decreasing* ∇_{ij} values. If ties occur, select an arbitrary ordering within each set of tied propositions. Let the superscript (b) index the rank of a cell's ∇_{ij}-value in this order, so that $\nabla^{(b)} \geqslant \nabla^{(b+1)}$. This procedure yields the ordered series

$$\nabla^{(1)} \geqslant \nabla^{(2)} \geqslant \ldots \geqslant \nabla^{(b)} \geqslant \ldots \geqslant \nabla^{(RC)}$$

As an example consider the following table:

	x_1	x_2	x_3	
y_1	.000	.340	.110	.450
y_2	.380	.000	.070	.450
y_3	.020	.060	.020	.100
	.400	.400	.200	1.000

Computing the ∇_{ij} values and ordering, we obtain

$$\nabla^{(1)} = \nabla_{11} = \quad 1.000; \; U^{(1)} = .180$$

$$\nabla^{(2)} = \nabla_{22} = \quad 1.000; \; U^{(2)} = .180$$

$$\nabla^{(3)} = \nabla_{31} = \quad .500; \; U^{(3)} = .040$$

$$\nabla^{(4)} = \nabla_{23} = \quad .222; \; U^{(4)} = .090$$

$$\nabla^{(5)} = \nabla_{33} = \quad .000; \; U^{(5)} = .020$$

$$\nabla^{(6)} = \nabla_{13} = \; -.222; \; U^{(6)} = .090$$

$$\nabla^{(7)} = \nabla_{32} = \; -.500; \; U^{(7)} = .040$$

$$\nabla^{(8)} = \nabla_{12} = \; -.889; \; U^{(8)} = .180$$

$$\nabla^{(9)} = \nabla_{21} = -1.111; \; U^{(9)} = .180$$

After ordering the cells, find the *boundary* cell $\mathcal{C}^{(B)}$ defined by

$$\sum_{b=1}^{B-1} U^{(b)} < U^* \leqslant \sum_{b=1}^{B-1} U^{(b)} + U^{(B)}$$

where $U^{(b)} = P_{i.}P_{.j}$ for the cell (i,j) that has rank b. Then an optimal proposition is defined by

$\omega_{ij} = 1$ for all cells ranking "higher" (earlier in the series) than B

$$= \frac{U^* - \sum_{b=1}^{B-1} U^{(b)}}{U^{(B)}} \quad \text{for the cell with rank } B$$

$= 0$ for all cells ranking "lower" (later in the series) than B

Stripping away the notation we find that this procedure is just a simple serial process that assigns $\omega_{ij} = 1$ to cells with high ∇ values. The exception is the last, boundary cell to be included. This cell must be assigned some error weight between 0 and 1 to satisfy the equality constraint in (13). Returning to our example, if $U^* = .41$, we find that $B = 4$ since $U^{(1)} + U^{(2)} + U^{(3)} = .180 + .180 + .040 = .400$, while adding $U^{(4)}$ to this total increases precision to .490. We must weight the fourth cell in the order $(.410 - .400)/.090 = .111$ to satisfy the U^* constraint exactly. Using this weight and (11), the maximizing ∇ value is

$$(1/.41)\left[.180(1.000) + .180(1.000) + .040(.500) + (.111).090(.222)\right] = .932$$

Except for this boundary problem our solution shows that *weighted errors have no role in ex post analysis* unless the investigator further constrains the maximization problem such that certain cells must have weights. Alternatively the boundary weighting problem can be avoided by increasing or decreasing U^*. For instance, considering parsimony and the fact that a solution to (13) will involve at most one weighted cell suggests that we define optimality to require unweighted errors. This suggests the maximization problem:

$$\text{Max } \nabla_{\mathcal{P}}$$

subject to (14)

$$\sum_i \sum_j \omega_{ij} P_{i.} P_{.j} \geqslant U^*$$

$$\omega_{ij} = 1 \text{ or } 0$$

However, a solution to this problem may fail to have the serial processing property of the solution to (13). Although the first $B-1$ cells would be included, the boundary cell might have to be replaced with some other cell or set of cells with a lower ∇-value. In the example,

$$U^{(1)} + U^{(2)} + U^{(3)} + U^{(4)} = .490 > U^{(1)} + U^{(2)} + U^{(3)} + U^{(5)} = .420$$

$$> U^* = .410$$

Now define $\mathcal{P}$ and $\mathcal{P}'$ so that

$$\mathcal{E}_{\mathcal{P}} = \mathcal{E}^{(1)} \cup \mathcal{E}^{(2)} \cup \mathcal{E}^{(3)} \cup \mathcal{E}^{(4)} \quad \text{and} \quad \mathcal{E}_{\mathcal{P}'} = \mathcal{E}^{(1)} \cup \mathcal{E}^{(2)} \cup \mathcal{E}^{(3)} \cup \mathcal{E}^{(5)}$$

Then

$$\nabla_{\mathcal{P}} = (1/.49)\left[.18(1.000) + .18(1.000) + .04(.500) + .09(.222)\right] = .816$$

which is less than

$$\nabla_{\mathcal{P}'} = (1/.42)\left[.18(1.000) + .18(1.000) + .04(.500) + .02(.000)\right] = .905$$

The low U value of $\mathcal{E}^{(5)}$ allows the first three cells to have a greater relative contribution in $\mathcal{P}'$ and more than compensates for the zero value of $\nabla^{(5)}$. In this case $\mathcal{P}'$ is the solution to (14), whereas serial processing yields the "suboptimal" proposition $\mathcal{P}$.

This failure of serial processing is undesirable, since it leads to a perverse result: if $\mathcal{P}$ is optimal according to (14) for U^* and $\mathcal{P}'$ is optimal for U^{**}, where $U^{**} > U^*$, we *cannot* infer that $\mathcal{E}_{\mathcal{P}}$ is contained in $\mathcal{E}_{\mathcal{P}'}$. However it would be useful were the optimality criterion to have the inclusion property: $\mathcal{E}_{\mathcal{P}} \subseteq \mathcal{E}_{\mathcal{P}'}$.

In order to achieve this property (and to simplify computations) we propose further constraining the basic maximization problem. The new constraint requires that ∇_{ij} for every error cell contained in the error set $\mathcal{E}_{\mathcal{P}*}$ for the optimal proposition is greater than ∇_{ij} for every cell not in $\mathcal{E}_{\mathcal{P}*}$.

The bivariate prediction logic proposition $\mathcal{P}$ is defined to be U^*-optimal for a population if and only if it is a solution to the following maximization problem:

$$\underset{\mathcal{P}}{\text{Max}}\ \nabla_{\mathcal{P}}$$

subject to

$$\sum_i \sum_j \omega_{ij} P_{i.} P_{.j} \geqslant U^* \tag{15}$$

$$\omega_{ij} = 1 \text{ or } 0$$

$$\nabla^{(b)} > \nabla^{(b')} \qquad \text{if } \omega^{(b)} = 1 \text{ and } \omega^{(b')} = 0$$

The solution to (15) is identical to that of (13) except that the boundary cell is included as an unweighted error cell. A further exception occurs in the unusual case in which cells with a lower rank than the boundary cell $(b > B)$ have $\nabla^{(b)} = \nabla^{(B)}$. All such cells must also be included in $\mathcal{E}_{\mathcal{P}*}$.

The $\mathcal{P}*$ solution is *uniquely* $U*$-optimal for the population. Note that, in general, this proposition also is $U*$-optimal for a range of values for $U*$. We adopt the convention of reporting the *maximum* value of $U*$ within this range for which the proposition $\mathcal{P}*$ is $U*$-optimal. Thus in the example the first four error cells in the order would define the optimal proposition for $U* = .49$. The solution has the desired inclusion property: *if the proposition $\mathcal{P}*$ is $U*$-optimal and $\mathcal{P}**$ is $U**$-optimal, where $U** > U*$, then $\mathcal{E}_{\mathcal{P}*} \subseteq \mathcal{E}_{\mathcal{P}**}$.*

Some investigators may wish to reject certain $U*$-optimal solutions and conclude that no proposition of precision $U \geqslant U*$ appropriately describes the data. For example, one might reject any solutions whose error set contains cells for which $\nabla_{ij} \leqslant 0$. Rejecting solutions with such error cells guarantees that one will accept only propositions that identify no entire row or column as errors; hence only strong propositions will be selected. Moreover if ∇_{ij} must exceed zero for every error cell, then no solution for any $U*$ will be accepted when the variables are statistically independent.

4.4.3 ∇-Optimality

Finding a $U*$-optimal proposition involves maximizing the value of $\nabla_{\mathcal{P}}$ subject to a constraint establishing the minimum level of prediction precision that is acceptable. Some investigators may wish to reverse the roles of objective function and constraint, seeking the most precise proposition that meets a required level of prediction success. The analogue to (13) is

$$\underset{\mathcal{P}}{\text{Max }} U_{\mathcal{P}}$$

subject to

$$\nabla_{\mathcal{P}} = \nabla* \tag{16}$$

The analogue to (15) is

$$\underset{\mathcal{P}}{\text{Max }} U_{\mathcal{P}}$$

subject to

$$\nabla_{\mathcal{P}} \geqslant \nabla* \tag{17}$$

$$\omega_{ij} = 1 \text{ or } 0$$

$$\nabla^{(b)} > \nabla^{(b')} \text{ if } \omega^{(b)} = 1 \text{ and } \omega^{(b')} = 0$$

The solutions are also analogous. Using the same ordering of cells as before, find the boundary cell $\mathcal{E}^{(B)}$ satisfying a $\nabla*$ (rather than $U*$)

constraint. Let the proposition $\mathcal{B}^-$ identify the first $B-1$ cells as unweighted errors. As a solution for (16) the error weights are

$$\omega_{ij} = 1 \quad \text{for all cells ranking higher than } B$$

$$= \frac{U_{\mathcal{B}} - [\nabla_{\mathcal{B}} - - \nabla^*]}{U^{(B)} [\nabla^* - \nabla^{(B)}]} \quad \text{for the cell with rank } B$$

$$= 0 \quad \text{for all cells ranking lower than } B$$

The following error weights specify the *unique* solution of (17):

$$\omega_{ij} = 1 \text{ for all cells ranking no lower than } B \, (b \leqslant B) \text{ and}$$
$$\text{for all lower ranking cells } (b > B) \text{ having } \nabla^{(b)} = \nabla^{(B)}$$
$$= 0 \text{ for all other cells}$$

There is an equivalence between the solutions of (15) and (17):

> Let $\mathcal{P}^*$ be U^*-optimal, and let U^{**} denote the maximum value of U^* for which $\mathcal{P}^*$ is U^*-optimal. Also, let ∇^* be the value of $\nabla_{\mathcal{P}}$ for the proposition $\mathcal{P}^*$. Then $\mathcal{P}^*$ is both U^{**}-optimal and ∇^*-optimal. (18)

Table 4.8   Number of candidates nominated by two parties in six types of election districts, Illinois General Assembly Elections, 1902–1954[a]

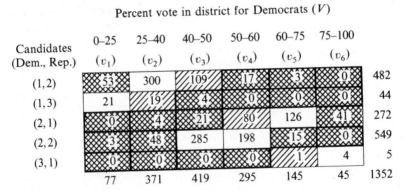

Percent vote in district for Democrats (V)

Candidates (Dem., Rep.)	0–25 (v_1)	25–40 (v_2)	40–50 (v_3)	50–60 (v_4)	60–75 (v_5)	75–100 (v_6)	
(1,2)	53	300	109	17	3	0	482
(1,3)	21	19	4	0	0	0	44
(2,1)	0	4	21	80	126	41	272
(2,2)	3	48	285	198	15	0	549
(3,1)	0	0	0	0	1	4	5
	77	371	419	295	145	45	1352

[a]Three error structures are shown. Heavy borders surround 18 error cells for the *ex post* proposition that is U^*-optimal for $U^* = .473$ and ∇^*-optimal for $\nabla^* = .820$. Double hatching indicates error cells for the game-theoretic model, and any (double or single) hatching indicates error cells for Sawyer and MacRae's predictions. Source: Data from Sawyer and MacRae (1962, p. 938). Omitted rows contained no observation (see Table 3.5).

4.4.4 Example: *Ex Post* Analysis of Candidate Nomination in Illinois

We now illustrate procedures for *ex post* analysis by examining once more Sawyer and MacRae's (1962) data concerning the population of 1352 two-party races for the Illinois General Assembly during the period 1902–1954. Table 4.8 contains all the rows of Table 3.5 with at least one observation. Hence it contains all the admissible cells. Each cell (dr,j) of Table 4.8 denotes an election for which the Democrats nominated d candidates and the Republicans nominated r candidates in the district, and the Democrats' share of the two-party vote in the district is described by column category v_j.

Category Definitions and Ex Post Analysis

These column categories are equivalence classes defined by the two models of the Sawyer-MacRae paper. That is, within each category there are no distinctions of predictive significance for these two models. However, as such distinctions might well arise in alternative models, a general *ex post* analysis would either treat the vote share as a continuous variable or at least use a large number of discrete categories. The former approach is beyond the scope of this book, and the latter is not possible since the Sawyer-MacRae paper displays the data only in cross-classified form. Consequently our *ex post* analysis is restricted to the six categories of vote divisions shown in Table 4.8.

U*-Optimality and a Decision Rule

Heavy borders in the table surround the 18 error cells for the U^*-optimal proposition when $U^* = .473$. By (18) this same proposition $\mathcal{P}^*$ is ∇^*-optimal for $\nabla^* = .820$.

Assume, as do Sawyer and MacRae, that both parties are able to forecast correctly their shares of the two-party vote (at least to a level of accuracy such that they correctly predict which state v_j will obtain). Then the proposition $\mathcal{P}^*$ asserts that *both* parties tend to act as if they obey the following decision rule: Field one candidate if the party is expected to win less than 40% of the vote, one or two candidates if between 40 and 50%, two if between 50 and 60%, and two or three if more than 60%.

Symmetric Processing

Although this proposition was selected through a serial processing of the $\nabla_{dr,j}$'s for individual cells, it treats the parties symmetrically: the 18 error cells for the proposition can be partitioned into nine pairs of cells so that if (dr,j) is an error event, then so is $(rd, 7-j)$. For example, both the cells

$(12,6)$ and $(21,1)$ are errors for the proposition $\mathcal{P}^*$. Suppose that instead of processing the $\nabla_{dr,j}$'s individually, we had processed symmetric *pairs* of cells, $\{(dr,j),(rd,7-j)\}$. Processing these pairs ensures that any proposition thereby selected will treat the parties in an identical fashion, recognizing only differences in their expected share of the two-party vote. Table 4.9 shows the rank b of each symmetric pair of cells in the serial ordering of values.

Table 4.9 *Ex post* analysis of data shown in Table 4.8 for candidate nominations in Illinois, 1902–1954

Rank in ordering, b	Symmetric pair of error cells, $\mathcal{E}^{(b)}$	Rule U error, $U^{(b)}$	$\nabla^{(b)}$	Cumulative rule U error, $\sum_{b'=1}^{b} U^{(b')}$	Overall ∇ for $\mathcal{E}_{\mathcal{P}} = \bigcup_{b'=1}^{b} \mathcal{E}^{(b)}$
1	$\{(12,6),(21,1)\}$	.023	1.000	.023	1.000
2	$\{(13,4),(31,3)\}$	.008	1.000	.032	1.000
3	$\{(13,5),(31,2)\}$	.004	1.000	.036	1.000
4	$\{(13,6),(31,1)\}$	.001	1.000	.037	1.000
5	$\{(12,5),(21,2)\}$	.093	.945	.131	.960
6	$\{(22,1),(22,6)\}$	.037	.940	.167	.956
7	$\{(12,4),(21,3)\}$	.140	.799	.308	.885
8	$\{(13,3),(31,4)\}$	.011	.728	.318	.879
9	$\{(22,2),(22,5)\}$	.155	.699	.473	.820
10	$\{(12,3),(21,4)\}$	.154	.095	.628	.642
11	$\{(13,2),(31,5)\}$	.009	$-.586$	.637	.624
12	$\{(22,3),(22,4)\}$	.214	$-.666$	.851	.299
13	$\{(12,1),(21,6)\}$	.027	-1.575	.878	.242
14	$\{(12,2),(21,5)\}$	.119	-1.639	.998	.016
15	$\{(13,1),(31,6)\}$	.002	-8.163	1.000	.000

The first nine pairs are errors for the proposition described above that is U^*-optimal for $U^* = .473$, with $\nabla = .820$. The first ten pairs are errors for the proposition that has $\nabla = .642$ and, subject to the constraint that only symmetric pairs are processed, is U^*-optimal for $U^* = .628$. This second proposition can be represented as the decision rule for both parties: Field one candidate if the party is expected to win less than 40% of the vote, two candidates if between 40 and 60%, and two or three if more than 60%. These two propositions may be compared to the game-theoretic model explored in Chapter 3. Its error set consists of the first nine pairs plus the thirteenth pair. Although the thirteenth pair has a very poor ∇ value, its low (.027) precision permits the overall ∇ value of .690 for the game-theoretic model to fall between the values of the two "optimal" propositions. Thus there is no strong evidence for favoring the *ex post* simple decision rule models over the game-theoretic model. Moreover the simple decision rules are specific representations of the proposition, "The greater the expected number of votes, the greater the number of candidates." Game theory thus provides an explanation of why the specific rules developed *ex post* appear optimal as against other rules such as: Field one candidate if less than 25%, two if between 25 and 75%, and three if more than 75%.

Although poorly behaved, the thirteenth pair does suggest ways in which the model could be amended. It shows that, in contrast to the game-theoretic model, each party is unwilling to field three candidates when it has a commanding vote share in excess of 75%. This result suggests that the model's assumptions should be changed to allow for duopolistic collusion within—or seat trading across—election districts.

Single-Outcome Prediction

The general approach just applied can and should be modified to serve the purposes of a particular investigation. For example, suppose the investigator wished to select, *ex post*, the pure strategy proposition for Table 4.8 that maximizes ∇ subject to the constraints that it not only predicts a single joint-strategy pair (row) for each election district but also treats the two parties symmetrically. This proposition can be identified by processing the ordered pairs of cells shown in Table 4.9 as follows: For each pair of columns, $\{x_j, x_{7-j}\}$, treat as *successes* the pair of events ranking last in the order. This procedure selects pairs 12, 14, and 15. All other pairs represent error events. The corresponding proposition makes predictions, *ex post*, that are identical to those of Sawyer and MacRae's augmented model discussed in Chapter 3. Consequently Sawyer and MacRae's model is an optimal proposition in the sense described in this paragraph. Although this proposition has greater precision (.664) than the U^*-optimal propositions identified earlier, it achieves this by adding three pairs of error cells ($b = 10$, 11, and 13) for which $\nabla < 0$. Consequently the increase in precision is purchased through a decrease in proportionate reduction in error; $\nabla = .534$.

4.4.5 Example: *Ex post* Analysis for Actuarial Models

If the optimality criteria we considered earlier are adopted, then actuarial theories will not be selected *ex post*. The solutions to (13) and (16) show that even if one allows for weighted errors, a weight other than one or zero will be assigned to at most one event, the boundary cell. Because of the equivalence (6) between weighted errors and mixed strategies, the solutions also tell us that one would use a mixed strategy for at most one independent variable state. Moreover comparison of the solution of (15) to that of (13) or of (17) to (16) indicates that the solution constrained to unweighted errors and serial processing will be "close" to the weighted solution. Hence the ∇ value for the U^*-optimal proposition will usually be only slightly less than the ∇ value when weights are permitted. Similarly the U value for the ∇^*-optimal proposition will be only slightly less than the corresponding weighted U value.

Other perspectives, however, may lead researchers to examine mixed strategies *ex post*. In model comparison, for example, one can design the *ex post* analysis so that the selected proposition is U^*-optimal for the level of precision U found for some *a priori* actuarial proposition. However, for mixed strategy predictions that always predict a single state, the condition $\sum_{i=1}^{R} q_{ij} = 1$ implies that the implicit error weights sum to $R-1$ in each column ($C-1$ in each row if Y is the independent variable). This condition

Table 4.10 *Ex post* analysis of data shown in Table 3.8 for coalition formation in the triad

Rank in ordering, b	Description of error events, by *t*ype and two-person coalition	$U^{(b)}$	$\nabla^{(b)}$	Cumulative rule U error, $\sum_{b'=1}^{b} U^{(b')}$	Overall ∇ for $\mathcal{E}_{\mathcal{P}} = \bigcup_{b'=1}^{b} \mathcal{E}^{(b')}$
1	t_1; no coalition	.047	.875	.047	.875
2	t_5; no coalition	.049	.758	.096	.816
3	t_2; no coalition	.047	.705	.143	.779
4	t_6; $(AB),(AC),(BC)$	.121	.626	.264	.709
5	t_4; $(AB),(AC),(BC)$	.117	.477	.381	.638
6	t_3; no coalition	.048	.423	.428	.614
7	t_5; (AB)	.040	.319	.468	.588
8	t_2; $(AB),(AC)$	.078	.271	.546	.543
9	t_5; (AC)	.041	.081	.587	.511
10	t_3; (BC)	.039	-.047	.627	.476
11	t_3; $(AB),(AC)$	.080	-.229	.707	.396
12	t_1; $(AB),(AC),(BC)$	.118	-.349	.825	.289
13	t_4; no coalition	.047	-1.195	.872	.210
14	t_2; (BC)	.039	-1.282	.910	.146
15	t_5; (BC)	.041	-1.294	.951	.084
16	t_6; no coalition	.048	-1.567	1.000	.000

will generally not be met by the U^*-optimal proposition or by a solution to (13); thus such a proposition is not interpretable as this type of actuarial proposition. To suggest how this condition might be incorporated in the analysis, we return to the coalition formation data of Table 3.8 that were analyzed in Section 4.2.

In order to simplify the *ex post* analysis, as in the previous example, it is useful to combine events into the sets shown in Table 4.10 before they are processed serially. These sets identify the minimum number of classifications that make all distinctions necessary to use these data for evaluating the models considered earlier.

The distinctions eliminated by this condensation, however, do have predictive significance for some propositions that might have been considered *ex post*. As discussed in the previous section, if it is anticipated that the data will be analyzed *ex post*, distinctions that *might* have theoretical significance should not be eliminated in forming the cross classification.

In the following we permit the *ex post* proposition to have weighted errors. As shown in Section 4.2.4, Chertkoff's (1967) actuarial model $\mathfrak{D}$ achieves $\nabla_{\mathfrak{D}} = .320$ with a rule U error rate of .746 in these data. Applying the solution procedure for (13) to the ordered sets shown in Table 4.10 rather than to individual cells, we can identify the weighted error proposition that solves (13) for $U^* = .746$, the precision of Chertkoff's model:

$$\omega^{(b)} = 1 \qquad \text{if } b < 12$$

$$\omega^{(12)} = \frac{.746 - .707}{.118} = .331$$

$$\omega^{(b)} = 0 \qquad \text{if } b > 12$$

This *ex post* proposition enjoys greater prediction success than Chertkoff's model since it achieves a greater value of $\nabla (.357)$ at the same level of precision ($U = .746$).

However this *ex post* prediction cannot be interpreted as an actuarial proposition. This interpretation cannot be used since the proposition identifies every cell in t_3 as an error cell. Consequently the error weights in the rows for this type sum to 4, as against the $C - 1 = 3$ required for an actuarial theory. Similarly in the rows for t_1 the sum of the error weights equals $[1.0 + 3(.331)] = 1.993 < 3.0$.

In order that the *ex post* proposition might have an actuarial interpretation, one might further constrain the procedure to require that the error weights sum to 3.0 in each row. If the procedure yielded a solution then, like Chertkoff's model, the *ex post* actuarial proposition (through a mixed strategy) would predict a single outcome for each observation. We can attempt to satisfy this additional constraint by modifying the *ex post* selection procedure as follows. Set $\omega^{(b)} = 1$ for all $b = 1, \ldots, 10$, thereby accounting for a total rule U error rate of .627. Since the criterion level $U = .746$ is not yet satisfied, the set of events $b = 11\,[t_3; (AB), (AC)]$ should be added as errors. However, since the other two events in t_3 ["no coalition" and "(BC)"] have already been assigned an error weight of one, we must make the weights sum to 3.00 by setting $\omega^{(11)} = 1/2$. The error weights established to this point account for a total rule U error of .667, still .079 short of the criterion level, so the next set of events $(b = 12)$ in the order must be admitted as errors. In this case we must set $\omega^{(12)} = 2/3$; therefore the event set $b = 12$ adds $2/3(.118) = .079$ weighted rule U error. With this addition it turns out that in this case the U^* constraint is satisfied exactly: total rule U error $= .746 = U^*$. Therefore the *ex post* optimal proposition that satisfies the additional constraint uses the stated error weights for $b = 1, \ldots, 12$ and sets the weights for all events having $b \geqslant 13$ to zero.

These error weights establish the *ex post* actuarial model $\mathcal{Q}' = \{q'_{ij}\} = \{1 - \omega_{ij}\}$. $\mathcal{Q}'$ disagrees with Chertkoff's model shown in Table 4.2 only in strength type t_5. Given t_5, the *ex post* proposition predicts coalition (BC) with conditional probability 1, whereas Chertkoff predicts (AC) with probability $1/3$ and (BC) with $2/3$. The optimal *ex post* proposition $\mathcal{Q}'$ dominates Chertkoff's model for these data since it achieves a greater proportionate reduction in error ($\nabla_{\mathcal{P}} = .351$ versus .320) at the same level of prediction precision ($U = .746$).

The differences between the unconstrained *ex post* model with a weighted boundary cell and, on the other hand, $\mathcal{Q}'$ or Chertkoff's model are not as striking as one might expect given our earlier remarks that pure strategies will always dominate mixed strategies *ex post*. The reason is that both $\mathcal{Q}'$ and the Chertkoff model are only partly actuarial. Chertkoff's model makes predictions with probabilities other than zero or one only in three of the six types; there are only two mixed strategy types for $\mathcal{Q}'$. For both models, in all six types at least one event is assigned zero probability. Greater differences between pure and mixed strategy results will occur when all the q_{ij} are greater than zero and less than one.

4.4.6 RE-optimality

It should be clear by now that there are a variety of ways in which one can constrain the *ex post* selection of a proposition. Bartlett (1974) has developed an additional approach that should be especially useful when the investigator (*a*) has a large number of categories on the independent variable, (*b*) has no or little *a priori* theory for the bivariate relation, or (*c*) wishes to develop a parsimonious *ex post* description in terms of a smaller number of categories that represent combinations of the original categories.

The Bartlett procedure, rather than setting U^* or ∇^*, treats as error events all cells with ∇_{ij} greater than a prespecified level l, the minimum value of ∇_{ij} for any cell to be included as an error cell. Again there is an equivalence: the proposition $\mathcal{P}$ selected is also U^*-optimal and ∇^*-optimal for $U^* = U_{\mathcal{P}}$ and $\nabla^* = \nabla_{\mathcal{P}}$.

He used reduction in error (RE) as well as PRE to assess prediction effectiveness. RE is defined by

$$RE = \text{rule } U \text{ error rate} - \text{rule } K \text{ error rate} = U\nabla$$

One reason for using RE is to ensure that the selected proposition has substantial precision. Since $RE = U\nabla$, to have relatively large RE requires that both U and ∇ be reasonably large. Choosing the minimal ∇_{ij} value $l = 0$ maximizes the value of RE. In the absence of other compelling strategies, predicting those cells with positive ∇_{ij} values is an appealing approach to *ex post* analysis.

The Bartlett procedure was developed for *ex post* multivariate analysis and is discussed in that context in Section 7.8. The procedure involves several refinements, including a means for combining states with similar, though not identical, predictions at intermediate stages of the algorithm, and also methods for approximating statistical properties of the procedure (as sketched in Sec. 6.7.3).

4.4.7 Conclusion: Optimality of *ex post* Propositions

In this section we have considered at length alternative but strongly related procedures for *ex post* selection. Although these procedures have some convenient properties, they are optimal only in the sense that they are solutions to the corresponding maximization problems. They might not be optimal in a scientific sense. In illustrating applications of these methods we have suggested that the basic *ex post* procedures be modified in the context of particular research problems. For example, one quite rightly may be willing to relinquish some prediction success in order to select a proposition having greater apparent heuristic value than that identified by the *ex post* procedures defined earlier. Chapter 6, in presenting statistical theory for making inferences about populations from samples, discusses further questions related to optimal *ex post* prediction. Until then, our primary focus returns to evaluating *a priori* propositions with data describing an entire population.

4.5 PREDICTION ANALYSIS WHEN CERTAIN EVENTS ARE EXCLUDED

There are two types of research situations in which investigators may wish to eliminate the effects of certain cells when measuring prediction success. First, if certain cells are excluded from the domain of the prediction being analyzed, then they are irrelevant events and their effects should be eliminated. For example, we showed in Section 2.6.1 that the domain of the predictions underlying Yule's Q excludes those observation pairs that are tied on either variable. In that case the effects of cells outside the domain were eliminated simply by considering the subpopulation that contains no ties. Second, logically (hence empirically) impossible events lead to "hole" cells that have zero probability. The effects of holes should be recognized.

Consider the following hypothetical example. Suppose that in studying ·* behavior in sequences of three-person games, the investigator develops a model that predicts which two-person coalition will form on a *defection trial*, that is, on a trial G whose coalition did not form on the previous trial, $G-1$. As Table 4.11 indicates, if the analytic domain is limited to defection trials, then we must exclude coalitions that form on both trials $G-1$ and G, for otherwise no defection occurs on trial G. Consequently the events in the main diagonal of Table 4.11 contain no data.

Table 4.11   Coalition formation on defection trial G, hypothetical data[a]

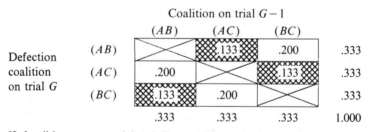

		Coalition on trial $G-1$			
		(AB)	(AC)	(BC)	
Defection	(AB)		.133	.200	.333
coalition	(AC)	.200		.133	.333
on trial G	(BC)	.133	.200		.333
		.333	.333	.333	1.000

[a]Infeasible events are deleted. Error cells are shaded.

Specifying the Domain

In the analysis of this type of cross classification, a complete statement of a prediction logic proposition must include a universal quantifier that specifies which events lie within the domain of the prediction. In the example the universal quantifier would be of the form "given that a defection occurs" or "given that a different coalition $(IJ) \in \{(AB),(AC),(BC)\}$ forms on trial G than on $G-1$." Symbolically the proposition in the example could be written as the prediction logic state-

ment, $\mathscr{P}:[(IJ)_G \neq (IJ)_{G-1}][(AB)_{G-1} \rightsquigarrow (AC)_G, \ (AC)_{G-1} \rightsquigarrow (BC)_G, \ \&$
$(BC)_{G-1} \rightsquigarrow (AB)_G]$. If no cell is to be excluded, then a prediction logic
proposition can be unambiguously specified by the set of error events.
However if certain events in the cross classification are to be excluded,
then we need to identify holes as well as errors. The error structure for the
proposition consists of the set of falsifying events in the corresponding
truth table which disregards events that lie outside the proposition's
domain.

The Need for a Correction

Unless corrective steps are taken, effects of holes produced in the bivariate
distribution by the impossibility of certain events or the exclusion of
irrelevant events can result in unjustified conclusions about the success of
a proposition. For example, suppose that in evaluating the proposition $\mathscr{P}$
shown in Table 4.11, we were to ignore effects attributable to the holes in
the main diagonal. Then, since $\nabla_{\mathscr{P}} = -.200$, the results would indicate that
the proposition fails in these data. However further analysis indicates that
this conclusion would be unjustified. The value of (uncorrected) ∇_{ij} is
$-.200$ in every error cell of the proposition, and equals $-.800$ in each of
its success cells. Thus if we continue to exclude the main diagonal cells
from the analytic domain, (uncorrected) $\nabla_{\mathscr{P}} = -.200$ is the "best" that can
be attained in these data. Despite this, the proposition should be judged to
achieve some success: Inspection of the table reveals that, of the two
possible coalition outcomes in each column, the coalition predicted by the
proposition is the more likely occurrence. The overall rule K error rate is
$3 \times 13.3\% = 40\%$. On the basis of the uniform marginals it would be
reasonable to assert as the null model that the two possible coalitions in
each defection trial are equally likely, leading one to a 50% error rate in
applying the proposition without knowledge of X. Then the proportionate
reduction in error for the proposition would equal $1 - (.400/.500) = +.200$,
rather than the $-.200$ value computed earlier.

The problem arises in the way rule U error is calculated. The negative
value for $\nabla_{\mathscr{P}}$ was computed earlier with the usual formula for the rule U
error rate: $\sum_i \sum_j \omega_{ij} P_{i.} P_{.j} = .333$. Yet we have argued that .500 seems to be a
reasonable value for the denominator of $\nabla_{\mathscr{P}}$. The problem with the usual
formula is, of course, that it does not recognize that the cells on the main
diagonal are necessarily empty. Given only the marginals and the assumption of statistical independence, one would expect $1/3$ of the cases in each
of the three sets of cells—successes, errors, and holes. By this procedure
only $2/3$ of the cases are assigned to relevant (error or success) outcomes.
Of these, exactly half are expected in the set of error cells. Thus one
approach to correcting for the effects of holes is to calculate the proportion

of relevant outcomes that are expected in the set of error cells. This is the approach taken in the next subsection.

4.5.1 Global Accounting for Holes

Let $\mathcal{H}$ be the set of cells that are to be treated as *holes*. Let the hole indicator $\phi_{ij} = 1$ if (y_i, x_j) belongs to $\mathcal{H}$; otherwise $\phi_{ij} = 0$. If the cells to be treated as irrelevant in the original probability distribution $\{P_{ij}\}$ have nonzero probability (as, in the example, if the original data for Table 4.11 included coalitions forming on both defection and nondefection trials), then they should be set to zero, and the remaining probabilities should be normalized:

$$P'_{ij} = \frac{(1 - \phi_{ij}) P_{ij}}{\sum_i \sum_j (1 - \phi_{ij}) P_{ij}}$$

so that

$$\sum_i \sum_j \phi_{ij} P'_{ij} = 0 \tag{19}$$

$$\sum_i \sum_j P'_{ij} = \sum_i P'_{i.} = \sum_j P'_{.j} = 1$$

Thus we assume throughout that $P'_{ij} = 0$ for every cell in the set of holes $\mathcal{H}$.

The next task is to modify the expressions for rule U error and $\nabla_{\mathcal{P}}$ so that effects of the set of holes are accounted for in evaluating the success of a proposition $\mathcal{P}$ in a probability distribution satisfying (19). No correction is necessary under rule K: Given a relevant observation with $X = x_j$, predict the observation's Y-state lies in the success set $\mathcal{S}(x_j)$ rather than in the error set. Then, as before, total rule K error is

$$K = \sum_i \sum_j \omega_{ij} P'_{ij}$$

If no adjustment is made for holes, then uncorrected rule U error is

$$U = \sum_i \sum_j \omega_{ij} P'_{i.} P'_{.j} \tag{20}$$

and the uncorrected PRE measure is, as before,

$$\nabla_{\mathcal{P}} = 1 - \frac{K}{U} \tag{21}$$

For Table 4.11 the uncorrected values for the proposition $\mathcal{P}$ are $U = 1/3$ and $\nabla_{\mathcal{P}} = -.200$.

Although the rule K error rate calculated in the usual way is correct, the rule U error rate computed according to expression (18) is too low, since it fails to take the holes into account. As a result, the uncorrected value of $\nabla_{\mathcal{P}} = 1 - (K/U)$ also is too low.

As a solution to this problem, we propose to replace U as defined by (20) with the proportion of all *relevant* (successes and errors) outcomes that are expected to be errors under rule U. The solution is based on the following procedure. Rule U requires that, without knowing the X-state of each randomly drawn observation, we predict with probability $P'_{.j}$ that its Y-state not only lies in the set of success events, but also does *not* lie in the set of error events for the $\mathcal{S}(x_j)$ prediction. Thus if y_i is neither an error nor a success for the $\mathcal{S}(x_j)$ prediction, then y_i is an irrelevant outcome for this prediction for all $NP'_{.j}$ observations to which $\mathcal{S}(x_j)$ is applied. Considering now the various $\mathcal{S}(x_j)$ predictions for all $j = 1, \ldots, C$, the overall probability that y_i is *labeled* an irrelevant event under rule U equals $\sum_j \phi_{ij} P'_{.j}$. Since cases are randomly drawn, $Y = y_i$ with the unconditional probability $P'_{i.}$. Thus, overall, the probability that the success or failure of the predictions for randomly selected observations cannot be determined equals

$$H = \sum_i \sum_j \phi_{ij} P'_{i.} P'_{.j}$$

$$= \sum_j P'_{.j} H_{.j} \qquad \text{where} \quad H_{.j} = \sum_i \phi_{ij} P'_{i.} \qquad (22)$$

This is just the total expected proportion in the set of all holes under statistical independence. The expected proportion of relevant outcomes for evaluating the predictions under rule U equals $(1 - H)$. Consequently the proportion of the relevant outcomes that can be expected under rule U to represent prediction errors for the proposition $\mathcal{P}$ equals

$$U' = \frac{U}{1 - H} \qquad (23)$$

Interpreting the Global Correction as Sequential Elimination of Irrelevant Outcomes

Further insight is provided by developing (23) from the following sequential prediction procedure. To simplify the exposition consider a population of N observations. (The results generalize readily to infinite populations.)

Select each observation randomly for prediction. Predict each $\mathcal{S}(x_j)$ with probability P'_j. The overall result is NU expected errors and NH expected irrelevant outcomes. To reduce the expected number of irrelevant outcomes, again draw randomly from the entire population, selecting each observation with probability H for a second round of predictions. For each observation for which a prediction is to be made in the second round, again predict each $\mathcal{S}(x_j)$ with probability P'_j. The second round predictions are expected to produce NUH errors and NH^2 irrelevant outcomes. Proceed to successive rounds, drawing cases with probability $H^{(t-1)}$ for the tth round of predictions, and expecting $NUH^{(t-1)}$ errors and NH^t irrelevant outcomes in the tth round. As the total number of rounds T for which this process is continued grows indefinitely large, then, by the formula for the sum of a geometric series, the expected total number of rule U' errors for all rounds approaches

$$\lim_{T \to \infty} \sum_{t=1}^{T} NUH^{(t-1)} = \frac{NU}{1-H} = NU'$$

By a similar use of the series formula, the expected number of relevant outcomes approaches N and the expected number of relevant outcomes for which $\mathcal{S}(x_j)$ was the prediction rule approaches $NP'_j(1-H_{\cdot j})/(1-H)$.

Prediction Success Using the Global Correction for Holes

Overall, adjusting in this way for the effects of holes, define the PRE measure

$$\nabla'_{\mathcal{P}} = 1 - \frac{\text{rule } K \text{ error}}{\text{rule } U' \text{ error}}$$

$$= 1 - \frac{K}{U'} = 1 - \frac{K}{U/(1-H)} \tag{24}$$

For Table 4.11, $K = .400$ and $H = U = 1/3$. Therefore, using (23) and (24), the corrected values are $U' = 1/2$ and $\nabla' = +.200$. Recall that we obtained these same corrected values in the earlier analysis of this example. Consequently in this case the rule U' procedure appears to make the appropriate correction for holes.

Properties of the Global Correction

In fact, this correction procedure has a number of appealing properties. With one exception (to be discussed in Sec. 4.5.3), it satisfies the design criteria specified in Section 2.1. Perhaps most importantly, $\nabla'_{\mathcal{P}}$ maintains the logical equivalence result (20) of Chapter 3. In particular, $\nabla'_{\mathcal{P}}$ has the same value for all logically equivalent propositions, regardless of which variable is treated as dependent. The global correction also preserves the relation used in decomposing prediction success into components:

$$\nabla'_{\mathcal{P}} = \sum_i \sum_j \frac{\omega_{ij} U'_{ij}}{U'} \nabla'_{ij} \tag{25}$$

where $U'_{ij} = P'_{i.} P'_{.j}/(1-H)$, and $\nabla'_{ij} = 1 - (P'_{ij}/U'_{ij})$. Moreover $\nabla'_{\mathcal{P}}$ has a simple relation to the uncorrected measure:

$$\nabla'_{\mathcal{P}} = H + (1-H)\nabla_{\mathcal{P}} \tag{26}$$

Using (26) it is easy to show that for any admissible proposition $\mathcal{P}$,

If $H = 0$ or $\nabla_{\mathcal{P}} = 1$,   then $\nabla'_{\mathcal{P}} = \nabla_{\mathcal{P}}$;   otherwise, $\nabla'_{\mathcal{P}} > \nabla_{\mathcal{P}}$ (27)

4.5.2   Example: Analyzing Feasible Outcomes of Status Games

Laing and Morrison (1974) used the global adjustment procedure and associated ∇' measure described earlier to account for effects of impossible outcomes in their analysis of a laboratory study of students playing a sequence of three-person games. The players know that the sequence of games ends without warning at a randomly chosen time, and that each player's share of a $15 payment is a specified, increasing function of his final *rank* based on the total points won over the entire sequence. A player finishing in sole possession of first place, for example, wins $10. At the beginning of each game G in the sequence players know each other's current total score and also know how many points can be won in game G by whatever two-person coalition (IJ) forms. Play of any game G involves bargaining among the three players until some coalition (IJ) forms and the points won by that coalition are divided between the two partners so that each wins a nonnegative, whole number of points, and the third player wins no points in game G.

At one level of analysis the outcome of game G can be described solely in terms of which coalition (IJ) forms and what ranks are achieved by the three players at the end of game G. Table 4.12 displays one column for each two-person coalition and one row for each of the thirteen ways in which three persons can be ranked. Thus there are $3 \times 13 = 39$ distinct outcomes. However for a particular game G some of these outcomes are not feasible. For the particular game shown in Table 4.12, any coalition (IJ) can win 100 points, and at the time this game begins, players 1 and 3 are tied for first place with total scores of 50 points each, and player 2 is in last place with zero points. If coalition (13) forms in this game, then player 2 must remain in sole possession of third place ($r_2 = 3$) in the game's outcome. Consequently all outcomes in which $r_2 \neq 3$ are infeasible when coalition (13) forms in this instance; those infeasible outcomes are deleted by $\times$ in the second column of Table 4.12. In fact, considering now all three coalitions, fully 24 of the 39 outcome descriptions cannot be achieved in this game.

Laing and Morrison compare the success of alternative models in predicting the outcomes observed in a total of 256 laboratory games. Their dependent variable is a composite formed by the Cartesian product of the

Table 4.12   Coalition and rank outcome of a status game[a]

Description of rank outcome	Vector of ranks, (r_1, r_2, r_3)	Coalition (IJ)		
		(12)	(13)	(23)
All tied:	(2,2,2)	✕	✕	✕
Two tied for first:	(1.5,1.5,3)	0	✕	✕
	(1.5,3,1.5)	✕	0	✕
	(3,1.5,1.5)	✕	✕	0
No ties:	(1,2,3)	0	✕	✕
	(1,3,2)	0	0	✕
	(3,1,2)	✕	✕	0
	(2,1,3)	0	✕	✕
	(2,3,1)	✕	0	0
	(3,2,1)	✕	✕	1
Two tied for last:	(1,2.5,2.5)	0	✕	✕
	(2.5,1,2.5)	0	✕	0
	(2.5,2.5,1)	✕	✕	0

[a] At the beginning of this game the players' total point scores are (50,0,50) and any coalition (IJ) wins 100 points. Infeasible outcomes are deleted. Shaded cells are error events identified by Laing and Morrison's (1974) heuristic model for this game. The three success cells are shown with bold borders.

13 alternative rank distributions with the 3 alternative coalitions. Each *cell* of Table 4.12 describes just one of the 39 mutually exclusive states of this composite dependent variable Y. The underlying independent variable is a typology of 293 distinct game situations described in terms of the total scores at the beginning of any game G and the point value of the alternative coalitions in that game. Each distinct class in this typology predicts a unique set of feasible outcomes. The 39 events of Table 4.12 form just one column in $Y \times X$; the single case shown there is the only observation in this column.

The entire 39×293 matrix is sparsely populated, averaging less than one observation per column. Moreover, as suggested by Table 4.12, many of the outcomes are infeasible and therefore must be treated as holes. In fact, for these data fully 46% of the cases are expected in the holes under statistical independence ($H = .464$). Consequently a large correction must be made in order to account for effects of the infeasible events.

We present results for two models analyzed by Laing and Morrison. Let $\mathcal{P}$ denote their heuristic model of coalition behavior and $\mathcal{M}$ denote their application of Aumann and Maschler's (1964) bargaining set, a solution concept from the theory of n-person games.[5] Table 4.12 shows the error events identified by $\mathcal{P}$ for just 1 of the 293 game situations.

Since both models necessarily must recognize the infeasibility of the events in the holes, the same correction must be made for each. Using the rule U' error rate given by (23), $U' = U/(1 - H)$, model $\mathcal{P}$ is more precise than $\mathcal{M}$ for these data ($U'_{\mathcal{P}} = .619$, $U'_{\mathcal{M}} = .531$) although both models attain a substantial level of precision. Because of the prevalence of infeasible events in the 39×293 cross classification, it is important that the results be based on measures that are corrected for holes, for otherwise one would reach an unjustified conclusion that the models fail in predicting these outcomes: the uncorrected values are $\nabla_{\mathcal{P}} = .094$ and $\nabla_{\mathcal{M}} = -.166$. However using the relation $\nabla' = H + (1 - H)\nabla$ given in (26) and $H = .464$, the corrected measures are $\nabla'_{\mathcal{P}} = .515$ and $\nabla'_{\mathcal{M}} = .375$. Thus, although $\mathcal{P}$ dominates $\mathcal{M}$ in these data, both models achieve at least modest success in predicting the observed outcomes. This success would not have been detected had the investigators made no correction for the effect of holes.

4.5.3 Set-by-Set Accounting for Holes

The global correction procedure used earlier has one disadvantage. As shown in Section 4.5.1, the total number of $\mathcal{S}(x_j)$ predictions for which

[5] $\mathcal{M}$ corresponds to model $\mathcal{M}_{\text{PPO}}$ in Laing and Morrison (1974), and the results given in the text are based on the combined data from the three laboratory studies they analyze separately. See Morrison (1974b) for additional details.

relevant outcomes are expected approaches $NP'_{.j}(1 - H_{.j})/(1 - H)$, where $H_{.j} = \sum_i \phi_{ij} P'_{i.}$. For $P'_{.j} \neq 0$, this total equals $NP'_{.j}$ only in the special case that $H_{.j} = H$. Yet criterion 3 given in Chapter 2 prescribes that appropriate recognition be given to a prediction's scope in evaluating its success. This objective would be accomplished more fully if exactly $NP'_{.j}$ relevant outcomes were expected for the prediction $\mathcal{S}(x_j)$ when X is unknown.

Sequential Elimination of Irrelevant Outcomes Using the Set-by-Set Procedure

The following alternative procedure achieves this goal by making set-by-set corrections for the effects of holes. The second procedure, called rule U'', again involves a series of predictions for a diminishing number of cases. Rule U'' is defined as follows: on the tth round of predictions, predict $\mathcal{S}(x_j)$ with probability $P'_j H_j^{(t-1)}$, $t = 1, \ldots, T$. Contrast this with rule U' where the prediction probability was simply $P'_j H^{(t-1)}$. For U' the number of relevant outcomes is increased by reapplying $\mathcal{S}(x_j)$ to the global proportion of irrelevant cases expected for *all* $\mathcal{S}(x_j)$ in the previous round. Under rule U'', $\mathcal{S}(x_j)$ is reapplied just to the expected proportion of irrelevant cases associated with that $\mathcal{S}(x_j)$ prediction in the previous round.

Under rule U'' on the tth round we expect to observe $NP'_{.j}[(1 - H_{.j})H_{.j}^{(t-1)}]$ relevant outcomes and $NU_{.j} H_{.j}^{(t-1)}$ errors for $\mathcal{S}(x_j)$, where $U_{.j} = \sum_i \omega_{ij} P'_{i.} P'_{.j}$. As the number of rounds T grows indefinitely large, the number of relevant outcomes expected for each $\mathcal{S}(x_j)$ prediction approaches the desired number, $NP'_{.j}$, and the total number of rule U'' errors approaches $NU''_{.j} = NU_{.j}/(1 - H_{.j})$. Hence summing these terms for all states of the independent variable X and dividing by N, we obtain the total expected rule U'' error rate:

$$U''_{\mathcal{P}_{YX}} = \sum_j U''_{.j} = \sum_j \frac{U_{.j}}{1 - H_{.j}} \tag{28}$$

Finally, accounting in this set-by-set way for the effects of holes, define the PRE measure for the proposition $\mathcal{P}_{YX}$:

$$\nabla''_{\mathcal{P}_{YX}} = 1 - \frac{\text{rule } K \text{ error}}{\text{rule } U'' \text{ error}}$$

$$= 1 - \frac{K}{U''} = 1 - \frac{K}{\sum U_{.j}/(1 - H_{.j})} \tag{29}$$

Properties of the Set-by-Set Correction

Note that, unlike $\nabla_{\mathcal{P}}$ and $\nabla'_{\mathcal{P}}$, the measure (29) is asymmetric in the sense that it is designed only for a bivariate proposition in which X is the independent variable. If, on the other hand, Y is the independent variable, then, by an analogous development,

$$\nabla''_{\mathcal{P}_{XY}} = 1 - \frac{K}{\sum\limits_{i} U_{i.}/(1 - H_{i.})} \qquad (29')$$

Thus ∇'' depends on which variable is treated as independent:

Logical equivalence of $\mathcal{P}_{YX}$ and $\mathcal{P}_{XY}$ does *not* imply that $\nabla''_{\mathcal{P}_{YX}} = \nabla''_{\mathcal{P}_{XY}}$.

However, the ∇'' measure satisfies our design criteria. It provides a distinct PRE measure for each logically distinct proposition and is unaffected by transformations of the variable states that have no predictive significance for the proposition under investigation. It seems to provide a more satisfactory accounting for prediction scope and precision than does ∇'. It also preserves the useful relation for decomposing prediction success into additive components:

$$\nabla''_{\mathcal{P}_{YX}} = \sum_{i} \sum_{j} \frac{\omega_{ij} U''_{ij}}{U''} \nabla''_{ij} \qquad (30)$$

where $U''_{ij} = P'_{i.} P'_{.j}/(1 - H_{.j})$ and $\nabla''_{ij} = 1 - P'_{ij}/U''_{ij}$.

Obviously if $\nabla = 1$ or if no hole is identified by $\mathcal{P}_{YX}$ in any column having $U_{.j} > 0$, then $\nabla'' = \nabla$. On the other hand, omitting degenerate situations, $\nabla'' > \nabla$. Thus, ignoring exceptional cases both ∇' and ∇'' typically are greater than the uncorrected measure. Moreover it is easy to construct hypothetical examples demonstrating that there is no necessary ordering of ∇' and ∇'' when they are not equal. If $H = H_{.j}$ for all x_j having $U_{.j} > 0$, as in the earlier example of Table 4.11, then $U' = U''$ and $\nabla' = \nabla''$.

Both ways of correcting ∇ for effects of holes have their advantages. The symmetric measure ∇' is identical for all logically equivalent propositions, unlike ∇'', whereas the asymmetric measure appears to provide a more effective accounting for prediction scope and precision in evaluating prediction success. The fact that we have indicated both types of correction reflects our view that the nature of an appropriate correction remains an open question. However, the examples presented in this section provide compelling testimony that it can be extremely important for investigators to account for the presence of holes.

4.5.4 Eliminating Main-Order Effects in Evaluating Supplementary Theory

For some investigations it may be useful to evaluate the success of one proposition after eliminating the effects of another proposition's error cells. Suppose one theory identifies the set of *error* cells $\mathcal{K}$, and a second proposition, $\mathcal{P}$, refines the predictions of the first by simply adding the set of error cells $\mathcal{E}_+$, so that, overall, the error set for the proposition $\mathcal{P}$ is $\mathcal{K} \cup \mathcal{E}_+$. Recall that in predicting candidate nominations in Illinois, as discussed in Sections 3.2 and 4.4, Sawyer and MacRae's model $\mathcal{P}$ attempts to refine the predictions of the game-theoretic model in exactly this way. From *ex post* comparisons of these two models, we discovered in Section 4.4.4 that Sawyer and MacRae's refinements are optimal in the sense that they maximize ∇ subject to the constraint that a single strategy pair is to be predicted for the two parties in each kind of district. However, in order to increase prediction precision Sawyer and MacRae added to the error set cells for which $\nabla_{ij} < 0$.

One way to evaluate the success of a model that refines the predictions of a successful theory in this way is to place holes in the error cells of the original theory, thus removing from the distribution its main-order effects. Then, treating the original error cells as holes and eliminating their influence, one can use ∇' and ∇'' to evaluate the success of the supplementary model in predicting second-order effects.

In Table 4.13, all error cells for the game-theoretic model have been deleted, so that all cells that are not deleted are successes for that theory. The shaded cells are error events that are added by Sawyer and MacRae's supplementary model. Given that zero probability is assigned to every hole in Table 4.13, the game-theoretic model $\mathcal{K}$ commits no rule K error in this distribution.

In the P'_{ij} distribution the overall success of Sawyer and MacRae's model is a weighted average of its success in $\mathcal{K}$ and $\mathcal{E}_+$:

$$\nabla_{\mathcal{K} \cup \mathcal{E}_+} = \frac{H}{H + U_+} \nabla_{\mathcal{K}} + \frac{U_+}{H + U_+} \nabla_{\mathcal{E}_+} \tag{31}$$

$$= \frac{.454}{.631}(1) + \frac{.177}{.631}(-.031) = .710$$

Thus, as before, we find that Sawyer and MacRae's model increases precision by adding cells $\mathcal{E}_+$ where ∇ is slightly negative, thus trading a decrease in PRE for greater precision.

In order to evaluate Sawyer and MacRae's success in predicting second-order effects in these data, we now use the procedures for discount-

Table 4.13   Candidate nominations in Illinois (1902–1954), omitting events that are errors for game-theoretic model[a]

Percent vote in district for Democrats (V)

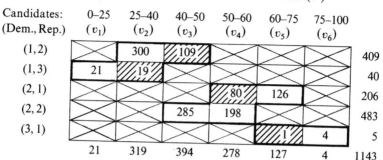

Candidates: (Dem., Rep.)	0–25 (v_1)	25–40 (v_2)	40–50 (v_3)	50–60 (v_4)	60–75 (v_5)	75–100 (v_6)	
(1, 2)		300	109				409
(1, 3)	21	19					40
(2, 1)				80	126		206
(2, 2)			285	198			483
(3, 1)					1	4	5
	21	319	394	278	127	4	1143

[a] Shaded events are errors for Sawyer and MacRae's (1962) model. Irrelevant events for evaluating second-order effects are deleted.

ing effects of the holes in Table 4.13. The supplementary model has a rule K error rate of .183 in the P_{ij}' distribution. Using the global procedure to discount the effect of the game-theoretic error cells, $U' = .325$ and $\nabla' = .437$. Using the set-by-set procedure $U'' = .258$ and $\nabla'' = .292$. Thus by either procedure Sawyer and MacRae's supplementary model achieves modest success in predicting second-order effects in these data.

APPENDIX 4.1   ∇-EQUIVALENT PURE AND MIXED STRATEGIES

This appendix establishes parts (b) and (c) of theorem (6) found in Section 4.2.3.

Given any proposition $\mathscr{P}$ with error weights $\mathscr{W}$, a ∇-equivalent, mixed strategy proposition $\mathscr{Q}$ with unweighted errors can be defined as follows:

1. Without loss of generality normalize the ω_{ij} such that the largest error weight is one.

2. If, in column j, ω_{ij} is the tth largest error weight, let $f_j(i) = t$ and $\omega_j^{(t)} = \omega_{ij}$. (If two or more error weights are identical in value, they may be ordered arbitrarily.) Further define $\omega_j^{(0)} = 1.0$ and $\omega_j^{(R+1)} = 0.0$.

3. Now in $\mathscr{Q}$, for $t = 1$ to $R + 1$, let

$$\omega_{ij}^{(t)} = 1 \text{ if } f_j(i) < t; \quad 0, \text{ otherwise}$$

$$q_{*j}^{(t)} = \omega_j^{(t-1)} - \omega_j^{(t)}$$

Some straightforward algebra then shows that $\mathcal{P}$ and $\mathcal{Q}$ are ∇-equivalent, establishing (b). Part (c) follows immediately by (i) using part (a) to find a ∇-equivalent pure strategy; (ii) using the logical equivalence property of bivariate ∇ to find an equivalent pure strategy with Y as the independent variable; (iii) applying part (b) to this pure strategy.

5

BIVARIATE PREDICTION ANALYSIS FOR PAIRS OF EVENTS

A proposition's "degree" is one of the important distinctions made in Section 1.2's discussion of alternative styles of prediction. Chapters 3 and 4 focused entirely on degree-1 predictions, that is, on event predictions for each observation singly. This chapter considers degree-2 propositions, which make an event prediction for each pair of observations jointly. Although the prediction logic and associated methods can be extended to event predictions for three or more observations, degree-2 receives nearly all this chapter's attention. We continue to ignore sampling considerations until Chapter 6. All the extensions of the basic model for degree-1 predictions, as developed in Chapter 4, may be applied to degree-2 predictions.

Interest in pair predictions has perhaps been strongest in research dealing with variables such as status that involve a relative comparison of individuals. These interindividual comparisons are frequently limited to those of relative order. That is, we can say that one individual has greater, equal, or less status than another, but not by how much they differ. From this viewpoint a pair of individuals can be ordered on a variable in one of three ways: (1) the first person's state is greater than the second's; (2) they are equal; (3) the first's is less than the second's. In fact, most past bivariate analysis of degree-2 prediction has been restricted to predictions involving 3×3 cross classifications of these condensed categories.

159

However we show in Section 5.1 that prediction logic also can be used for a larger class of degree-2 predictions pertaining to an $R^2 \times C^2$ cross classification of observation pairs that is formed from the original $R \times C$ table of single observations. In Section 5.2, this cross classification is condensed into the form described previously. This 3×3 table is employed implicitly by standard measures of "association" based on degree-2 predictions. We show in Section 5.3 that these measures are special cases of $\nabla_{\mathcal{P}}$. After developing the general strategy of degree-2 prediction analysis and illustrating some specialized applications in the measurement of ordinal association, we turn in Section 5.4 to comment on appropriate conditions for using degree-2 predictions. Section 5.5 investigates comparisons of degree-1 and degree-2 predictions. Such comparisons are facilitated by a basic result, namely, that every degree-1 prediction has a logically equivalent degree-2 representation. Thus the results presented in this chapter permit a unified analysis of both degree-1 and degree-2 cross classifications of a population.

The concluding section applies methods developed in the chapter to the analysis of Goldberg's (1969) data on political party identification.

5.1 DEGREE-2 PREDICTION ANALYSIS OF EVENT PAIRS IN EXTENSIVE FORM

We now develop and analyze a cross classification of paired observations from an original $Y \times X$ table of single observations from Ransford's (1972) study based on a survey of 455 suburban whites in the Los Angeles area. As shown in Table 5.1, the independent variable E describes the extent of each respondent's formal education by one of four states, ordered from high to low education: $E = \{e_1, e_2, e_3, e_4\}$. Note that we follow the standard convention of using low numerical indices to indicate high ordinal values. The dependent variable A uses three categories also ordered from high to low to summarize the extent to which the respondent gave answers expressing antagonism toward protests of blacks: $A = \{a_1, a_2, a_3\}$. Thus the 4×3 cross classification in Table 5.1 locates each respondent in one cell of $E \times A$. This table presents the data in a form appropriate for prediction analysis of degree-1 propositions.

Notation for the Degree-2 Extensive Form

In order to address degree-2 predictions, we first need to describe each pair of observations on the variables. Consider the variable $A = \{a_1, a_2, a_3\}$ in the example. These three states can be paired in $3 \times 3 = 9$ ways: $A^{*2} = \{(a_1,a_1),(a_1,a_2),(a_1,a_3),(a_2,a_1),(a_2,a_2),(a_2,a_3),(a_3,a_1),(a_3,a_2),(a_3,a_3)\}$. Now

Table 5.1   Degree-1 cross classification of survey data for white respondents, by education and antagonism toward black protest[a]

Education, E:		Antagonism, A			
		High (a_1)	Medium (a_2)	Low (a_3)	
College graduate	(e_1)	.066	.070	.081	.218
Some college	(e_2)	.152	.099	.070	.321
High school graduate	(e_3)	.195	.079	.056	.330
Less than high school	(e_4)	.101	.020	.011	.132
		.514	.268	.218	1.000
					$(N = 455)$

Source: Ransford (1972, p. 339).
[a]The entries in the table were computed from conditional row proportions given to only two decimals in the source. Consequently they may differ somewhat from the original data because of rounding error.

let A_I and A_{II} denote the respective states of the first and second observations in a pair. Then we may describe the pair on the variable by assigning it to the corresponding state of A^{*2}, namely, (A_I, A_{II}). For example, if the first observation has $A = a_3$ and the second has $A = a_1$, then the pair has $A^{*2} = (a_3, a_1)$. Sometimes we will omit the comma, writing, for example, $(a_3 a_1)$.

Since the set of nine states of A^{*2} includes all pairs of A-states, we call this the *extensive form* of A^{*2}. Similarly define the extensive form of E^{*2} as the set of $4^2 = 16$ pairs (E_I, E_{II}) of the four E-states. Then each pair of respondents in Ransford's study may be located in the 16×9 cross classification of $E^{*2} \times A^{*2}$ shown in Table 5.2. In the general bivariate situation degree-2 predictions relating $Y = \{y_1, y_2, \ldots, y_i, \ldots, y_R\}$ and $X = \{x_1, x_2, \ldots, x_j, \ldots, x_C\}$ can be applied and evaluated with respect to the $R^2 \times C^2$ cross classification in extensive form, $Y^{*2} \times X^{*2}$. In Section 2.6 we developed this form for the special case of two dichotomies to identify the degree-2 prediction underlying Yule's Q.

Computing Degree-2 Probabilities

The rules for computing the entries in the extensive form of $Y^{*2} \times X^{*2}$ from the original $Y \times X$ cross classification are simple, although their application without the aid of a computer can be tedious. As in previous

Table 5.2   Extensive form of degree-2 cross classification of survey data shown in Table 5.1[a]

$A^{*2} = \{(A_{\mathrm{I}}A_{\mathrm{II}})\}$, where A_{I} and A_{II} are antagonism to black protests of first & second respondents in pair

$E^{*2} = \{(E_{\mathrm{I}}E_{\mathrm{II}})\}$, where E_{I} and E_{II} are education of first and second respondents in pair

	a_1a_1	a_1a_2	a_1a_3	a_2a_1	a_2a_2	a_2a_3	a_3a_1	a_3a_2	a_3a_3	
e_1e_1	.004	.005	.005	.005	.005	.006	.005	.006	.007	.048
e_1e_2	.010	.007	.005	.011	.007	.005	.012	.008	.006	.070
e_1e_3	.013	.005	.004	.014	.006	.004	.016	.006	.005	.072
e_1e_4	.007	.001	.001	.007	.001	.001	.008	.002	.001	.029
e_2e_1	.010	.011	.012	.007	.007	.008	.005	.005	.006	.070
e_2e_2	.023	.015	.011	.015	.010	.007	.011	.007	.005	.103
e_2e_3	.030	.012	.009	.019	.008	.006	.014	.006	.004	.106
e_2e_4	.015	.003	.002	.010	.002	.001	.007	.001	.001	.042
e_3e_1	.013	.014	.016	.005	.006	.006	.004	.004	.005	.072
e_3e_2	.030	.019	.014	.012	.008	.006	.009	.006	.004	.106
e_3e_3	.038	.015	.011	.015	.006	.004	.011	.004	.003	.109
e_3e_4	.020	.004	.002	.008	.002	.001	.006	.001	.001	.044
e_4e_1	.007	.007	.008	.001	.001	.002	.001	.001	.001	.029
e_4e_2	.015	.010	.007	.003	.002	.001	.002	.001	.001	.042
e_4e_3	.020	.008	.006	.004	.002	.001	.002	.001	.001	.044
e_4e_4	.010	.002	.001	.002	.000⁺	.000⁺	.001	.000⁺	.000⁺	.017
	.264	.138	.112	.138	.072	.058	.112	.058	.048	1.000

$(N^2 = 207{,}025$ pairs)

[a]Shading denotes error set for proposition $e_1e_4 \rightsquigarrow A_{\mathrm{I}} < A_{\mathrm{II}}$, $e_4e_1 \rightsquigarrow A_{\mathrm{I}} > A_{\mathrm{II}}$.

chapters we consider a finite population; the resulting developments extend directly to infinite populations. It is possible to construct N^2 ordered pairs from a population of N observations. The ordering corresponds to which observation is listed first in the pair: for example, (Flo, Joe) and (Joe, Flo), are identified separately. Note that this construction also pairs each observation with itself. For example, (Flo, Flo) is a pair. Thus there are $N^2 = 455^2 = 207{,}025$ pairs of respondents in Ransford's data. To illustrate how to compute the entries of Table 5.2, suppose that the first observation in the pair has $E = e_1$ and $A = a_1$. Referring back to Table 5.1, there are $NP_{11} = 30$ respondents in the (e_1, a_1) cell. Suppose also that the second respondent in the pair has $E = e_4$ and $A = a_3$. There are $NP_{43} = 5$ such respondents in Ransford's data. Thus there are $30 \times 5 = 150$ pairs in which the first respondent lies in cell (e_1, a_1) of Table 5.1 and the second lies in (e_4, a_3). Dividing this number by the total number of pairs yields the probability, $150/N^2 = .001$, that a pair of respondents is assigned to the $(e_1 e_4, a_1 a_3)$ cell of Table 5.2. Note that we could have obtained the same results by multiplying probabilities rather than numbers of cases: $(NP_{11})(NP_{43})/N^2 = P_{11} P_{43} = .001$. By symmetry, reversing the order of the two respondents in the pairs, this same probability $[(P_{43} P_{11}) = (P_{11} P_{43})]$ appears in cell $(e_4 e_1, a_3 a_1)$.

Returning now to the general bivariate situation, suppose the first observation in a pair has (y_g, x_h) and the second has (y_i, x_j), where $g = 1, \ldots, R,\ i = 1, \ldots, R,\ h = 1, \ldots, C,\ j = 1, \ldots, C$. Denote the location of this pair in the extensive form of $Y^{*2} \times X^{*2}$ as the cell (gi, hj). Then the probability of such a pair in a random draw is the product of bivariate population probabilities for single observations:

$$P_{gihj} = P_{gh} P_{ij} \tag{1}$$

Similarly, the marginals of $Y^{*2} \times X^{*2}$ in extensive form may be computed from the $Y \times X$ probabilities:

$$P_{gi..} = P_{g.} P_{i.}$$
$$P_{..hj} = P_{.h} P_{.j} \tag{2}$$

where

$$\sum_{g=1}^{R} \sum_{i=1}^{R} P_{gihj} = P_{..hj}$$

$$\sum_{h=1}^{C} \sum_{j=1}^{C} P_{gihj} = P_{gi..}$$

All these probabilities for degree-2 prediction analysis must, of course, satisfy the usual conditions for a cross classification:

$$\sum_{g=1}^{R} \sum_{i=1}^{R} \sum_{h=1}^{C} \sum_{j=1}^{C} P_{gihj} = \sum_{g=1}^{R} \sum_{i=1}^{R} P_{gi..} = \sum_{h=1}^{C} \sum_{j=1}^{C} P_{..hj} = 1 \tag{3}$$

Symmetry and Its Implications for Prediction Statements

Because each set of two observations contributes two ordered observation pairs [(Flo, Joe) and (Joe, Flo)], there is a symmetry in the probability structure for the extensive form of $Y^{*2} \times X^{*2}$:

$$P_{gihj} = P_{gh}P_{ij} = P_{ij}P_{gh} = P_{igjh}$$

$$P_{gi..} = P_{g.}P_{i.} = P_{i.}P_{g.} = P_{ig..} \qquad (4)$$

$$P_{..hj} = P_{.h}P_{.j} = P_{.j}P_{.h} = P_{..jh}$$

Any degree-2 event prediction relating X and Y can be represented in the extensive form of $Y^{*2} \times X^{*2}$. Pure strategy degree-2 predictions may be given in prediction logic statements of the form $\mathcal{P}_{Y^{*2}X^{*2}} : \{(x_h x_j) \rightsquigarrow \mathcal{S}(x_h x_j)\}$, where $\mathcal{S}(x_h x_j)$ is a set of Y^{*2}-states. One such degree-2 statement for Ransford's data shown in Table 5.2 predicts that if the first respondent is a college graduate and the second has less than a high school education, then the first will be less antagonistic to black protest than the second:

$$(e_1 e_4) \rightsquigarrow \{(a_3 a_1), (a_3 a_2), (a_2 a_1)\}$$

In stating degree-2 predictions, however, one should recognize the necessary symmetries in $Y^{*2} \times X^{*2}$ given by (4) that result from the double counting of each set of two observations in the pairs. Consider any observation pair that is an error for the proposition given at the end of the last paragraph. When the observations are reordered, this pair contributes an observed error for the symmetric prediction: $(e_4 e_1) \rightsquigarrow \{(a_1 a_3), (a_2 a_3), (a_1 a_2)\}$. Consequently it is appropriate to require that any degree-2 prediction obey the following condition:

$$(gi, hj) \text{ is an error cell if and only if } (ig, jh) \text{ is an error cell.} \qquad (5)$$

Moreover each event in such a symmetric pair of cells should be given the same error weight:

$$\omega_{gihj} = \omega_{igjh} \qquad (5')$$

The ∇ Measure for Degree-2 Predictions in Extensive Form

Once the probability structure of $Y^{*2} \times X^{*2}$ in extensive form has been computed and the error weights are specified, methods of prediction analysis paralleling those developed earlier for degree-1 predictions apply.

Thus the prediction success of a degree-2 prediction may be measured with:

$$\nabla_{\mathscr{P}} = 1 - \frac{\sum_g \sum_i \sum_h \sum_j \omega_{gihj} P_{gihj}}{\sum_g \sum_i \sum_h \sum_j \omega_{gihj} P_{gi..} P_{..hj}} \tag{6}$$

In the example the proposition we have been considering identifies only those unweighted error events that are shaded in rows $(e_1 e_4)$ and $(e_4 e_1)$ of Table 5.2. This proposition is modestly successful for these data $(\nabla_{\mathscr{P}} = .402)$ although its predictions are quite imprecise $(U = .040)$.

5.2   DEGREE-2 PREDICTION ANALYSIS OF EVENT PAIRS IN CONDENSED FORM

Although the extensive form of $Y^{*2} \times X^{*2}$ can be used for any degree-2 prediction, this $R^2 \times C^2$ table can be extremely large. Fortunately the extensive form provides more information than is needed to analyze the most frequently used form of prediction relating two ordinal variables: "The greater is X, the greater Y tends to be." For such propositions we can simplify matters by converting from the extensive form to the *condensed ordinal form* of $Y^{*2} \times X^{*2}$. The condensed form has been used widely in the literature pertaining to the measurement of ordinal association discussed in the next section.

Creating the Condensed Ordinal Form

Suppose that for Ransford's data we predicted that, if two respondents differ in education level, the one with *more* education tends to be *less* antagonistic toward black protest and that people with the same amount of education hold the same views toward protest. Note that our prediction involves only the relative order of one person with respect to another on a given variable, without specifying how much the two persons differ. Consequently use the condensed ordinal state $E_I > E_{II}$ to describe all pairs for which the education level of the first observation is greater than that of the second: $E_I > E_{II} = \{(e_1, e_2), (e_1, e_3), (e_1, e_4), (e_2, e_3), (e_2, e_4), (e_3, e_4)\}$. Similarly, define $E_I = E_{II}$, $E_I < E_{II}$, and the corresponding states of A^{*2}. Then this degree-2 proposition in condensed ordinal form can be stated in prediction logic as: $E_I > E_{II} \rightsquigarrow A_I < A_{II}$, $E_I = E_{II} \rightsquigarrow A_I = A_{II}$, & $E_I < E_{II}$

$\sim\!\!\rightarrow A_I > A_{II}$. (The last component is implied from the first by the symmetry requirement.) More compactly this may be written

$$E_I \left\{ \begin{matrix} \geq \\ = \\ < \end{matrix} \right\} E_{II} \sim\!\!\rightarrow A_I \left\{ \begin{matrix} \leq \\ = \\ > \end{matrix} \right\} A_{II} \qquad (7)$$

Notice that this proposition requires only ordinal comparisons of the two observations, regardless of their particular states on the two variables.

Consequently to analyze this proposition we can transform Table 5.2 into the condensed ordinal form of $E^{*2} \times A^{*2}$ shown as Table 5.3. For example, compute the entry for $(E_I = E_{II}, A_I = A_{II})$ as $P(e_1e_1, a_1a_1) + P(e_1e_1, a_2a_2) + \cdots + P(e_4e_4, a_3a_3) = .004 + .005 + \cdots + .000 = .112$. Assigning all error cells (shaded) for the proposition given by (7) a weight of one, the proposition achieves a PRE of only $\nabla_{\mathscr{P}} = .138$ for these data, although it is highly precise ($U = .671$). Thus the probability structure shown in Table 5.3 can be subjected to the same kind of prediction analysis that we developed earlier for any bivariate distribution.

Table 5.3   Condensed ordinal form of survey data in Table 5.2[a]

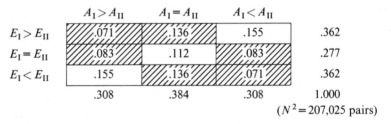

	$A_I > A_{II}$	$A_I = A_{II}$	$A_I < A_{II}$	
$E_I > E_{II}$	.071	.136	.155	.362
$E_I = E_{II}$	.083	.112	.083	.277
$E_I < E_{II}$	.155	.136	.071	.362
	.308	.384	.308	1.000

($N^2 = 207{,}025$ pairs)

[a] Shading denotes error set for $E_I > E_{II} \sim\!\!\rightarrow A_I < A_{II}$, $E_I = E_{II} \sim\!\!\rightarrow A_I = A_{II}$, & $E_I < E_{II} \sim\!\!\rightarrow A_I > A_{II}$.

In general, the condensed ordinal form of $Y^{*2} \times X^{*2}$ is the 3×3 cross classification given as Table 5.4. Note that, for example, $E_I > E_{II} \sim\!\!\rightarrow A_I < A_{II}$ may be written in the extensive form as the logically equivalent proposition $[(e_1e_2), (e_1e_3), (e_1e_4), (e_2e_3), (e_2e_4),$ or $(e_3e_4)] \sim\!\!\rightarrow \{(a_3a_1), (a_3a_2), (a_2a_1)\}$. In fact, it is easy to prove the general result that

> Every degree-2 proposition in condensed ordinal form has a logical equivalent in extensive form.[1] (8)

[1] The converse is not true: not every extensive form proposition has a logical equivalent in condensed form.

Table 5.4   Simplified notation for the condensed ordinal form of $Y^{*2} \times X^{*2}$

	$X_{\mathrm{I}} > X_{\mathrm{II}}$	$X_{\mathrm{I}} = X_{\mathrm{II}}$	$X_{\mathrm{I}} < X_{\mathrm{II}}$	
$Y_{\mathrm{I}} > Y_{\mathrm{II}}$	$\frac{1}{2}P(\mathbf{C})$	$\frac{1}{2}P(\mathbf{T}_{X\bar{Y}})$	$\frac{1}{2}P(\mathbf{D})$	$\frac{1}{2}[1 - P(\mathbf{T}_Y)]$
$Y_{\mathrm{I}} = Y_{\mathrm{II}}$	$\frac{1}{2}P(\mathbf{T}_{\bar{X}Y})$	$P(\mathbf{T}_{XY})$	$\frac{1}{2}P(\mathbf{T}_{\bar{X}Y})$	$P(\mathbf{T}_Y)$
$Y_{\mathrm{I}} < Y_{\mathrm{II}}$	$\frac{1}{2}P(\mathbf{D})$	$\frac{1}{2}P(\mathbf{T}_{X\bar{Y}})$	$\frac{1}{2}P(\mathbf{C})$	$\frac{1}{2}[1 - P(\mathbf{T}_Y)]$
	$\frac{1}{2}[1 - P(\mathbf{T}_X)]$	$P(\mathbf{T}_X)$	$\frac{1}{2}[1 - P(\mathbf{T}_X)]$	

Event Symbol	
C	"concordance": $X_{\mathrm{I}} > X_{\mathrm{II}}$ & $Y_{\mathrm{I}} > Y_{\mathrm{II}}$, or $X_{\mathrm{I}} < X_{\mathrm{II}}$ & $Y_{\mathrm{I}} < Y_{\mathrm{II}}$
D	"discordance": $X_{\mathrm{I}} > X_{\mathrm{II}}$ & $Y_{\mathrm{I}} < Y_{\mathrm{II}}$, or $X_{\mathrm{I}} < X_{\mathrm{II}}$ & $Y_{\mathrm{I}} > Y_{\mathrm{II}}$
	"ties":
$\mathbf{T}_Y$	$Y_{\mathrm{I}} = Y_{\mathrm{II}}$
$\mathbf{T}_X$	$X_{\mathrm{I}} = X_{\mathrm{II}}$
$\mathbf{T}_{XY}$	$X_{\mathrm{I}} = X_{\mathrm{II}}$ & $Y_{\mathrm{I}} = Y_{\mathrm{II}}$
$\mathbf{T}_{X\bar{Y}}$	$X_{\mathrm{I}} = X_{\mathrm{II}}$ & $Y_{\mathrm{I}} \ne Y_{\mathrm{II}}$
$\mathbf{T}_{\bar{X}Y}$	$X_{\mathrm{I}} \ne X_{\mathrm{II}}$ & $Y_{\mathrm{I}} = Y_{\mathrm{II}}$

This result demonstrates that we may condense the degree-2 cross classification by combining states (and adding their probabilities) whenever the predictions analyzed are based solely on ordinal comparisons. The cell entries in Table 5.4 are expressed as probabilities of events designated by symbols that are commonly used in discussing methods for analyzing ordinal data. These event symbols are explained in the glossary accompanying Table 5.4. Notice that the cell probabilities in this table are radially symmetric about the center cell; thus the probability in each outer cell equals the probability in the outer cell on the opposite side of the center cell. The "outer" marginals also are equal on either variable. This radial symmetry is a direct consequence of (4), which reflects the double counting of each set of two observations in forming the ordered pairs. Therefore equal error weight should be given to each symmetric pair of events in the condensed ordinal form. Using the event symbols defined in Table 5.4, at most five distinct error weights

$$\omega_{\mathbf{C}}, \omega_{\mathbf{D}}, \omega_{\mathbf{T}_{Y\bar{X}}}, \omega_{\mathbf{T}_{\bar{Y}X}}, \omega_{\mathbf{T}_{YX}}$$

should be employed with a degree-2 proposition in condensed ordinal form.

Given the radial symmetry, the expressions shown in Table 5.4, and the requirement that the cell entries sum to the respective marginals, the probabilities of the various events shown in Table 5.4 may be calculated from the original probabilities of $Y \times X$:

$$P(\mathbf{T}_Y) = \sum_{i=1}^{R} P_{i.}^2$$

$$P(\mathbf{T}_X) = \sum_{j=1}^{C} P_{.j}^2$$

$$P(\mathbf{T}_{XY}) = \sum_{i=1}^{R} \sum_{j=1}^{C} P_{ij}^2 \qquad (9)$$

$$P(\mathbf{C}) = 2 \sum_{g=1}^{R-1} \sum_{h=1}^{C-1} \sum_{i=g+1}^{R} \sum_{j=h+1}^{C} P_{gh} P_{ij}$$

$$P(\mathbf{T}_{X\bar{Y}}) = P(\mathbf{T}_X) - P(\mathbf{T}_{XY})$$

$$P(\mathbf{T}_{\bar{X}Y}) = P(\mathbf{T}_Y) - P(\mathbf{T}_{XY})$$

$$P(\mathbf{D}) = 1 - P(\mathbf{T}_Y) - P(\mathbf{C}) - P(\mathbf{T}_{X\bar{Y}})$$

We have developed the extensive form and condensed ordinal form of cross classifications for evaluating degree-2 predictions. In the case of nominal (or ordinal) variables, we could similarly create the condensed *nominal* form of $X^{*2} = \{X_I = X_{II}, \ X_I \neq X_{II}\}$. This variable retains only relative nominal information about the paired comparison. Moreover in some degree-2 applications it may be useful to cross-classify the extensive form of one variable with the condensed ordinal form of another. This can be useful in analyzing degree-2 predictions when the first variable is nominal and the second is ordinal. Any such transformation is allowed if it permits the proposition to be stated in logically equivalent form.

In working with the condensed form special attention should be given to the fact that:

> The condensed form probabilities are affected by combi-
> nation or subdivision[2] of the states of Y or X. In general,
> $\nabla_{\mathcal{P}}$ measures for degree-2 predictions in condensed form   (10)
> can be affected by changes in the ordinal category defini-
> tions.

[2] Morris (1970) develops the latter point for the condensed ordinal form in a somewhat different context.

In the example, if one respondent is a college graduate and another only had some college, then this pair of observations is strongly ordered on the education variable. However if the original states are combined into the category "more than high school," then this transformation treats the two respondents as having equal education. This combination is *necessary*, however, if the investigator wishes to use the condensed ordinal form to evaluate a proposition that in the extensive form attributes no predictive significance to the "college graduate" versus "some college" distinction, but uses the ordering of "more than high school" and the other states of the education variable.

5.3 MEASURES OF ORDINAL ASSOCIATION

This section interprets various standard measures of ordinal "association" as special applications of $\nabla_{\mathscr{P}}$. The main results are given in Table 5.7.

One persistent topic of the literature on measuring ordinal association is how ties should be treated in counting prediction errors. This is a consequence of attempts to identify a single summary measure of ordinal association. Rather than adopting this objective, we argue that measures should indicate the success attained by a specific prediction. Which events should be counted as prediction errors clearly must depend on what prediction is being evaluated.

Goodman and Kruskal's (1954) solution to the problem of ties is to discard them—they eliminate all ties before measuring association. In effect, they delete the second row and second column of the condensed ordinal form shown in Table 5.4 and normalize the remaining probabilities so that they sum to one. This produces Table 5.5A. $P(\overline{\text{T}})$ is the probability of no ties: $P(\overline{\text{T}}) = P(\text{C}) + P(\text{D}) = 1 - P(\text{T}) = 1 - P(\text{T}_{XY}) - P(\text{T}_{\overline{X}Y}) - P(\text{T}_{X\overline{Y}})$. (Note that the radial symmetry in the condensed ordinal form causes the marginal probabilities of Table 5.5 to equal one-half.) Table 5.5B shows the condensed ordinal form of Ransford's data, omitting all ties.

Goodman and Kruskal's measure of ordinal association is

$$\gamma = \frac{P(\text{C}) - P(\text{D})}{P(\text{C}) + P(\text{D})} = \frac{P(\text{C}) - P(\text{D})}{P(\overline{\text{T}})} = P(\text{C}|\overline{\text{T}}) - P(\text{D}|\overline{\text{T}}) \qquad (11)$$

where $P(\text{C}|\overline{\text{T}})$ and $P(\text{D}|\overline{\text{T}})$ are the conditional probabilities that a pair is concordant and discordant, respectively, given no ties. Thus γ is simply the probability that an observation pair falls on the major diagonal minus the probability it falls on the minor diagonal of Table 5.5A.

Table 5.5   Condensed ordinal form for pairs, omitting ties

(*A*) The General Case

	$X_{\mathrm{I}} > X_{\mathrm{II}}$	$X_{\mathrm{I}} < X_{\mathrm{II}}$	
$Y_{\mathrm{I}} > Y_{\mathrm{II}}$	$\frac{1}{2}\left[\dfrac{P(\mathbf{C})}{P(\overline{\mathbf{T}})}\right]$	$\frac{1}{2}\left[\dfrac{P(\mathbf{D})}{P(\overline{\mathbf{T}})}\right]$	$\frac{1}{2}$
$Y_{\mathrm{I}} < Y_{\mathrm{II}}$	$\frac{1}{2}\left[\dfrac{P(\mathbf{D})}{P(\overline{\mathbf{T}})}\right]$	$\frac{1}{2}\left[\dfrac{P(\mathbf{C})}{P(\overline{\mathbf{T}})}\right]$	$\frac{1}{2}$
	$\frac{1}{2}$	$\frac{1}{2}$	1.0

(*B*) The Survey Data Example, Showing Error Cells of
Prediction for Which $\nabla_{\mathcal{P}} = \gamma$

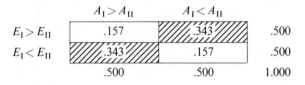

	$A_{\mathrm{I}} > A_{\mathrm{II}}$	$A_{\mathrm{I}} < A_{\mathrm{II}}$	
$E_{\mathrm{I}} > E_{\mathrm{II}}$	.157	.343	.500
$E_{\mathrm{I}} < E_{\mathrm{II}}$	.343	.157	.500
	.500	.500	1.000

Let us identify a prediction logic statement for which γ is the $\nabla_{\mathcal{P}}$ measure. For the domain of pairs tied on neither variable, let $\mathcal{P}_{\gamma}$ be $(X_{\mathrm{I}} > X_{\mathrm{II}}) \longleftrightarrow (Y_{\mathrm{I}} > Y_{\mathrm{II}})$. Then

$$\text{Rule } K \text{ error} = P\left(Y_{\mathrm{I}} < Y_{\mathrm{II}} \ \& \ X_{\mathrm{I}} > X_{\mathrm{II}} | \overline{\mathbf{T}}\right) + P\left(Y_{\mathrm{I}} > Y_{\mathrm{II}} \ \& \ X_{\mathrm{I}} < X_{\mathrm{II}} | \overline{\mathbf{T}}\right)$$

$$= \frac{1}{2}\frac{P(\mathbf{D})}{P(\overline{\mathbf{T}})} + \frac{1}{2}\frac{P(\mathbf{D})}{P(\overline{\mathbf{T}})} = \frac{P(\mathbf{D})}{P(\overline{\mathbf{T}})}$$

and

$$\text{Rule } U \text{ error} = P\left(Y_{\mathrm{I}} < Y_{\mathrm{II}} | \overline{\mathbf{T}}\right) P\left(X_{\mathrm{I}} > X_{\mathrm{II}} | \overline{\mathbf{T}}\right) + P\left(Y_{\mathrm{I}} > Y_{\mathrm{II}} | \overline{\mathbf{T}}\right) P\left(X_{\mathrm{I}} < X_{\mathrm{II}} | \overline{\mathbf{T}}\right)$$

$$= \tfrac{1}{2}\left(\tfrac{1}{2}\right) + \tfrac{1}{2}\left(\tfrac{1}{2}\right) = \tfrac{1}{2}.$$

Therefore

$$\nabla_{\mathcal{P}_{\gamma}} = 1 - \frac{P(\mathbf{D})/P(\overline{\mathbf{T}})}{1/2} = \frac{P(\overline{\mathbf{T}}) - 2P(\mathbf{D})}{P(\overline{\mathbf{T}})} = \frac{P(\mathbf{C}) - P(\mathbf{D})}{P(\overline{\mathbf{T}})} = \gamma \quad (12)$$

In the Ransford example,

$$\gamma = \frac{P(\mathbf{C}) - P(\mathbf{D})}{P(\mathbf{T})} = \frac{.142 - .310}{.452} = -.372$$

It is easy to prove that $-1 \leqslant \gamma \leqslant +1$. If $\gamma \geqslant 0$, then the value of γ indicates the proportionate reduction in error afforded by the *a priori*, degree-2 prediction $X_I > X_{II} \longleftrightarrow Y_I > Y_{II}$ given no ties. If $\gamma < 0$, then the absolute value of the measure equals the proportionate *increase* (negative decrease) in prediction error committed by the latter prediction. If, on the other hand, the prediction is $\mathcal{P}_{-\gamma} : (X_I > X_{II}) \longleftrightarrow (Y_I < Y_{II})$, then $\nabla_{\mathcal{P}_{-\gamma}} = -\gamma$ is the appropriate measure. The prediction $(E_I > E_{II}) \longleftrightarrow (A_I < A_{II})$ is logically equivalent to $\mathcal{P}_{-\gamma}$, and achieves modest success in Ransford's survey data: $\nabla_{\mathcal{P}_{-\gamma}} = -\gamma = +.372$.

There are some alternative prediction logic interpretations of γ where ties on only one variable are excluded from the domain. These are given, without further discussion, in Table 5.7.

Somers' (1962) d_{yx} is also based on the condensed ordinal form, but ties are not discarded from the analysis. Consider the proposition

$$\mathcal{P}_{d_{yx}} : X_I > X_{II} \rightsquigarrow Y_I > Y_{II} \ \& \ X_I < X_{II} \rightsquigarrow Y_I < Y_{II}$$

with the error weights shown in Table 5.6. Then

$$\text{Rule } K \text{ error} = 2\left[\omega_{\mathbf{D}} \tfrac{1}{2} P(\mathbf{D}) + \omega_{\mathbf{T}_{\overline{XY}}} \tfrac{1}{2} P(\mathbf{T}_{\overline{XY}}) \right]$$

$$= P(\mathbf{D}) + \tfrac{1}{2} P(\mathbf{T}_{\overline{XY}})$$

Table 5.6 Error weights for Somers' d_{yx}

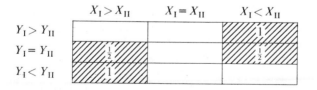

	$X_I > X_{II}$	$X_I = X_{II}$	$X_I < X_{II}$
$Y_I > Y_{II}$			1
$Y_I = Y_{II}$	$\frac{1}{2}$		$\frac{1}{2}$
$Y_I < Y_{II}$	1		

[a]Numbers in shaded error cells are error weights. Unshaded cells have zero weight.

$$\text{Rule } U \text{ error} = 2\left[\tfrac{1}{2}\left[1 - P(\mathbf{T}_X)\right]\left\{\omega_\mathbf{D}\tfrac{1}{2}\left[1 - P(\mathbf{T}_Y)\right] + \omega_{\mathbf{T}_{XY}}P(\mathbf{T}_Y)\right\}\right]$$

$$= \left[1 - P(\mathbf{T}_X)\right]\left\{\tfrac{1}{2}\left[1 - P(\mathbf{T}_Y)\right] + \tfrac{1}{2}P(\mathbf{T}_Y)\right\}$$

$$= \tfrac{1}{2}\left[1 - P(\mathbf{T}_X)\right] = \tfrac{1}{2}P(\overline{\mathbf{T}}_X)$$

and

$$\nabla_{\mathscr{P}_{d_{yx}}} = 1 - \frac{P(\mathbf{D}) + \tfrac{1}{2}P(\mathbf{T}_{X\overline{Y}})}{\tfrac{1}{2}P(\overline{\mathbf{T}}_X)} = \frac{P(\mathbf{C}) - P(\mathbf{D})}{P(\overline{\mathbf{T}}_X)} = d_{yx} \qquad (13)$$

[Note that since $P(\overline{\mathbf{T}}_X) \geqslant P(\overline{\mathbf{T}})$, $|d_{yx}| \leqslant |\gamma|$.]
For Ransford's data shown in Table 5.3,

$$d_{yx} = \frac{P(\mathbf{C}) - P(\mathbf{D})}{P(\overline{\mathbf{T}}_X)} = \frac{.142 - .310}{.616} = -.273$$

The preceding development of d_{yx} closely follows the prediction scheme originally used by Somers. An alternative $\nabla_{\mathscr{P}}$ interpretation of d_{yx}, similar to that developed by Kim (1971), uses the error weights of Table 5.6 for the logical equivalent of $\mathscr{P}_{d_{yx}}$,

$$\mathscr{P}' : Y_\mathrm{I}\left\{\begin{matrix}\geqslant\\<\end{matrix}\right\}Y_\mathrm{II} \longleftarrow X_\mathrm{I}\left\{\begin{matrix}\geqslant\\\leqslant\end{matrix}\right\}X_\mathrm{II}$$

Then $\nabla_{\mathscr{P}', \mathscr{U}} = d_{yx}$. That is, Kim's statistic for Y as the independent variable is identical to Somer's statistic for X as the independent variable, and vice versa. Thus, by exchanging the roles of independent and dependent variables from those originally assigned by Somers, one can interpret d_{yx} as a ∇ measure of a degree-2 proposition that makes a nonvacuous prediction for observation pairs tied on the independent variable.

In addition to interpreting Somers' measure, Kim proposed his own "symmetric" measure d. As shown in Table 5.7 this also has a $\nabla_{\mathscr{P}}$ interpretation. Table 5.7 summarizes the $\nabla_{\mathscr{P}}$ interpretation of a variety of conventional measures of ordinal association. This table identifies the proposition $\mathscr{P}$ and associated error weights such that $\nabla_{\mathscr{P}, \mathscr{U}}$ is equivalent to the indicated measures. Note that, as shown in the first and last rows of Table 5.7, the quadrant measure is based on a degree-1 prediction whereas Spearman's ρ_s is developed as a ∇ measure for a degree-3 prediction. The remaining measures, including Kendall's τ_b^2, are based on degree-2 predict-ions.[3]

The prediction logic framework not only provides a unified interpreta-tion of the various ordinal measures of association but also serves to point out their strengths and weaknesses. Kendall's τ_b^2 is restricted to an *ex post* mixed strategy proposition. The quadrant measure requires *ex post* knowl-edge of the median states before the otherwise *a priori* prediction can be applied. The other measures are *a priori*, and they do translate such statements as "X and Y have a monotonic relation" or "The greater the X, the greater the Y" into event predictions.

If one of these measures had to be chosen as an all-purpose screening device in exploratory research, we would prefer one of the Somers–Kim measures. Their interpretation is more direct than those of Kendall's τ_b^2 and ρ_s, and, unlike γ and ρ_s, they do not discard ties from the domain of the prediction.

On the other hand, this set of conventional measure ignores many possibilities that might be important in certain research contexts. For example, when no ties are excluded from the domain, none of these measures is designed for the degree-2 proposition,

$$X_{\mathrm{I}} \left\{ {> \atop <} \right\} X_{\mathrm{II}} \rightsquigarrow Y_{\mathrm{I}} \left\{ {\geqslant \atop \leqslant} \right\} Y_{\mathrm{II}}$$

In contrast, the $\nabla_\mathscr{P}$ approach provides a distinct measure for each distinct degree-2 proposition for either the extensive or condensed form, whether stated *a priori* or *ex post* as a pure or mixed strategy prediction.

---

[3]Given that $R^2 = \beta_{yx}\beta_{xy}$ in bivariate linear regression and that $\tau_b^2 = d_{yx}d_{xy}$, Somers (1962, p. 804) suggests that his measure is analogous to a regression slope. Hawkes (1971) amplifies and extends this interpretation. If $X_{\mathrm{I}} > X_{\mathrm{II}}$ is scored $+1$, $X_{\mathrm{I}} = X_{\mathrm{II}}$ is scored 0, $X_{\mathrm{I}} < X_{\mathrm{II}}$ is scored -1, and Y^{*2} is scored similarly, and if the scores for Y^{*2} are regressed on the X^{*2} scores, then Kendall's $\tau_b^2 = R^2$ and Somers' $d_{yx} = \beta_{yx}$.

This interpretation shares some advantages with the prediction logic interpretation of Kendall's τ_b^2 given in Table 5.7. It provides a strategy for developing measures based on τ_b^2 and d_{yx} that indicate the partial contribution of various subsets of independent variables in predicting the dependent variable. Thus this approach permits the development of an analytic scheme for multivariate predictions.

However the linear model interpretation of Kendall's τ_b^2 is strained. In a linear model, the predicted value of the dependent variable is normally interpreted as the value that would be found for the observation were it possible to remove the influence of all the random effects that are not included in the model. Unless $\tau_b^2 = d_{yx} = 1.0$ and there are no random effects or $\tau_b^2 = d_{yx} = 0.0$ and there is no interest in prediction, this interpretation cannot apply to the condensed form. When $X^{*2} = +1$, for example, the linear model predicts $+d_{yx}$, a score value that no observation can possibly exhibit. [Hawkes (1971, p. 914) recognizes this difficulty, and proposes that d_{yx} be interpreted as the average score of Y^{*2} for positive score values of X^{*2}.]

Table 5.7 Measures of ordinal association[a]

Units of analysis	Measure of bivariate association	Prediction domain	Prediction	Error events	Error weights
Single observations	Quadrant measure = $1 - 2[\Pr(X > x^\circ \,\&\, Y \leq y^\circ) + \Pr(X \leq x^\circ \,\&\, Y > y^\circ)]$	There must exist median values y° and x° such that $\Pr(Y > y^\circ) = \Pr(Y \leq y^\circ) = \Pr(X > x^\circ) = \Pr(X \leq x^\circ) = 1/2$	$X > x^\circ \longleftrightarrow Y > y^\circ$	$Y \leq y^\circ \,\&\, X > x^\circ$ $Y > y^\circ \,\&\, X \leq x^\circ$	1 1
Observation pairs	Goodman & Kruskal's $\gamma = \dfrac{P(\mathbf{C}) - P(\mathbf{D})}{P(\mathbf{C}) + P(\mathbf{D})}$	Condensed ordinal form, given either (a) $X_I \neq X_{II} \,\&\, Y_I \neq Y_{II}$ (b) $Y_I \neq Y_{II}$ or (c) $X_I \neq X_{II}$	(a) or (b): $X_I \{\substack{>\\<}\} X_{II} \longrightarrow Y_I \{\substack{>\\<}\} Y_{II}$ (c): $X_I \{\substack{>\\<}\} X_{II} \longrightarrow Y_I \{\substack{>\\<}\} Y_{II}$	$Y_I < Y_{II} \,\&\, X_I > X_{II}$ $Y_I > Y_{II} \,\&\, X_I < X_{II}$	1 1
	Somers' $d_{yx} = \dfrac{P(\mathbf{C}) - P(\mathbf{D})}{P(\bar{\mathbf{T}}_X)}$	Condensed ordinal form	$X_I > X_{II} \longrightarrow Y_I > Y_{II}$ $X_I < X_{II} \longrightarrow Y_I < Y_{II}$	$Y_I < Y_{II} \,\&\, X_I > X_{II}$ $Y_I = Y_{II} \,\&\, X_I \neq X_{II}$ $Y_I > Y_{II} \,\&\, X_I < X_{II}$	1 1/2 1
	Kim's d (symmetric) = $\dfrac{2[P(\mathbf{C}) - P(\mathbf{D})]}{P(\bar{\mathbf{T}}_X) + P(\bar{\mathbf{T}}_Y)}$	Condensed ordinal form	$X_I > X_{II} \longrightarrow Y_I > Y_{II}$ $X_I = X_{II} \longrightarrow Y_I = Y_{II}$ $X_I < X_{II} \longrightarrow Y_I < Y_{II}$	$Y_I < Y_{II} \,\&\, X_I > X_{II}$ $Y_I = Y_{II} \,\&\, X_I \neq X_{II}$ $Y_I \neq Y_{II} \,\&\, X_I = X_{II}$ $Y_I > Y_{II} \,\&\, X_I < X_{II}$	1 1/4 1/4 1

			Given the state of X^{*2}, predict each state of Y^{*2} with its conditional probability (ex post mixed strategy)	If prediction is $Y_I \left\{ \begin{matrix} \geq \\ = \\ < \end{matrix} \right\} Y_{II}$ then use

Observation pairs

Kendall's $\tau_b^2 = \dfrac{[P(C) - P(D)]^2}{P(\bar{T}_X)\,P(\bar{T}_Y)}$

[Note: This measure is simply a weighted error version of τ (unsquared) as developed in Section 2.5.2]

Condensed ordinal form

$Y_I > Y_{II}$
$Y_I = Y_{II}$
$Y_I < Y_{II}$

	$(>)$	$(=)$	$(<)$
$(>)$	0	1/4	1
$(=)$	1/4	0	1/4
$(<)$	1	1/4	0

Deuchler's $\mathfrak{R} =$ Wilson's $e = \dfrac{P(C) - P(D)}{1 - P(T_{XY})}$

Condensed ordinal form. Cell $(X_I = X_{II}, Y_I = Y_{II})$ defined as a "hole". Set-by-set normalization (see Section 4.5)

$X_I > X_{II} \longrightarrow Y_I > Y_{II}$
$X_I = X_{II} \longrightarrow \emptyset$
$X_I < X_{II} \longrightarrow Y_I < Y_{II}$

$Y_I < Y_{II}$ & $X_I > X_{II}$	1
$Y_I = Y_{II}$ & $X_I \neq X_{II}$	1/2
$Y_I \neq Y_{II}$ & $X_I = X_{II}$	1/2
$Y_I > Y_{II}$ & $X_I < X_{II}$	1

Observation triples

Spearman's $\rho_s = 3 - 6[\Pr\{X_I > X_{II}$ & $Y_I < Y_{III}\} + \Pr\{X_I < X_{II}$ & $Y_I > Y_{III}\}]$

Condensed ordinal form omitting all ties

$X_s > X_t > X_u \longrightarrow$
$[(Y_s > Y_t \ \& \ Y_s > Y_u)$ or $(Y_t > Y_s > Y_u)]$,
where s, t, u index distinct observations in the triple

$[(Y_s < Y_t \ \& \ Y_s < Y_u)$ & $(X_s > X_t > X_u)]$	1
$[(Y_u > Y_s > Y_t)$ & $(X_s > X_t > X_u)]$	1

[a] Sources for earlier interpretations of measures: quadrant measure, ρ_s, Kruskal (1958); γ, Costner (1965); d_{xy}, Somers (1962); d, Kim (1971); τ_b^2, Wilson (1973); $\mathfrak{R}$, Deuchler (1914); e, Wilson (1974).

5.4 A COMMENT ON MODES OF PREDICTION ANALYSIS FOR QUALITATIVE DATA: SINGLE OBSERVATIONS VERSUS OBSERVATION PAIRS

Reflecting on our review (in the preceding section and Chapter 2) of various measures of association for qualitative variables, we encounter an interesting paradox. With the exception of Yule's Q, all the prediction-related measures of association for *nominal* variables analyze the probability distribution in $Y \times X$ for single observations rather than pairs. On the other hand, of the measures of *ordinal* association only the quadrant measure is based on predictions about single observations. The rest deal with observation pairs or triples.[4]

Should the Analysis of Ordinal Variables Be Restricted to Pair Predictions?

This strange correlation in modes of analysis is perhaps simply the perpetuation of an historical accident by intellectual tradition. In measuring ordinal association should one indeed devote primary attention to comparisons within pairs? Kim (1971, p. 895), for example, has argued that "'order' is a relative concept, and its measurement value in itself has no specific meaning except in comparisons." Somers (1974, p. 231) has asserted this position even more strongly:

> ... the restriction to pairs in the case of ordinal statistics is compatible with the inherently comparative nature of ordinal data. And in a fundamental sense, the empirical foundation of science is comparison and contrast rather than the study of a single, isolated observation, *in vacuo*.

Yet each ordinal variable contains two specific forms of information. First, it identifies a set of equivalence classes for describing each single observation. For example, one respondent might "strongly approve" of the President's conduct in office. This opinion can be voiced regardless of the opinions of other respondents. Second, the categories of an ordinal variable also are ranked according to the "amount" or "degree" of that variable they represent. In some research applications, the statement that a person "strongly approves" is of more predictive significance than the statement that one person "has a higher degree of approval than another."

---

[4]Following the work by Kruskal (1958), we have interpreted Spearman's ρ_s to be based on *a priori* prediction about the condensed ordinal form of $Y^{*3} \times X^{*3}$ for observation triples. In contrast, Costner (1965, p. 347n) suggests using Spearman's ρ_s to predict single observations. Yet ρ_s can be taken as predicting single observations only in an *ex post* model where errors are measured as the squared difference between "predicted" and "actual" rank.

For the first of these statements attention must be given to the location of single observations in the ordered classes of the variable. In the second situation the investigator should compare observations to each other.[5] To be sure, the investigator should select methods in response to the nature of the prediction to be investigated. We have designed $\nabla_{\mathscr{P}}$ to permit this custom designing to a specific prediction, whether the proposition is chosen to predict events for single observations or observation pairs. Scientific "comparison and contrast" is provided by the comparison of conditional and unconditional probabilities in the ∇ measure; pair predictions are not necessary. Consequently choice of a prediction's degree is an important step in the research process.

Prediction Reliability: An Example of Contrasting Degree-1 and Degree-2 Predictions

The following example highlights some important differences in measuring prediction success with single observations rather than observation pairs. Suppose a doctor summarizes prognoses for 100 patients by assigning each patient to one of the expected life span categories shown as column headings in Table 5.8. Suppose further that two years after the last patient was classified, a follow-up study discovers that the actual life spans of these patients are as shown in the table.[6] The optimistic doctor overestimated the life span of all patients except those living less than one month or more than two years.

How might one assess the prediction reliability of the doctor?[7] Were we to base the analysis on comparisons among observation pairs using the conventional measures of ordinal association, we would judge the doctor to be a successful prognosticator: $\gamma = 1$, $d_{yx} = d = .842$ for the data shown in Table 5.8.

[5]For related comments see Wilson (1974), pp. 248–249. Wilson makes the important point that it may be far more natural to formulate theory in terms of individual observations than in terms of pairs.

[6]The investigator might wish to treat these data as ordinal even if instead of making set predictions the doctor had predicted life spans to, say, the nearest day. For example, a one month's error may be considered more important the shorter is the predicted life span. Consequently the squared error approach of linear models might be deemed inappropriate in this research setting. By the result (10), different categorizations of the life span variable will affect the values of the degree-2 measures given in the text. Similarly degree-1 analysis can be affected.

[7]The reliability problem was posed by Goodman and Kruskal (1954, pp. 756–758). They raise it in another substantive context, that of evaluating competing methods of nominal assignment, such as two psychological tests. Our discussion of reliability, which provides alternative analyses to those offered by Goodman and Kruskal but identical to those of Cohen (1960, 1968), also applies to the assignment problem.

Table 5.8   Hypothetical data relating predicted to actual life span[a]

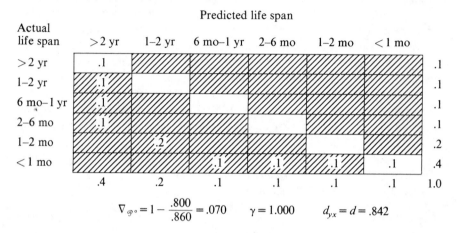

Actual life span	Predicted life span						
	>2 yr	1–2 yr	6 mo–1 yr	2–6 mo	1–2 mo	<1 mo	
>2 yr	.1						.1
1–2 yr	.1						.1
6 mo–1 yr	.1						.1
2–6 mo	.1						.1
1–2 mo		.2					.2
<1 mo			.1	.1	.1	.1	.4
	.4	.2	.1	.1	.1	.1	1.0

$$\nabla_{\mathcal{P}^\circ} = 1 - \frac{.800}{.860} = .070 \qquad \gamma = 1.000 \qquad d_{yx} = d = .842$$

[a]Shading shows errors set for $\mathcal{P}^\circ : \{x_j \leadsto y_j\}$

In contrast, as a standard of perfect prediction suppose we adopted the proposition $\mathcal{P}^\circ = \{x_j \leadsto y_j\}$ shown in this table. Based on the difficult standard of this precise prediction ($U = .860$), the doctor's success rate is, at best, unimpressive: $\nabla_{\mathcal{P}^\circ} = .070$.

Weighting "Extreme" Errors

One important difference between $\nabla_{\mathcal{P}^\circ}$ for single observations and conventional measures of ordinal association for observation pairs is the greater sensitivity of the latter measures to "extreme" prediction errors. This can be illustrated by comparing results for (A) and (B) of Table 5.9. In (A) no observation has been predicted correctly; $\nabla_{\mathcal{P}^\circ}$ is correspondingly negative. In (B), 40% of the population has been predicted correctly, and $\nabla_{\mathcal{P}^\circ}$ is positive. In contrast, the ordinal measures based on observation pairs are greater in (A) than in (B). This anomaly occurs because the data tend to deviate less from the main diagonal in (A) than in (B). The more distant observations in (B) increase the proportion of pairs that are discordant and thereby reduce the value of γ and d. Placing more weight on "extreme" errors might be desirable. However the methods of the next section can be used to show that the degree-2 predictions underlying conventional measures of association do not have equivalent degree-1 predictions about $Y \times X$. This applies to any degree-2 measure (including $\nabla_{\mathcal{P}}$) that is defined for the condensed ordinal form. Consequently none of these ordinal measures can be based on error weights for single observations in $Y \times X$.

Table 5.9   Extreme errors in predicting life span

(A)

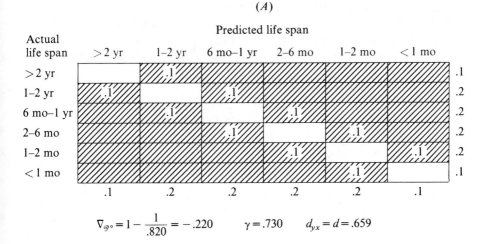

$$\nabla_{\mathscr{P}^\circ} = 1 - \frac{1}{.820} = -.220 \qquad \gamma = .730 \qquad d_{yx} = d = .659$$

(B)

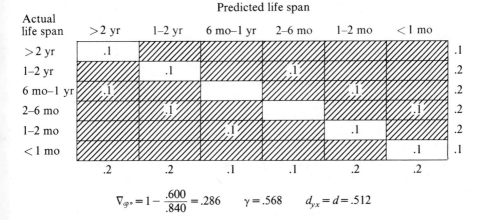

$$\nabla_{\mathscr{P}^\circ} = 1 - \frac{.600}{.840} = .286 \qquad \gamma = .568 \qquad d_{yx} = d = .512$$

5.5   COMPARATIVE PREDICTION ANALYSIS OF DEGREE-1 VERSUS DEGREE-2 PROPOSITIONS

Competition among alternative theories about the same phenomena is fundamentally important in science. Earlier in this book we discussed a variety of ways in which the success of alternative degree-1 propositions about the same cross classification can be compared. Moreover the methods of Section 5.2 can be used to translate degree-2 propositions in condensed form to their logical equivalents in extensive form, thus permitting their comparison with degree-2 propositions that have no representation in the condensed form. We now develop methods for comparative prediction analysis of degree-1 versus degree-2 propositions.

The following result (proved in Appendix 5.1) makes such comparisons possible:

Let $\mathcal{P}$ be a degree-1 proposition with error weights $\{\omega_{ij}\}$ in $Y \times X$. Let $\mathcal{P}'$ be a degree-2 proposition with error weights $\{\omega_{gihj}\} = \{\frac{1}{2}\omega_{gh} + \frac{1}{2}\omega_{ij}\}$ in the extensive form of $Y^{*2} \times X^{*2}$. Then

(14)

- (i) $\mathcal{P}'$ and $\mathcal{P}$ are logically equivalent
- (ii) $U_{\mathcal{P}'} = U_{\mathcal{P}}$
- (iii) $\nabla_{\mathcal{P}'} = \nabla_{\mathcal{P}}$

This result indicates how any degree-1 proposition can be translated into a logically equivalent degree-2 proposition (but not vice versa!) without changing the values of ∇ and rule U error.

To illustrate, assume $\omega_{12} = 1$, $\omega_{33} = 1$, and all other $\omega_{ij} = 0$. Then the equivalent degree-2 proposition has $\omega_{1323} = \omega_{3132} = \omega_{1122} = \omega_{3333} = 1$, $\omega_{1i2j} = \omega_{i1j2} = \omega_{3i3j} = \omega_{i3j3} = \frac{1}{2}$ for all $ij \neq 12$ or 33, otherwise $\omega_{gihj} = 0$. Note that, in accordance with (5'), $\omega_{gihj} = \omega_{igjh}$.

Once the degree-1 proposition $\mathcal{P}$ is restated as its degree-2 equivalent, $\mathcal{P}'$, then, in accordance with (14), it may be compared in this representation to any alternative degree-2 proposition $\mathcal{P}''$. Since by (14) the $\nabla_{\mathcal{P}}$ value is not changed, one does not actually need to carry out the $Y^{*2} \times X^{*2}$ extensive form representation of a degree-1 proposition in order to compare its $\nabla_{\mathcal{P}}$ value with that of a degree-2 proposition. However one may wish to create the extensive form in order to use the decomposition procedures of Section 4.2.5 for comparing propositions.

Thus all degree-1 and degree-2 propositions or their equivalents may be compared within the common domain, the extensive form of $Y^{*2} \times X^{*2}$. For example, Table 5.10 shows the degree-2 extensive form representation of the main diagonal degree-1 prediction $\mathcal{P}^\circ = \{x_j \rightsquigarrow y_j\}$ for a 3×3 table along with the extensive form representations of the degree-2 predictions underlying γ and d_{yx}. (Presumably comparisons of $\nabla_{\mathcal{P}^\circ}$ with γ would involve computing $\nabla_{\mathcal{P}^\circ}$ with tied rows and columns deleted.) Using Table 5.10 and equation (7) of Chapter 4, one can decompose $\nabla_{\mathcal{P}^\circ}$ into d_{yx}, ∇_+, and ∇_-.

Table 5.10 Extensive form representation of error sets for main diagonal prediction, gamma, d_{yx}, 3×3 table[a]

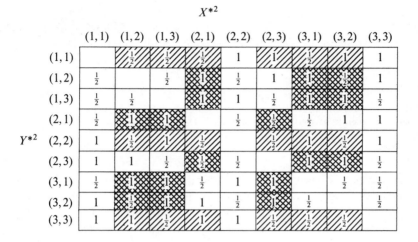

X^{*2}

[a]Numbers indicate nonzero error weights for main diagonal prediction. Cells with double hatching are error cells for γ and cells weighted 1.0 by d_{yx}. Rows and columns (1, 1), (2, 2), and (3, 3) are deleted for γ. Cells with single hatching are error cells weighted 0.5 by d_{yx}.

5.6 APPLICATION: DEFECTION FROM FATHER'S POLITICAL PARTY

Differences between degree-1 and degree-2 prediction analysis of ordinal variables can be illustrated by considering a cross classification from Goldberg's (1969) study of political party identification. Goldberg used several ordinal scales as alternative independent variables for predicting whether respondents defected from or remained loyal to their fathers' political party.[8] Table 5.11 shows one of his cross classifications.

Goldberg predicts that the proportion defecting will increase with the independent variable. He apparently was interested in degree-1 analysis, but faced with the degree-2 nature of the extant measures of ordinal association, he used Goodman and Kruskal's τ measure for nominal data. Finding very low values for τ in his data (on the order of .05), Goldberg's article is, in part, an apologia for the low explanatory power of his theory.

[8]Goldberg's independent and dependent variables are not measured independently as is required for application of $\nabla_{\mathscr{G}}$. However the confounding introduced by his techniques should be of small effect.

Table 5.11   Defection from father's party identification versus intergenerational strain $(X)^a$

	(High)										(Low)	
	x_1	x_2	x_3	x_4	x_5	x_6	x_7	x_8	x_9	x_{10}	x_{11}	
Defection	.001	.000	.002	.019	.056	.050	.073	.056	.024	.001	.004	.286
No defection	.002	.000	.002	.027	.060	.080	.148	.256	.119	.012	.008	.714
	.003	.000	.004	.046	.116	.130	.221	.312	.143	.013	.012	1.000

$(N = 915)$

Source: Goldberg (1969, Table 1, p. 9).
[a] Domain is restricted to respondents who are either Republican or Democrat identifiers and who report their fathers as either Republican or Democrat identifiers. Shading denotes error cells for *ex post* prediction discussed in text.

However τ is based on a prediction that has no relation to Goldberg's theory. As stated in Section 4.2.6, this measure equals $\nabla_{\mathscr{P}}$ for the mixed prediction strategy "If x_j, then predict y_i with probability $P_{ij}/P_{.j}$," the conditional probability of y_i given x_j *observed in the data*. This mixed prediction strategy is chosen *ex post*, and is determined solely by the observed distribution. It can, and does, result in predictions that contradict Goldberg's proposition. For an example of this, as X *increases* from x_5 to x_4 in Table 5.11, the proportion of respondents defecting from their father's party *decreases* from .48 to .41. Other reversals occur in these data. Since, as shown in Section 2.5, the predictions of τ follow the conditional probabilities, the predictions will not be consistent with Goldberg's prediction that defection increases with the X-state. Consequently the value of τ is not relevant to Goldberg's prediction.

Any degree-1 proposition that satisfies the following general model is consistent with Goldberg's proposition: Let $a < b$ be numbers between 0 and 12, and let j be the index of X states, $j = 1, \ldots, 11$. Then

If $j \leqslant a$ predict defection
If $a < j < b$ predict defection or no defection
If $j \geqslant b$ predict no defection

There are 78 distinct degree-1 predictions defined by this model, one for each pair of values for the parameters a and b. For example, the proposition selected *ex post* and shown in Table 5.11 is a particular instance of the general model when $a = 5$ and $b = 8$. In fact, following the definitions of Section 4.4, this proposition is U^*-optimal ($U = .258$) among the alternative predictions consistent with Goldberg's model.

This *ex post* proposition achieves modest success for these data: $\nabla_{\mathscr{P}}$ indicates more than 30% reduction in error. Despite the low precision value of .258 for this pure strategy proposition, its ∇ value contrasts markedly with the 5% reduction in error measured by Goodman and Kruskal's τ for the best-fitting (but irrelevant) *ex post* actuarial proposition. (Recall that Section 4.2.6 indicated that good fit does not imply a high level of prediction success.) This difference is impressive and could easily have changed the investigator's conclusions. What we have found for this table applies to other cross classifications in Goldberg's study. (See Hildebrand, Laing, and Rosenthal, 1976, for more detailed results.)

On the other hand, suppose that Goldberg had opted to operationalize his proposition, instead, as a degree-2 prediction in condensed ordinal form. (For this purpose, "defection" and "no defection" are treated as high and low categories.) One prominent degree-2 proposition that is consistent with Goldberg's prediction is

$$\mathcal{P}': X_\mathrm{I} \left\{ \genfrac{}{}{0pt}{}{\geq}{<} \right\} X_\mathrm{II} \rightsquigarrow Y_\mathrm{I} \left\{ \genfrac{}{}{0pt}{}{\geq}{<} \right\} Y_\mathrm{II}$$

For the data shown in Table 5.12 this proposition achieves little more than

Table 5.12   Degree-2 prediction analysis of Table 5.11[a]

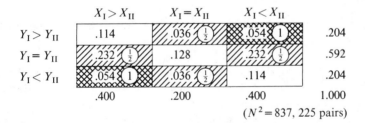

	$X_\mathrm{I} > X_\mathrm{II}$	$X_\mathrm{I} = X_\mathrm{II}$	$X_\mathrm{I} < X_\mathrm{II}$	
$Y_\mathrm{I} > Y_\mathrm{II}$	.114	.036 $(\frac{1}{2})$	.054 (1)	.204
$Y_\mathrm{I} = Y_\mathrm{II}$	.232 $(\frac{1}{2})$	.128	.232 $(\frac{1}{2})$	.592
$Y_\mathrm{I} < Y_\mathrm{II}$	.054 (1)	.036 $(\frac{1}{2})$	.114	.204
	.400	.200	.400	1.000

$(N^2 = 837, 225 \text{ pairs})$

Degree-2 measure		Effective U: proportion of N^2 pairs expected to be errors under rule U
$\nabla_{\mathcal{P}'}$	$= .103$	.719
$\nabla_{\mathcal{P}', \mathcal{U}}$	$= .146$	.441
$\nabla_{\mathcal{P}''}$	$= .334$	.164
γ	$= .353$	.169
d_{yx}	$= .149$	.400
d_{xy}	$= .291$	.204
d	$= .197$	.302
e	$= .136$	.436

[a] All single or double-hatched cells are unweighted errors for $\mathcal{P}'$. Circled numbers are error weights for $\nabla_{\mathcal{P}', \mathcal{U}}$ and e. Double-hatched cells are errors for $\mathcal{P}''$ and γ.

10% reduction in error, but its predictions are highly precise ($U = .72$). By introducing the error weights shown in Table 5.12 for this proposition, PRE is increased to 15%, but the weighted rule U error indicates a substantial loss in precision.

The measure $\nabla_{\mathscr{P}', \mathscr{W}}$ is very similar to Kim's d and Wilson's e measures. Kim's d equals the ∇ measure for proposition $\mathscr{P}'$ if we change each error weight that equals $\frac{1}{2}$ in the table to, instead, $\frac{1}{4}$, and otherwise use the same weights. Wilson's e is also a del measure for this proposition if we use the error weights shown in the table, remove the data from its center cell, and account for the hole thereby created by using the set-by-set procedure discussed in Section 4.5.3.

Finally consider the degree-2 prediction

$$\mathscr{P}'' : X_{\mathrm{I}} \left\{ \begin{array}{c} > \\ < \end{array} \right\} X_{\mathrm{II}} \rightsquigarrow Y_{\mathrm{I}} \left\{ \begin{array}{c} \geqslant \\ \leqslant \end{array} \right\} Y_{\mathrm{II}}$$

On the basis of $\nabla_{\mathscr{P}''} = .33$ this proposition is more successful than any of the others for all N^2 pairs. However it is the least precise: $U_{\mathscr{P}''} = .164$. Goodman and Kruskal's γ is based on the equivalent to this proposition except that it treats as irrelevant events those pairs of observations tied on one or both variables.

APPENDIX 5.1 LOGICAL EQUIVALENCE OF DEGREE-1 AND DEGREE-2 PROPOSITIONS

To prove result (14) of Chapter 5 let the degree-1 proposition $\mathscr{P}$ be defined for $Y \times X$ by the set of error weights $\{\omega_{ij}\}$ and $\mathscr{P}'$ be the degree-2 proposition in extensive form defined by $\{\omega_{gihj}\}$ where $\omega_{gihj} = \frac{1}{2}(\omega_{ij} + \omega_{gh})$.

(i) Two propositions are logically equivalent if, for every possible observation, the error weights they assign are always equal. More formally, $\mathscr{P}$ and $\mathscr{P}'$ with the respective sets of error weights $\mathscr{W}$ and $\mathscr{W}'$ are logically equivalent if for every observation in any population within the domain of either of the propositions, $\mathscr{P}$ assigns an error weight $\omega \in \mathscr{W}$, $\mathscr{P}'$ assigns $\omega' \in \mathscr{W}'$, and $\omega = \omega'$. We may use this to prove (i) of (14).

Consider any generic event (y_i, x_j). The NP_{ij} observations of this event contribute $\omega_{ij} P_{ij}$ to the rule K error rate for $\mathscr{P}$. Now consider $\mathscr{P}'$. The event (y_i, x_j) contributes $N^2 P_{ij}^2$ observations of the paired event $(y_i y_i, x_j x_j)$; this represents $\omega_{ij} P_{ij}^2$ rule K error for $\mathscr{P}'$. There also are $2N^2 P_{ij}(1 - P_{ij})$ pairs in which an observation in (y_i, x_j) is paired as first or second element with an observation *not* in (y_i, x_j); the contribution of the event (y_i, x_j) via such pairs to the rule K error of $\mathscr{P}'$ is $(\omega_{ij}/2)[2N^2 P_{ij}(1 - P_{ij})](1/N^2) = \omega_{ij} P_{ij}(1 - P_{ij})$. Therefore the overall contribution of (y_i, x_j) to the rule K

error rate of $\mathcal{P}'$ equals the sum

$$\omega_{ij} P_{ij}^2 + \omega_{ij} P_{ij} (1 - P_{ij}) = \omega_{ij} P_{ij}$$

This is identical to the contribution of (y_i, x_j) to the rule K error rate of $\mathcal{P}$.

Since this holds for every event (y_i, x_j) in the domain of $\mathcal{P}$ and thus for every event in the domain of $\mathcal{P}'$, and since this also obtains in every population, the two propositions with associated error weights are logically equivalent. Also note that the two propositions must have equal rule K error rates *overall*.

(ii)

$$U_{\mathcal{P}'} = \sum_g \sum_h \sum_i \sum_j \omega_{gihj} P_{gi..} P_{..hj}$$

$$= \sum_g \sum_h \sum_i \sum_j \tfrac{1}{2}(\omega_{ij} + \omega_{gh}) P_{g.} P_{i.} P_{.h} P_{.j}$$

$$= \tfrac{1}{2}\left[\sum_i \sum_j \omega_{ij} P_{i.} P_{.j} \sum_g \sum_h P_{g.} P_{.h} + \sum_g \sum_h \omega_{gh} P_{g.} P_{.h} \sum_i \sum_j P_{i.} P_{.j} \right]$$

$$= \tfrac{1}{2}\left[\sum_i \sum_j \omega_{ij} P_{i.} P_{.j} + \sum_g \sum_h \omega_{gh} P_{g.} P_{.h} \right]$$

$$= \sum_i \sum_j \omega_{ij} P_{i.} P_{.j} = U_{\mathcal{P}}$$

(iii) Since $K_{\mathcal{P}} = K_{\mathcal{P}'}$ and $U_{\mathcal{P}} = U_{\mathcal{P}'}$, $\nabla_{\mathcal{P}} = \nabla_{\mathcal{P}'}$.

6

BIVARIATE STATISTICAL INFERENCE

Until now we have assumed that predictions could be evaluated with observations that either represented an entire population or were sufficiently numerous that sampling error could be ignored. Although the population approach was a useful fiction in explaining the methods of prediction analysis, an investigator almost always wants to regard his subjects—be they people, counties, societies, rats, or whatever—as only a sample from a larger underlying population or process. Given the previous definition of $\nabla_{\mathcal{P}}$ as a population parameter, the primary problem then becomes one of *inference*—using sample data to determine something about the value of $\nabla_{\mathcal{P}}$ for the whole population.

For instance, the Udy data presented in Table 3.2 certainly do not encompass all societies organized since the beginning of recorded time. Even if that were the case one would want to postulate laws applying not only to past societies, but also to future ones. Viewed this way the Udy data represent a "sample slice" taken from an ongoing process. As a sample the data may or may not be representative. The basic task of statistics is to determine the probable degree of "nonrepresentativeness" —in other words, how much the data might be lying. In this chapter we employ statistical theory to find methods for using sample data to estimate population (or process) $\nabla_{\mathcal{P}}$ values, to determine probable ranges for these values, and to assess the possibility that an apparently positive value may have arisen purely by chance.

Some readers may be helped by a brief sketch of the procedures followed in establishing statistical inference methods. The first step is to consider the way in which the data have been gathered and to examine possible biases in that data collection. If the data collection process is believed to be reasonably free from bias, which usually requires some form of random sampling assumption, the next step is to amalgamate the sample data into an estimate of the true population quantity. Usually this step is done by direct analogy—one estimates the population parameter by the corresponding sample value. When estimating the population mean value, one uses the sample mean. (This rule is by no means inviolate; there are several situations where it fails badly. Nonetheless it is usually a good procedure.) The result of this estimating process—technically the value of the *estimating statistic*, or *estimator*—is a single number, which is just a guess and gives no indication of the probable degree of precision of that guess. To evaluate the probable precision of an estimator, an investigator needs the estimator's *sampling distribution*.

To illustrate the idea of a sampling distribution within a $\nabla_{\mathscr{P}}$ context, an artificial example may help. Imagine an urn containing a population of 200 marbles, some black (x), the rest white ($\bar{x}$), some iron (y), the rest glass ($\bar{y}$). A sample of five is drawn at random. (As each marble is drawn it is not replaced in the urn.) If the specified prediction is that black marbles tend to be iron, we can estimate $\nabla_{\mathscr{P}}$ as one minus the ratio of the fraction of black glass marbles to the product of the fraction of black marbles times the fraction of glass marbles. The result of this estimate varies with the actual sample selected (at random), and there are 2,535,650,040 possible samples. Given any assumed distribution of marble types in the population, it is possible to count how many of these samples will lead to any particular estimated $\nabla_{\mathscr{P}}$ value.[1] For example, suppose our population consists of 50 marbles of each of the 4 possible types (black glass, black iron, white glass, white iron), so that population

$$\nabla_{\mathscr{P}} = 1 - \frac{50/200}{(100/200)(100/200)} = 0$$

There are 591,600,030 samples containing no black glass marbles. Despite the presence of zero prediction errors, estimated $\nabla_{\mathscr{P}}$ is not always 1.0 for these samples. In 148,456,280 of the samples a zero marginal occurs for either the category "black" or the category "glass" or both categories; the

---

[1]The various calculations of the numbers of possible samples are made using the combinatorial analysis methods found in introductory probability texts. See, for example, Freund (1962).

result is an undefined estimated $\nabla_{\mathscr{P}}$ of $1 - 0/0$. Only in the remaining 443,143,750 samples does estimated $\nabla_{\mathscr{P}} = 1.0$; the probability of this result is therefore $443,143,750/2,535,650,040$, or about .175. In principle, it is possible to calculate the probability for each possible estimated $\nabla_{\mathscr{P}}$ value when sampling from this particular population, that is, to calculate the probability distribution, or *sampling distribution*, of the estimated $\nabla_{\mathscr{P}}$ value.

The sampling distribution will reflect the population distribution. To continue our example, if the population instead has only 2 black glass marbles and 66 of each of the other 3 types in the population so that

$$\nabla_{\mathscr{P}} = 1 - \frac{2/200}{(68/200)(68/200)} = .913$$

then the probability that estimated $\nabla_{\mathscr{P}}$ equals 1.0 is much higher, namely, .710. This result illustrates a more general one; sample estimates close to the true population value will have relatively high probability. Furthermore the larger the sample size, the more closely will the sampling distribution clump near the true value.

Once one has this sampling distribution, it is easy to find the probability that a sample estimate will fall within a specified distance of the true population value, or, reversing the viewpoint, that the true value will be within a specified distance of the sample value. This calculation specifies a *confidence interval*—a range of values within which the true value will fall with specified probability. For instance, data might allow a researcher to say, "With 90% confidence, the true $\nabla_{\mathscr{P}}$ value is between .45 and .65." Alternatively the sampling distribution allows one to calculate the probability of a large deviation of the estimated value from the true value. With this information one can test the *hypothesis* that the true value is some given number (often chosen as a "nothing-doing" number, such as $\nabla_{\mathscr{P}} = 0$ error reduction); if the estimated value is improbably far from the hypothesized value, the hypothesis should be rejected. As a slight variant, if, according to the sampling distribution, there is a small probability of observing at least as large a discrepancy between observed and hypothesized values as one actually finds, then the hypothesis is highly suspect and should be rejected. (In much published empirical research this probability is referred to either as the *p*-value or the attained significance level.)

In many cases, including those arising in this work, the sampling distribution approximates a bell-shaped, *normal* curve. Using a normal approximation yields particularly simple confidence interval and hypothesis testing procedures. All that is needed for normal-curve methods is a specification of the mean and variance that apply to the particular sampling distribution. [Definitions of the mean and variance for finite popula-

tions were provided in Section 3.2.7. Similar, but slightly different definitions are used for a normally distributed random quantity such as the
sample value of ∇_φ. The mean locates the middle of the sampling distribution; the variance, or average squared deviation from the mean (or its
square root, the standard deviation), measures the probable degree of
variation or "spread-out-ness" from that center.] A basic theorem of
mathematical statistics, the central limit theorem, states that if one draws a
sample of "large" size n from a population having mean μ and standard
deviation σ, the sampling distribution of the sample mean $\bar{x}$ is approximately normal. The mean of the sampling distribution is the same as the
population mean μ; the standard deviation is $\sigma/\sqrt{n}$. One further result of
normal theory is that, given an observed sample mean $\bar{x}$, the unknown
population mean μ falls in the interval

$$\bar{x} - \frac{1.96\sigma}{\sqrt{n}} \leqslant \mu \leqslant \bar{x} + \frac{1.96\sigma}{\sqrt{n}}$$

with probability .95. If $\sigma = 12$, $n = 100$, and $\bar{x}$ in a particular sample is 56.4,
then

$$56.4 - \frac{(1.96)(12)}{\sqrt{100}} \leqslant \mu \leqslant 56.4 + \frac{(1.96)(12)}{\sqrt{100}}$$

or

$$54.048 \leqslant \mu \leqslant 58.752$$

is a 95% confidence interval. One may also want to test the null hypothesis
that $\mu = 50$ against the alternative that $\mu \neq 50$. If the true mean is 50 the
probability is .95 that $\bar{x}$ falls in the acceptance region

$$50 - \frac{1.96\sigma}{\sqrt{n}} \leqslant \bar{x} \leqslant 50 + \frac{1.96\sigma}{\sqrt{n}}$$

or

$$47.648 \leqslant \bar{x} \leqslant 52.352$$

Since $\bar{x} = 56.4$ falls outside this region (and inside the critical or rejection
region), we reject the null hypothesis at a $1 - .95 = .05$ significance level.
Alternatively the discrepancy of $6.4 = 56.4 - 50$ is $5\frac{1}{3}$ times the standard
deviation (a standard deviation being $\sigma/\sqrt{n} = 1.2$). Since the probability
of this large or larger a discrepancy—the p-value—is less than .00001, the
hypothesis that $\mu = 50$ is untenable.

Confidence intervals and hypothesis tests, despite the apparent dissimilarities, are closely related. For the preceding it is easy to verify that any particular null hypothesis value of μ between 54.048 and 58.752 could not be rejected at the .05 significance level, and any value outside that interval would be rejected. In general, a 95% confidence region can always be interpreted as the set of nonrejectable (significance level .05) values.

A one-sided test, which accepts (fails to reject at the .05 level) the null hypothesis that μ equals some μ_0 when

$$\bar{x} \leqslant \mu_0 + \frac{1.645\sigma}{\sqrt{n}}$$

is equivalent to a one-sided 95% confidence interval,

$$\mu \geqslant \bar{x} - \frac{1.645\sigma}{\sqrt{n}}$$

In prediction analysis one-sided tests are usually appropriate because the investigator typically is concerned with whether an estimated ∇ is significantly *greater* than zero; however a corresponding one-sided confidence interval would only provide a lower bound for ∇. In most situations an investigator wants both lower and upper bounds, or a two-sided confidence interval. Since a one-sided test and a two-sided interval are not equivalent, it is possible that estimated $\hat{\nabla}$ may be significantly greater than zero while the confidence interval includes zero.

These procedures require the assumption that the only unknown parameter is the population mean μ. It was necessary to assume that the "nuisance" parameter σ was known to be 12. In many situations this is an unreasonable assumption. The population standard deviation must be estimated from the data. As is shown in most elementary statistics texts [for example, Hays (1973, Ch. 9)], this requires using the t distribution rather than the normal. However even for moderate sample sizes the numerical difference between t and normal probabilities is quite small. With the sampling distributions used in this chapter it is possible to replace "nuisance" parameters by estimates without materially affecting confidence or significance probabilities.

Another aspect of statistical inference, one neglected by virtually all basic texts, is the issue of how one selects the appropriate parameter for confidence interval estimation or hypothesis testing.

In the previous chapters we have seen that there are many, many ways to select a parameter (measure) for a cross classification. Unfortunately, very often in practice the selection procedure has been essentially the following: one examines the data for striking patterns, selects a particularly impressive pattern for testing, and then tests the hypothesis that this pattern is attributable to chance, using a statistical theory that assumes that the pattern or theory to be tested has been selected in advance! The inevitable result is the overstating of the strength of statistical evidence, or, in other words, excessive capitalizing on chance effects. For one example,[2] it is widely (but not universally) assumed that leukemia is a noncommunicable disease, which implies that multiple cases in a given geographical region at a given time are attributable only to chance. Yet if one searches records long enough one will find an "epidemic"of leukemia somewhere; not suprisingly there is a case in a Michigan city of a one-year leukemia rate 15 times the expected rate. We have made quite a fuss over the distinction between *a priori* and *ex post* selection of prediction rules. This point becomes most crucial in the sample-to-population inference process; it is distressingly easy to use *a priori* statistical procedures to analyze an *ex post* theory. In the leukemia case this error would lead to a conclusion that the results were overwhelmingly significant, whereas a simulation indicates that the probability of observing such an "epidemic" *somewhere* in the records is just about $1/2$!

Another aspect of statistical theory is the distinction between exact and asymptotic theories. Exact theories are valid for any sample size. In principle asymptotic theories hold only for large samples. Yet in fact asymptotic theories often are the most useful, even for modest sample sizes. A classic example is the standard chi-square test, which, although formally based on large-sample theory, is often appropriate for modest sample sizes. The statistical methods in this chapter are asymptotic, but they can be used with sample sizes that are about as small as most investigators are likely to use.

Methods for sample-based inferences concerning the population value of $\nabla_{\mathscr{P}}$ are presented in Section 6.3. As a necessary preface to these methods Section 6.1 considers several possible sampling schemes that might be found in practice. Section 6.2 then develops estimates of $\nabla_{\mathscr{P}}$ for various sampling schemes. This section also investigates the probability distribution of these estimates. The problem is complex enough that we generally cannot find exact results; our results are mainly large-sample approximations. Fortunately the numerical studies presented in Section 6.4 indicate that only modest sample sizes are required for the approximations to attain reasonable accuracy. Sections 6.5 and 6.6 briefly present the sampling distributions for degree-1 predictions with holes and degree-2 predictions

---

[2] Communicated by Dr. Herbert A. Wilf.

in extensive and condensed forms. In Section 6.7 we take up the problem of *ex post* selection of a prediction rule; we present some limited results that essentially say that *ex post* selection of a "statistically significant" prediction rule is equivalent to "explaining" a statistically significant chi-square test. Finally in Section 6.8 we examine the contaminating effects of classification errors in handling data.

6.1 THREE SAMPLING MODELS

In this section we indicate three models for the process of sampling data from a population. They are not the only ones that could conceivably be used, but they seem to be the basic ones. We believe these cases will cover most applications.

Sampling schemes differ in at least two dimensions. The first dimension concerns the investigator's knowledge of the marginal proportions or probabilities in the *population* (rather than sample). There are four possible cases.

A1. The marginal probabilities $(P_{.1}, P_{.2}, \ldots, P_{.C})$ are known only for the independent variable.

A2. The marginal probabilities $(P_{1.}, P_{2.}, \ldots, P_{R.})$ are known only for the dependent variable.

A3. The marginal probabilities are known for both variables.

A4. The marginal probabilities are unknown for both variables.

The second dimension indicates how the investigator has controlled the sample marginal totals:

B1. The marginal totals $(n_{.1}, n_{.2}, \ldots, n_{.C})$ are fixed by the investigator only for the independent variable.

B2. The marginal totals $(n_{1.}, n_{2.}, \ldots, n_{R.})$ are fixed by the investigator only for the dependent variable.

B3. Both sets of marginal totals are fixed.

B4. Neither set is fixed.

The investigator's knowledge of marginal distributions is described on the first dimension (A). What types of circumstances affect his knowledge?

In certain situations (A3) an investigator may have accurate estimates of marginal proportions on both variables but relatively little information on the cell-by-cell cross tabulation. One obvious example would be a retrospective study of ticket splitting in an election. The election returns would specify the exact (barring clerical error or vote-counting chicanery) vote totals for each candidate for each office; however, published results would not indicate, for example, how many of the electorate voted for both the Democratic candidate for President and the Republican candidate for governor. This information would have to be secured separately, probably via a sample survey.

In other situations (A1 or A2) one set of marginal proportions is known whereas the other is not. For example, in the Ransford study discussed in Chapter 5, the investigator might have had good information on the educational mix in the population studied but no notion of the distribution of attitudes on black protest.

In still other, perhaps most, cases (A4) neither set of marginal totals is known. In the preceding example if the population of interest to the researcher were one for which the distribution of education is unknown (such as for construction workers in Pittsburgh), he would not know either set of marginal proportions.

The available information, as well as sheer convenience, affects the investigator's choice of sampling scheme. In the preceding example of educational level versus protest attitudes the investigator may choose a random sample of 400 from the entire population, or he may *stratify* the sample by choosing, say, 100 individuals from each of four educational levels. This stratification may make the sample more "representative," and thereby increase the precision (reduce the variance) of the estimator; the cost is that there are no data on the population proportions falling in the various classes. Since [by equation (17) of Chapter 3] the ∇ measure is a weighted average of the prediction success attained by the various predictor-variable categories, one must have weights to assign to the several predictor categories. Thus if an investigator can identify population proportions in advance (in situation A1, A2, or A3), sample fractions can be controlled to advantage (using, respectively, sampling controls B1, B2, or B3); otherwise, in situation A4, fractions in various predictor categories are essential data to be gathered and no marginal controls (B4) should be imposed. It is not possible to handle prediction problems without a sense of when the relevant prediction may be applied.

There is one further dimension on which sampling designs may differ—whether one samples with replacement or without. In taking the sample of 5 balls from our urn of 200 we could replace a ball each time it was drawn, in which case we would draw 1 of 200 balls on each draw. Alternatively we could set aside each ball as it is drawn, thus randomly selecting the next ball from those not yet drawn from the urn. Typically the "without replacement" procedure is followed in real-world data collection. However it is technically a little easier to assume that sampling is done with replacement, so that an individual in the population could theoretically be sampled twice or more. The convenient "with replacement" assumption leads to a slight overstatement of the variability of a statistic (specifically, of its sampling variance).[3] Unless the sample is a substantial fraction of the whole population, the effect is negligible, and we neglect it here.

In this chapter we develop estimation techniques for the following three sampling conditions:

S1. The investigator *knows both* sets of marginal proportions (A3) and *fixes neither* set of sample totals (B4).
S2. The investigator *knows neither* set of marginal proportions (A4) and *fixes neither* set of sample totals (B4).
S3. The investigator *knows only one* set of marginal proportions (A1 or A2) and fixes the sample total for the corresponding variable (B1 or B2).

(The ordering of the sampling schemes is of no importance. We believe S2 will be the most common one.) S1–S3 should cover the great majority of research applications. Clearly these are not the only situations; one could have both sets of marginal probabilities known and fix one set of sample totals, or indeed—in principle—have any of the 16 possible combinations of the two dimensions. However seven of the combinations do not yield enough information to estimate ∇. One particular combination that does not allow exact estimation of ∇ is the situation in which the independent variable is an experimental variable with sample (marginal) totals set by the experimenter. In this case the population probability of a state of the variable is a purely hypothetical construct or at best is unknown. Therefore such situations cannot be evaluated by ∇, unless the investigator is willing to assume some more or less arbitrary marginal probabilities and perform an "as if" analysis. As these probabilities enter the definition of ∇ only as the weights of a weighted average, the choice of these probabilities is not always critical; it should be recognized, though, that the use of ∇ in such situations requires an arbitrary specification of marginal probabilities.

6.2 ESTIMATES OF $\nabla_{\!\mathcal{P}}$ IN THE THREE SAMPLING CONDITIONS

We next indicate how the true (population) value of $\nabla_{\!\mathcal{P}}$ can be estimated from sample data. For each sampling scheme we seek an appropriate way of combining any known marginal probabilities with sample evidence regarding other probabilities to yield an effective estimation procedure.

---

[3] See, for example, Freund (1962, p. 183).

After developing estimation procedures we next specify the degree to which the random variability in the sample data results in random variability of the estimates; that is, we find the probability distributions of the estimates.

We use the same notation as in the previous chapters. The true (population) proportion having $Y=y_i$ and $X=x_j$ is P_{ij}. The number (not proportion) of observations in the sample having $Y=y_i$ and $X=x_j$ is denoted n_{ij}. The sample proportion (fraction) with $Y=y_i$ and $X=x_j$ is f_{ij}. Of course $f_{ij}=n_{ij}/n$, where n is the sample size. We treat the general case of weighted errors. We assume for now that a particular prediction $\mathcal{P}$ is specified *a priori*;[4] therefore we want to estimate

$$\nabla_{\mathcal{P}} = 1 - \frac{\sum_i \sum_j \omega_{ij} P_{ij}}{\sum_i \sum_j \omega_{ij} P_{i.} P_{.j}} \tag{1}$$

We use the notation $\hat{\nabla}_{\mathcal{P},S1}$, $\hat{\nabla}_{\mathcal{P},S2}$, and $\hat{\nabla}_{\mathcal{P},S3}$ to denote, respectively, the estimators under the three sampling conditions, dropping either subscript when the meaning is clear. The "hat" symbol $\hat{}$ always indicates *estimate of*.

In condition S1 both the $P_{i.}$ and the $P_{.j}$ proportions are known (to a high degree of accuracy). Therefore the denominator of the second term is known, and only the numerator need be estimated. The most obvious way[5] to estimate the true proportion P_{ij} is by the corresponding fraction f_{ij}. This leads to the S1 estimate

$$\hat{\nabla}_{S1} = 1 - \frac{\sum_i \sum_j \omega_{ij} f_{ij}}{\sum_i \sum_j \omega_{ij} P_{i.} P_{.j}} \tag{2}$$

Thus if in a study of ticket splitting the data had been as in Table 6.1, with the actual vote fractions indicated in the margins, we would estimate ∇ (for the prediction that voters do not split) by

$$\hat{\nabla}_{\mathcal{P},S1} = 1 - \frac{(40+60)/400}{(.52)(.44)+(.48)(.56)} = .498$$

Notice that the sample marginal fractions do not match the population

---

[4] Statistical theory for *ex post* statements in the prediction logic is presented in Sec. 6.7.
[5] Here we construct our estimators by analogy with the population measures without regard to their "optimality"; a more extensive discussion of the estimators is found in Appendix 6.1.

Table 6.1 Hypothetical split-ticket voting[a]

		Senator Dem.	Senator Rep.	Total	Sample proportions	Actual election proportions
Governor	Dem.	180	60	240	.60	.56
	Rep.	40	120	160	.40	.44
	Total	220	180	400		
Sample proportions		.55	.45			
Actual election proportions		.52	.48			

[a] Table shows actual number of "survey respondents." Shading shows error cells for the prediction that voters do not split.

marginals and that we do not attempt any corrections for that fact; it is possible, but more complicated, to make the adjustment.

In condition S2 we have no information and must estimate both numerator and denominator. We estimate P_{ij} by f_{ij}, as in condition S1, and, by analogy, estimate $P_{i.}$ and $P_{.j}$ by $f_{i.}$ and $f_{.j}$, respectively. The resulting estimate of $\nabla_{\mathscr{G}}$ is

$$\hat{\nabla}_{S2} = 1 - \frac{\sum_i \sum_j \omega_{ij} f_{ij}}{\sum_i \sum_j \omega_{ij} f_{i.} f_{.j}} \tag{3}$$

Condition S3 is a little trickier. Recall that the population measure $\nabla_{\mathscr{G}}$ does not depend on which variable, X or Y, is labeled *independent* (see Sec. 3.3). Similarly the S3 estimation and inference methods are not affected by whether the marginal totals are fixed for the independent or the dependent variable. However the estimate for $\nabla_{\mathscr{G}}$ if the X marginal totals are fixed (be X independent or dependent) will in general differ from that if the Y totals are fixed.

In order to develop the estimate for this case we assume, without loss of generality, that the X marginals have been fixed. We first write the measure in a more usable form. Using the relations

$$P_{ij} = P_{.j} P(y_i | x_j), \qquad P_{i.} = \sum_{j'=1}^{C} P(y_i | x_{j'}) P_{.j'}$$

we find:

$$\nabla_{\mathscr{S}} = 1 - \frac{\displaystyle\sum_{i=1}^{R}\sum_{j=1}^{C} \omega_{ij} P_{.j} P(y_i|x_j)}{\displaystyle\sum_{i=1}^{R}\sum_{j=1}^{C} \omega_{ij} P_{.j} \sum_{j'=1}^{C} P(y_i|x_{j'}) P_{.j'}}$$

where, since we have to take two different sums over the column subscript, we denote one summation index by j and the other by j'. The only unknowns in this formulation of $\nabla_{\mathscr{S}}$ for S3 are the conditional probabilities $P(y_i|x_j)$. Since the natural (and maximum likelihood) estimator of $P(y_i|x_j)$ is $n_{ij}/n_{.j}$ or, equivalently, $f_{ij}/f_{.j}$, the natural estimator of $\nabla_{\mathscr{S}}$ is

$$\hat{\nabla}_{S3} = 1 - \frac{\displaystyle\sum_{j=1}^{C} P_{.j} \sum_{i=1}^{R} \omega_{ij}(f_{ij}/f_{.j})}{\displaystyle\sum_{j=1}^{C} P_{.j} \sum_{i=1}^{R} \omega_{ij} \sum_{j'=1}^{C} \left[(f_{ij'}/f_{.j'})P_{.j'}\right]} \tag{4}$$

For example, if we had known, for the Udy data shown in Table 3.2, that the fraction of permanent organizations was .75, and therefore the fraction of temporary organizations was .25, *and* that the fraction engaging in tillage, construction, animal husbandry, or manufacturing was .5 (so that we were in condition S1), we would estimate $\nabla_{x_1 \longleftrightarrow y_1}$ (with unweighted errors) by

$$\hat{\nabla}_{S1} = 1 - \frac{(.188)+(.020)}{(.75)(.50)+(.25)(.50)} = .584$$

More realistically, knowing neither set of marginal fractions we estimate the same ∇ by

$$\hat{\nabla}_{S2} = 1 - \frac{(.188)+(.020)}{(.718)(.450)+(.282)(.550)} = .565$$

Had we known only the .50, .50 activity marginals and stratified the sample fractions as .55, .45, we would estimate ∇ by

$$\hat{\nabla}_{S3} = 1 - \frac{(.5)\left[\dfrac{.020}{.550}\right] + (.5)\left[\dfrac{.188}{.450}\right]}{.5\left[\dfrac{.530}{.550}(.5) + \dfrac{.188}{.450}(.5)\right] + .5\left[\dfrac{.020}{.550}(.5) + \dfrac{.262}{.450}(.5)\right]} = .546$$

The S3 estimator in effect adjusts for the known marginal probabilities. An equivalent way to calculate $\hat{V}_{S3}$ is to adjust the cell frequencies so that the column sums of the adjusted frequencies equal (in this case) the known column probabilities, as shown in the table:

$.530\left(\dfrac{.5}{.550}\right)$	$.188\left(\dfrac{.5}{.450}\right)$
$.020\left(\dfrac{.5}{.550}\right)$	$.262\left(\dfrac{.5}{.450}\right)$
.5	.5

With the adjusted frequencies we can proceed as in condition S2. In fact this method applies in a case we have ignored—known row probabilities (A1) and complete random sampling (B4). A similar but more complex adjustment, the Deming–Stephan (1940a, 1940b) algorithm, could be applied in condition S1. This adjustment would yield a more efficient estimate (one having a smaller variance), but the complexity of the adjustment makes the sampling distribution computation too unwieldy for our purposes.

Expressions (2), (3), and (4) each give a single numerical guess—formally termed a point estimate—as to the true value of $\nabla_{\mathscr{P}}$. Since, by assumption, the data are collected by random sampling, the estimates are subject to sampling error. As usual we want to consider the probable range of this error, by way of either a confidence interval or a significance test. To do so we need the probability distribution of each $\hat{V}$.

In describing these distributions we use the following notation:

$$U = \sum_i \sum_j \omega_{ij} P_{i.} P_{.j} = \text{rule } U \text{ error rate}$$

$$K = \sum_i \sum_j \omega_{ij} P_{ij} = \text{rule } K \text{ error rate}$$

$$B = \frac{K}{U} = 1 - \nabla_{\mathscr{P}} \tag{5}$$

$$\pi_{i.} = \sum_j \omega_{ij} P_{.j}$$

$$\pi_{.j} = \sum_i \omega_{ij} P_{i.}$$

For condition S1 it is an easy matter to find the probability distribution of $\hat{\nabla}_{S1}$. Since both sets of marginal probabilities are known, $\hat{\nabla}_{S1}$ is a linear function of K. In the unweighted case K has a binomial distribution. Although the general weighted case is slightly more complicated, the variance of $\hat{\nabla}_{S1}$ is found to be

$$\mathrm{Var}(\hat{\nabla}_{S1}) = \frac{1}{nU^2}\left[\left(\sum_i \sum_j \omega_{ij}^2 P_{ij}\right) - K^2\right] \tag{6}$$

In the unweighted case we can simplify this expression, noting that $\omega_{ij}^2 = \omega_{ij} \in \{1,0\}$. We obtain, *for unweighted errors only*,

$$\mathrm{Var}(\hat{\nabla}_{S1}) = \frac{K(1-K)}{nU^2} \tag{6'}$$

This variance (6 or 6') tends to zero as n becomes large, so $\hat{\nabla}_{S1}$ is close to $\nabla_{\mathscr{P}}$ (with high probability) for large n. In statistical jargon $\hat{\nabla}_{S1}$ (when applicable) is a *consistent estimator* of $\nabla_{\mathscr{P}}$; the other estimates also have this property (see Appendix 6.1). The variance also decreases with increases in the prediction precision factor U. For the case in hand, that of known marginals, this dependence is easy to understand. From (2) note that in the sample estimator of $\hat{\nabla}_{S1}$, the only unknowns are the cell fractions f_{ij} in rule K error. Fluctuations in rule K error, however, affect $\nabla_{\mathscr{P}}$ with a proportionality factor of $1/U$. Therefore if the prediction is highly precise, the large value of U dampens the effect of sample fluctuations in *observed* rule K error. Although the U term is unknown for $\hat{\nabla}_{S2}$ and $\hat{\nabla}_{S3}$, in these situations the variance also decreases with increased precision.

Ideally, we could treat conditions S2 and S3 as condition S1. That is, we could calculate the exact probability of each possible value of $\hat{\nabla}$ as a function of the sample size n, the fixed marginal totals (if any), and the known marginal proportions (if any). However this would require a monstrous amount of calculation. Moreover the result would be fairly uninformative because of its complexity. Therefore we will develop instead large sample approximations for the respective probability distributions. Section 6.4 indicates that these approximations are quite accurate even for modest sample sizes. The basic technique for finding this approximation is the Taylor series approximation or delta method (Wilks, 1962, p. 260), which can be paraphrased as follows: If a set of sample statistics has, for large sample sizes, approximately a normal distribution with small variance, any smooth[6] function of these statistics also has approximately a normal distribution.

This technique can be applied to our estimates, because each is a function of the f_{ij}, $f_{i.}$, and $f_{.j}$, whose joint distribution is approximately normal for large samples and whose variances all tend to zero as the sample size grows indefinitely large. We both verify that the conditions for applying the delta method are met and calculate means and variances of the approximating distributions in Appendix 6.2. In this chapter we state the results and concentrate on their interpretation.

The basic result is that $\hat{V}_{S2}$ and $\hat{V}_{S3}$ are subject (to a good approximation) to the normal probability distribution. In each case the mean of the approximating distribution equals the true population value $\nabla_{\mathcal{P}}$. That is, if $\hat{V}_{S2}$ and $\hat{V}_{S3}$ were in fact subject to a normal distribution, then the expected value of the estimate $E(\hat{V}_{S2}) = E(\hat{V}_{S3}) = \nabla_{\mathcal{P}}$. In condition S1 it is true that for the *exact* distribution of $\hat{V}_{S1}$, $E(\hat{V}_{S1}) = \nabla_{\mathcal{P}}$. The exact distribution mean is unlikely to be $\nabla_{\mathcal{P}}$ in conditions S2 and S3; we only know that the mean is approximately $\nabla_{\mathcal{P}}$. (In the jargon of statistics, $\hat{V}_{S1}$ is, when applicable, an unbiased estimator of $\nabla_{\mathcal{P}}$; but $\hat{V}_{S2}$ and $\hat{V}_{S3}$ are not necessarily unbiased.)

Since $\hat{V}_{S2}$ and $\hat{V}_{S3}$ are asymptotically normal, specifying their means and variances is sufficient to specify their asymptotic probability distribution. We have just seen that the means equal $\nabla_{\mathcal{P}}$. As shown in Appendix 6.2, the variance of $\hat{V}$ in the second sampling scheme is approximately

$$\text{Var}(\hat{V}_{S2}) \cong (n-1)^{-1} \left[\sum_i \sum_j a_{ij}^2 P_{ij} - \left(\sum_i \sum_j a_{ij} P_{ij} \right)^2 \right] \qquad (7)$$

where

$$a_{ij} = U^{-1} \left[\omega_{ij} - B(\pi_{i.} + \pi_{.j}) \right]$$

This variance indicates the amount of *probable deviation* of $\hat{V}_{S2}$ from the true value of $\nabla_{\mathcal{P}}$. The computation of this variance requires that the true population proportions be known. However, the variance itself may be estimated by replacing population values by sample values without affecting the asymptotic distribution. For instance, regarding the Udy data in Table 3.2 as a sample (with off-diagonal error weights both 1) we found an estimated ∇ of .565; hence B is $1 - .565 = .435$.

---

[6]*Smooth* basically means that first partial derivatives exist (and at least one is nonzero) at the mean of the multivariate distribution of these statistics.

To calculate the estimated π-values, we apply (5)

$$\hat{\pi}_{1.} = \omega_{11} f_{.1} + \omega_{12} f_{.2} = 0 \times .550 + 1 \times .450 = .450$$

$$\hat{\pi}_{2.} = \omega_{21} f_{.1} + \omega_{22} f_{.2} = 1 \times .550 + 0 \times .450 = .550$$

$$\hat{\pi}_{.1} = \omega_{11} f_{1.} + \omega_{21} f_{2.} = 0 \times .718 + 1 \times .282 = .282$$

$$\hat{\pi}_{.2} = \omega_{12} f_{1.} + \omega_{22} f_{2.} = 1 \times .718 + 0 \times .282 = .718$$

The sample size $n = 149$. Finally we have already calculated the estimated U value as follows:

$$(.718)(.450) + (.282)(.550) = .478$$

Therefore, factoring out U^{-2} in (7), the estimated S2 variance is

$$\frac{1}{148(.478)^2} \Big[\big\{ [0 - .435(.450 + .282)]^2 (.530)$$

$$+ [1 - .435(.450 + .718)]^2 (.188)$$

$$+ [1 - .435(.550 + .282)]^2 (.020)$$

$$+ [0 - .435(.550 + .718)]^2 (.262) \big\}$$

$$- \big\{ [0 - .435(.450 + .282)](.530)$$

$$+ [1 - .435(.450 + .718)](.188)$$

$$+ [1 - .435(.550 + .282)](.020)$$

$$+ [0 - .435(.550 + .718)](.262) \big\}^2 \Big] = .00425$$

This and other variance calculations are, evidently, rather tedious. The programming required to estimate these by computer is not difficult. An algebraic device that can slightly reduce the tedium of hand calculations is provided in Appendix 6.3.

The delta method can also be used to obtain the asymptotic variance of $\hat{V}_{S3}$. The derivations and simplifications are best deferred to Appendix 6.2. We simply state here that the variance in this case is

$$
\mathrm{Var}(\hat{V}_{S3}) \cong \left[(n-1)U^2\right]^{-1} \left\{ \sum_j \left(\eta_{.j}^2 f_{.j}\right) \left[\frac{\sum_i (\omega_{ij} - B\pi_{i.})^2 P_{ij}}{P_{.j}} \right. \right.
$$

$$
\left. \left. - \left(\frac{\sum_i (\omega_{ij} - B\pi_{i.}) P_{ij}}{P_{.j}} \right)^2 \right] \right\}
\tag{8}
$$

where $\eta_{.j} = P_{.j}/f_{.j}$.

In estimating this variance, set B to $1 - \hat{V}_{S3}$, U to the denominator in (4), P_{ij} to $f_{ij}\eta_{.j}$, and $\pi_{i.}$ to $\sum_j \omega_{ij} P_{.j}$.

In summary, the S2 and S3 estimators are approximately normally distributed with mean $V_{\mathscr{P}}$. Their respective variances are given by (7) and (8). Although we found the exact distribution for condition S1, the S1 estimator is, for large samples, also approximately normal with mean $V_{\mathscr{P}}$ and variance given by (6). Consequently, the inference methods of the next section apply to all three sampling conditions.

6.3   CONFIDENCE INTERVALS AND SIGNIFICANCE TESTS FOR $V_{\mathscr{P}}$

The normal-distribution approximations developed in the preceding section can be used to construct either a confidence interval or a test of significance for the underlying population value of $V_{\mathscr{P}}$.

We still assume random sampling according to one of the three schemes described before. We also assume, and the assumption becomes crucial here, that the prediction $\mathscr{P}$ is specified *a priori*, not on the basis of the sample data. *Use of these methods with data-generated propositions could lead to considerable distortions.* Later (Sec. 6.7) we present separate statistical methods for propositions that are generated from the sample *ex post*.

6.3.1   Confidence Intervals

In constructing a confidence interval for V, one seeks to describe the precision of the estimate. Finding that the 95% interval for V is $.5 \pm .02$

indicates a more precise estimate of $\nabla_{\mathcal{P}}$ than does a 95% confidence interval of $.5 \pm .4$. What is the interpretation of such intervals? A confidence interval of, say, 95% is not a fixed interval but a random variable, since it is computed with data from a random sample. If we drew many samples from a population, then approximately 95% of the computed intervals would contain the true value of $\nabla_{\mathcal{P}}$ for the population. Similarly, computed 99% intervals contain the true value of $\nabla_{\mathcal{P}}$ in approximately 99 of every 100 samples.

It is simple to construct confidence intervals for ∇ since they are based on a normal sampling distribution. In order to compute a $100(1 - \alpha)\%$ confidence interval one simply takes the point estimate and the positive square root (written here as the $\frac{1}{2}$ power) of the estimated variance obtained previously to form the interval:

$$\hat{\nabla} - Z_{\alpha/2} \left[\text{est. var}(\hat{\nabla}) \right]^{1/2} \leqslant \nabla \leqslant \hat{\nabla} + Z_{\alpha/2} \left[\text{est. var}(\hat{\nabla}) \right]^{1/2} \qquad (9)$$

where $Z_{\alpha/2}$ is the published normal-table value appropriate to the specified confidence level (1.645 for 90%, 1.960 for 95%, 2.576 for 99%).[7] [Later in this chapter we refer to tabled t- and F-values as well as tabled normal values. Virtually any elementary statistics text such as Hays (1973) contains these tables and explains their use.] The probability theory underlying this interval does involve a large-sample approximation; with modest sample sizes a better approximation can be had with a continuity correction discussed in Section 6.3.5.

6.3.2 Hypothesis Testing

A very similar procedure can be applied in hypothesis testing. Again assume the proposition $\mathcal{P}$ is selected *a priori*; the procedures of this section do not apply to testing the significance of $\hat{\nabla}$ when $\mathcal{P}$ is selected *ex post facto*. (They especially do not apply to testing the *largest* $\hat{\nabla}$ that can be calculated from the sample data.)

We assume the reader has some familiarity with testing procedures. Briefly, to review the major features, a procedure requires the following specifications:

1. A procedure requires a null hypothesis. The traditional null hypothesis is that X and Y are statistically independent. The appropriate

---

[7]Since the interval involves an estimated variance it is plausible that a t-distribution table might be more appropriate. For typical sample sizes that are used the difference is negligible. Use of t rather than Z did not improve the results of the Monte Carlo studies reported in Section 6.4.

null hypothesis using $\nabla_{\mathcal{P}}$ when $\mathcal{P}$ is specified *a priori* is $\nabla_{\mathcal{P}} = c$; usually c is set to zero so that the particular prediction rule is hypothesized to yield no error reduction. However a target greater than zero may be hypothesized. The choice of $\nabla_{\mathcal{P}} = 0$ as a null hypothesis is equivalent to statistical independence *only* in 2×2 tables.

2. A procedure also requires an alternative (research) hypothesis. The alternative of most interest, $\nabla_{\mathcal{P}} > c$, states that the prediction success attained in the population by the *a priori* proposition $\mathcal{P}$ is greater than the level specified under the null hypothesis.

3. The test procedure uses the sample data to produce a *test statistic*. Prior to computation of the statistic one also must specify a *critical region* such that the null hypothesis is to be rejected in favor of the alternative hypothesis if the statistic falls into that region. The region is chosen such that the statistic has a given probability, referred to as the size or *significance level* of the test, of falling into the region if the null hypothesis is true. That is, the significance level is the probability of accepting the alternative hypothesis when, in fact, the null hypothesis is true. Commonly used probabilities are .10, .05, .01, and .001. Alternatively one can calculate the probability of equaling or exceeding the actual sample value and then compare this probability to the desired significance level.

The procedures here are based on the one-sided alternative $\nabla_{\mathcal{P}} > c$. That the alternative should be one-sided (as opposed to the two-sided $\nabla_{\mathcal{P}} \neq c$) is really a consequence of the basic viewpoint of this book. We have tried to provide a useful way of specifying a type or direction of relation (via the prediction logic) as well as a way ($\nabla_{\mathcal{P}}$) of specifying the extent to which the bivariate distribution in the population offers support for the prediction. The research hypothesis of interest then should match this focus; it should indicate positive predictive success.

The problem is to test whether the sample $\hat{\nabla}_{\mathcal{P}}$ (corresponding to the prespecified $\mathcal{P}$) attains a value such that the value cannot plausibly be attributed to random sampling variation about a true value of c. The procedure follows the standard technique: find the estimate, subtract the hypothetical population value c (typically zero), divide by the square root of the estimated variance of the estimate, and compare the result to tables of the standard normal distribution. In general, the test statistic is given by

$$\hat{Z} = \frac{\hat{\nabla} - c}{\left[\text{est. var}(\hat{\nabla})\right]^{1/2}} = \text{``standardized''} \ \hat{\nabla}, \text{ or std. } \hat{\nabla} \qquad (10)$$

If the normal table gives Z_{α} as the critical one-tailed value for the α significance level, then the support offered the research hypothesis is significant at the α level if and only if $\hat{Z} \geqslant Z_\alpha$. As in the confidence interval problem, we already have the relevant formulas for the estimate and for the standard error (the square root of its estimated variance). Thus the actual hypothesis test is an easy adjunct of the previous ideas.

6.3.3 Comparing Two $\nabla_{\mathcal{P}}$ Values

These methods can be extended to the problem of testing for the equality of two ∇ values corresponding either to different predictions ($\mathcal{P}_1$ and $\mathcal{P}_2$, say) for the same data or to the same prediction for different data sets (or, as far as the mathematics goes, to different propositions on different data sets). However, before jumping into the mathematics, we must caution that the methods may not always lead to insightful results. Prediction analysis is many-faceted. It includes evaluating not only the degree of error reduction as measured by ∇, but also the degree of precision as measured by the denominator U, and even the scientific plausibility of the prediction as measured by whether a "story goes with it." A prediction that yields a large ∇ value may not necessarily be more valuable than one with a smaller ∇ value that also has much greater scope and precision. Thus the comparison of two propositions should be based not merely on the statistical significance of the difference in estimated ∇ values, but on a comparison of ∇, U, and "plausibility" values. The comparison of ∇ values for the same prediction applied to two different data sets (perhaps acquired under somewhat different circumstances) is more direct. The scientific plausibility of the prediction is presumably the same in both cases, and unless the marginal probabilities vary widely between the two data sets, the precision measures should be similar. Therefore the problem reduces more completely to a comparison of ∇ values.

The mathematics of the statistical comparison is a fairly direct extension of previous methods. When comparing estimated ∇ values for different, independently gathered, data sets, we need only recall that the variance of the difference of two statistically independent quantities is the sum of their individual variances (as proved in most elementary statistics books). So to test the null hypothesis of equal ∇ values for *independent samples*, compute the separate estimated values, say $\hat{\nabla}_1$ and $\hat{\nabla}_2$, and their respective estimated variances and form the test statistic

$$\hat{Z} = \frac{\hat{\nabla}_1 - \hat{\nabla}_2}{\left[\text{est. var}\left(\hat{\nabla}_1\right) + \text{est. var}\left(\hat{\nabla}_2\right)\right]^{1/2}} \tag{11}$$

This $\hat{Z}$-value should be compared to a normal-table value [$Z_{\alpha/2} = 1.96$ for a two-sided, 5% significance level test ($\alpha = .05$), for instance]. Large positive or negative $\hat{Z}$-values are improbable if the true ∇ values are equal; hence such values lead to rejection of that null hypothesis. That is, the research hypothesis of unequal ∇ values achieves the α level of significance if and only if $|\hat{Z}| \geqslant |Z_{\alpha/2}|$. A $100(1-\alpha)\%$ confidence interval for the difference can be established as usual by taking

$$
\hat{\nabla}_1 - \hat{\nabla}_2 - Z_{\alpha/2} \Big[\text{est. var}\big(\hat{\nabla}_1\big) + \text{est. var}\big(\hat{\nabla}_2\big) \Big]^{1/2}
$$

$$
\leqslant \nabla_1 - \nabla_2 \leqslant \hat{\nabla}_1 - \hat{\nabla}_2 + Z_{\alpha/2} \Big[\text{est. var}\big(\hat{\nabla}_1\big) + \text{est. var}\big(\hat{\nabla}_2\big) \Big]^{1/2}
\tag{12}
$$

If one wants to compare ∇ values for two predictions on the *same data*, the computations are a bit more complicated, since the two estimated ∇ values are statistically dependent. As indicated in Appendix 6.4, it still is possible to estimate the variance of the difference. Then, as usual, the statistical test of hypothesized equality of ∇ values[8] is based on a test statistic

$$
\hat{Z} = \frac{\hat{\nabla}_1 - \hat{\nabla}_2}{\Big[\text{est. var}\big(\hat{\nabla}_1 - \hat{\nabla}_2\big) \Big]^{1/2}}
\tag{13}
$$

The continuity correction discussed in a later section will improve the probability approximation.

6.3.4 Rules of Thumb for Using the Normal Approximation

The inference procedures developed in the past section are, in a strict sense, valid only for very large samples, where, paradoxically, the procedures are usually not needed. When can they be used safely with small- or moderate-sized samples? Fortunately there are well-known general principles (see Wallace, 1958) that can be applied to the normal approximation of the binomial distribution—that is, to condition S1. We use these results to establish a general rule of thumb. Later, in Section 6.4, we summarize studies that indicate that the same rules of thumb also apply well to condition S2.

The key to the use of general statistical principles is the numerator of the error ratio used in computing $\hat{\nabla}_{S1}$ and $\hat{\nabla}_{S2}$,

$$
\hat{K} = \frac{1}{n} \sum_i \sum_j \omega_{ij} n_{ij}
\tag{14}
$$

[8]If for some reason one wants to test the null hypothesis that $\nabla_1 - \nabla_2 = c$, merely subtract c from the difference in estimated values.

If each error weight ω_{ij} equals zero or one[9], the double sum simply counts the number of cases in the sample that are rule K errors; if each such case is called a failure, we have the standard setup for the binomial distribution, with n trials and probability of failure

$$K = \sum_i \sum_j \omega_{ij} P_{ij} = U(1 - \nabla) \qquad (15)$$

The normal approximation to the binomial is the best known of all statistical approximations. It is most accurate for $K = .5$, when samples as small as 10 or 12 give decent approximations. The general rule of thumb is that nK and $n(1 - K)$ should be at least 5 to get good results—roughly, that the true probability is within $\pm.01$ of the approximation. The rule of thumb may be written alternatively as follows:

$$\frac{5}{n} \leqslant U(1 - \nabla) \leqslant 1 - \frac{5}{n}$$

To apply this rule of thumb to our S1 estimate note that $\hat{\nabla}_{S1}$ is a linear function (with nonrandom coefficients) of the double sum in equation (14). In general, performing linear operations on a random quantity does not alter the quality of a normal approximation at all (since the standardized value of the new quantity is the same as, or the negative of, that of the old one). Consequently, the same rule of thumb applies to $\hat{\nabla}_{S1}$. The quantity $U(1 - \nabla)$ is small, requiring a large n for a good approximation, if either

1. The prediction has limited precision, so that U, the expected rate of rule U errors, is small.
2. The prediction is highly successful, so that the true value of $\nabla_{\mathcal{P}}$ is near one.

In particular, under the null hypothesis that $\nabla = 0$, $U(1 - \nabla) = U$ should be moderately large (ideally .5) to indicate a good approximation. For confidence intervals $U(1 - \nabla)$ is estimated by $U(1 - \hat{\nabla}) = \hat{K}$. Therefore one would want a large U and a modest $\hat{\nabla}$ as indications of a good approximation. With a large estimated ∇ the approximate confidence interval for ∇ can extend above one; such an interval would of course indicate a poor approximation. There is a conflict between the technical desirability of a modest-sized ∇ and the scientific desirability of a large ∇. We would hope that a researcher would welcome the finding of a large estimated ∇ and

---

[9]We have not done a detailed analysis of weighted-error approximations, but believe that the same general results will hold.

merely recognize that the normal approximation may be somewhat misleading.

As we have mentioned previously, the quality of the approximations depends on the *total* number (or rate) of rule K errors. This contrasts with the usual chi-square test, in which the approximation requires a certain minimum expected number (often taken as 5) in *each* cell. There is no such requirement for $\nabla_{\mathscr{P}}$; as long as the total error rate is large enough, the expected number in individual cells may even be zero (as in the Sawyer–MacRae data shown in Table 3.5). This follows because the ∇ measure treats data additively, whereas chi-square requires division by each cell expectation separately. This result suggests that the prediction analysis approach is especially useful in problems with a large number of cells but moderate sample sizes. Again we regard this technical value as secondary to the conceptual value of prediction analysis in stating and evaluating useful statements, particularly in complex large-table problems. It is comforting, though, that the technical properties of the method encourage its use in such problems.

A second rule of thumb, based on the results contained in Section 6.4, applies only to hypothesis testing. This rule of thumb says that the approximate p-value (α) corresponding to a given $\hat{Z}$ is most accurate when the exact p-value is in the range .05 to .001—the range of significance levels generally used in research. For high exact values, such as .25, the approximate value of the significance level is too large, hence, conservative. For extremely low values, such as 10^{-7}, the approximate p-value is too small.

6.3.5 Improving the Approximation—Continuity Correction

The quality of the normal approximation in any binomial situation can be improved by a *continuity correction*. The studies summarized in Section 6.4 also indicate that this correction makes a clear improvement in the S2 sampling situation and should also lead to improvement for S3.

The corrected expression for hypothesis testing is

$$\hat{Z} = \frac{\hat{\nabla}^+ - c}{\left[\text{est. var } (\hat{\nabla})^2\right]^{1/2}} \tag{10'}$$

whereas the corrected $100(1 - \alpha)\%$ confidence interval is

$$\hat{\nabla}^+ - Z_{\alpha/2}\left[\text{est. var}(\hat{\nabla})\right]^{1/2} \leqslant \nabla_{\mathscr{P}} \leqslant \hat{\nabla}^- + Z_{\alpha/2}\left[\text{est. var}(\hat{\nabla})\right]^{1/2} \tag{9'}$$

where

$$\hat{V}^+ = 1 - \frac{\hat{K} + D/n}{\hat{U}} \qquad (16)$$

$$\hat{V}^- = 1 - \frac{\hat{K} - D/n}{\hat{U}} \qquad (17)$$

$D = 1.0$ for a 2×2 table with diagonally opposed error cells or any table that can be shown to be logically equivalent to such a table by transformations of the variables

$\qquad = 0.5$ otherwise

Note that no change is made in the estimated variance. (Some readers may wish to skip the following discussion and turn directly to the numerical illustrations in the next subsection.)

The correction is motivated by the fact that $\hat{V}$ can take on only a limited, discrete number of values whereas the approximation uses the continuous normal curve to calculate probabilities. To perform the correction arithmetically we move the relevant $\hat{V}$ value half the distance to the next possible value given the sample marginals.[10] In most cases this means adding or subtracting one-half from the number of rule K errors; however in the case of a 2×2 table with diagonally opposed error cells (or any case that can be reduced to such a table by combining categories without changing the definition of error) the number of rule K errors must change in increments of 2 if the marginal totals are to remain constant. Thus with the Udy data of Table 3.2, if the number in the $(2,1)$ cell were to increase to 4, the $(2,2)$ cell would have to decrease to 38 to maintain a total of 42 in the y_2 row; in turn the $(1,2)$ cell would have to increase to 29 to maintain 67 in the x_2 column, for a net increase of 2 errors. Therefore in this case, and only this case, the continuity correction requires adding or subtracting one entire error to the rule K total to "split the difference."

6.3.6 Numerical Illustrations

Our numerical illustrations are all based on the assumption of condition S2 sampling, the situation that is undoubtedly most widespread in practice. We can begin with the Udy example, for we have already computed

---

[10]Since for different marginal totals there might be other $\hat{V}$ values within the range of correction, this argument is not entirely convincing. However our numerical studies indicate that this method improves the approximation substantially.

$\hat{V} = .565$ and est. var $(\hat{V}) = .0042$. Equations (16) and (17) are then used to compute $\hat{V}^+ = .551$ and $\hat{V}^- = .579$.

Using (9') we compute the 95% confidence interval as follows:

$$.551 - 1.96(.00425)^{1/2} \leqslant \nabla_\mathscr{P} \leqslant .579 + 1.96(.00425)^{1/2}$$

or

$$.423 \leqslant \nabla_\mathscr{P} \leqslant .707$$

Since $n\hat{U}(1 - \hat{V}) = 31$, this interval should be reliable. To test the null hypothesis that $\nabla_\mathscr{P} = 0.0$ versus the alternative hypothesis that $\nabla_\mathscr{P} > 0.0$, use (10') to form

$$\hat{Z} = \frac{.551}{(.00425)^{1/2}} = 8.44$$

From a normal table we find that the probability of finding this value of $\hat{Z}$ or a larger one is less than 10^{-7}. Since $n\hat{U} = 149(.478) = 71.23$, we are well within the safe range for application of the test. For this same data the conventional chi-square statistic is 54.205. Recalling from Section 3.1.2 that $\nabla_{x_1 \longleftrightarrow y_1} = 0$ if and only if statistical independence holds, chi-square can be used for the two-tailed test $\nabla \neq 0.0$ versus $\nabla = 0.0$. Compared to the 10^{-7} value for the one-tailed test based on $\hat{Z}$, the p-value for chi-square is only 2.0×10^{-4}. The true probability for the alternative $\nabla > 0.0$ actually should be exactly one-half the true probability for the alternative $\nabla \neq 0.0$. That this ratio fails to hold recalls that both χ^2 and $\hat{Z}$ are asymptotic test statistics. The very low p-value for $\hat{Z}$ also recalls our second rule of thumb, that the asymptotic test underestimates very low significance probabilities.

The Udy example is one where our methods and conventional *ex post* testing can both safely be applied. This is not the case for the Korteweg main diagonal hypothesis. [Recall that we are treating this "*ex priori*" hypothesis as if it were (totally) *a priori*. Since the results of Sec. 6.7 provide very conservative procedures for propositions that are totally *ex post*, we believe that, in the absence of procedures fully appropriate to "*ex priori*" analysis, our *a priori* procedures should be used.] From Table 3.3 we find there are fewer than three "expected" cases in an average cell. The standard rule of thumb of five "expected" in each cell would lead one to be wary of the chi-square test. In contrast, $n\hat{U} = 24 \times .5833 = 14$ indicates that the true p-level is well approximated by .0007, based on a $\hat{Z}$ statistic of 3.286. The very wide 95% confidence interval, $.195 \leqslant \nabla_\mathscr{P} \leqslant .805$, illustrates the general point that a "significant" rejection of the null hypothesis

$\nabla_{\mathcal{P}} = 0.0$ does not imply that the estimate of $\nabla_{\mathcal{P}}$ is very precise. In fact, since $n\hat{K} = 14 \times .50 = 7$ is not substantially greater than 5, the correct interval may well be somewhat wider than that given by the approximation.

As a final illustration of the inference methods for an *a priori* specified $\mathcal{P}$, consider again the 3×4 data matrix shown in Table 2.3, which cross-tabulates the positions of French deputies on two scales developed from their records on roll call votes. Although on inspection the non-monotonic relation in the table discussed in Chapter 2 may appear to be fairly strong, one cannot reject the null hypothesis of statistical independence (even at the .10 level) using the chi-square test. However suppose the research hypothesis predicted, *a priori*, an inverted U by specifying just one of the (unweighted) error sets shown in Table 6.2. The results vindicate the earlier judgment based on inspection of the data: The relation is strong and easily passes the .01 level of significance. Note that the two $\hat{Z}$ statistics are roughly equal despite the larger $\hat{\nabla}$ for $\mathcal{P}_2$. In effect $\mathcal{P}_2$ also has a larger estimated variance than $\mathcal{P}_1$, primarily because the prediction is less precise. The larger variance thus offsets the larger $\hat{\nabla}$. We return to these data in Section 6.7, where $\mathcal{P}_1$ and $\mathcal{P}_2$ are viewed as having been selected *ex post*.

Table 6.2 Error sets and test statistics for French roll call data in Table 2.3

Error set	$\mathcal{E}_1 = \{(2,2),(2,3),(3,2),(3,3)\}$	$\mathcal{E}_2 = \{(3,2),(3,3)\}$
$\hat{\nabla}_{S2}$	.4833	.6204
Sample $\hat{U}$	.1457	.0850
$n\hat{U}$	13.55	7.91
$Z = \text{std } \hat{\nabla}_{S2}$	3.1569	3.220
$p <$	.001	.001

6.4 ADEQUACY OF THE NORMAL APPROXIMATION

Our numerical computations based on S2 sampling relied not only on asymptotic theory but also on rules of thumb and correction factors related to the normal approximation to the binomial distribution. However since the exact S2 distribution is not binomial, it is appropriate to carry out a direct assessment of the accuracy of the S2 approximation.

Two techniques were used to study the approximation. First, we made comparisons with the exact distribution for the special case of 2×2 tables with $\nabla = 0$. Second, we performed Monte Carlo simulation studies by drawing repeated samples from a known population and constructing (to the simulation accuracy) the actual sampling distribution.

6.4.1 Numerical Comparisons with Exact Probabilities in 2×2 Tables

In this subsection we compare the nominal probabilities given by the normal approximation to exact probabilities in a special case. In the 2×2 table the hypothesis that $\nabla_{\mathcal{P}} = 0$ for any nontrivial proposition is equivalent to the hypothesis of statistical independence. The probabilities of various error cell entries, given the marginal totals, can be calculated directly; this is the basis of Fisher's *exact test*. Therefore it is possible to compare these probabilities with our normal approximation. Strictly speaking, the comparison is forced, since the exact probabilities are conditional on the marginal totals and the approximation is unconditional. However, inasmuch as unconditional probabilities are weighted averages of conditional ones, if we find that our approximate probabilities are generally close to the exact conditional ones, we can be sure that we have a good approximation. Our numerical studies for sample sizes of 25 and 100 indicate that the approximation is quite good except when marginal totals are badly skewed.

These computations compared the exact and approximate probabilities of obtaining a $\hat{\nabla}_{S2}$ value equaling or exceeding a specified value (p-value) for a prediction specifying only one error cell or two diagonally opposed cells, using the continuity correction (10′). The general principles revealed were as follows:

1. With respect to the effect of the rule U error rate, the approximation is best, and conservative in the sense of overstating error probabilities, when U is moderately large—perhaps slightly less than .5. This result parallels the S1 sampling result.

2. With respect to marginal distributions the approximation is typically best when the marginal totals are about equal. When the marginals are skewed by a ratio exceeding 4 : 1, fairly substantial errors in approximation can occur even when $n = 100$; these errors occur generally when the rule U error rate is small or very large.

These principles have been studied directly only in the very limited 2×2 case under the hypothesis of independence. They are consistent with the ideas in previous subsections and with the results of the next subsection, which gives us some confidence that they are valid more generally.

6.4.2 Monte Carlo Experiments

In the preceding subsection we took one step toward indicating that the asymptotic distribution gives an adequate approximation for use with small samples. That step does not establish the adequacy of the approximation in general, since it brings no evidence to bear on larger tables and, even in 2×2 tables, on confidence intervals and hypothesis tests in which $\nabla_{\mathcal{P}} > 0$. The case of $\nabla_{\mathcal{P}} > 0$ is especially important, as the actual distribution of $\hat{\nabla}$ becomes highly skewed as $\nabla_{\mathcal{P}}$ approaches its upper bound of 1.0. This skewness affects the accuracy of approximation by the normal distribution, which is symmetric rather than skewed.

Procedures

To study the adequacy of the normal approximation for condition S2 we conducted a variety of Monte Carlo experiments. In a Monte Carlo experiment a randomly selected sequence of numbers is used to simulate a sample that would be drawn from a fixed probability distribution. We investigated sample sizes of 25 and 100. In condition S2 the data should be generated by a multinomial distribution. In the 2×2 case, for example, an observation would fall in the $(1, 1)$ cell with probability P_{11}, in $(2, 1)$ with P_{21}, in $(1, 2)$ with P_{12}, and in $(2, 2)$ with $P_{22} = 1 - P_{11} - P_{12} - P_{21}$. These four probabilities are uniquely determined by specifying the population marginals, a proposition $\mathcal{P}$, and the value of $\nabla_{\mathcal{P}}$ for the population.

We investigated the quality of the approximation for

2×2 tables with a single error cell.

2×2 tables with two diagonally opposed error cells.

3×2 tables with two diagonally opposed error cells.

For each type of table a variety of specifications of marginal probabilities and true $\nabla_{\mathcal{P}}$ values were used to investigate the effects of asymmetric marginals and nonzero ∇ values. For each specification there were 1000 replications. The results for the 3×2 table were so similar to those for the 2×2 table that only a limited number of specifications needed to be run.

In each of the 1000 replications n (25 or 100) independent random numbers were generated by a uniform distribution on the closed interval $[0, 1]$. Thus a sequence of n thousand independent numbers was generated in each experiment. Each experiment generally had a unique sequence, guaranteeing that our results do not reflect a particular random sequence. By testing the random number against the specified values for the cell probabilities, an observation was "assigned" to a cell. For example, if $P_{11} = .1$ and $P_{12} = .3$, an observation was assigned to cell $(1, 2)$ if its random number was greater than .1 but less than or equal to .4.

It is quite possible, indeed highly probable for some cases with $n = 25$, that a column or row (rows in the 3×2 case) has no observation assigned to it. This will make one or more propositions for the data table inadmissible (in a sampling sense) and no estimator can be computed. We simply omitted these samples, which is what a practical investigator typically does when facing zero marginals on relevant variable states. Averages were based only on admissible samples.

Results

The results of the experiments may be summarized as follows:

1. The average estimated value of $\nabla_{\mathcal{P}}$ closely approximates the true value.

Although we know from analysis that our S2 estimator is asymptotically unbiased, the Monte Carlo studies must be used to infer the extent of bias in small samples. Bias is slight in 2×2 tables, even in samples as small as 25, except for very skewed marginals. Bias is always very small if $\nabla_{\mathcal{P}} < .2$. Because of the upper bound of 1.0, bias is less for $\nabla_{\mathcal{P}} = .9$ than for intermediate values. Bias is virtually eliminated (less than .01) for $n = 100$, except for very highly skewed marginals. Similarly in our specifications for the 3×2 table the greatest bias occurred for skewed marginals and intermediate values of $\nabla_{\mathcal{P}}$. The results for the 3×2 table generally resemble those for the 2×2 table; therefore we do not refer further to the 3×2 results.

One further remark concerning bias is in order. In our specifications the results show that the bias in the mean estimate of $\nabla_{\mathcal{P}}$ for a single error cell in the 2×2 table is more frequently positive than negative. However with but one exception the bias is negative (conservative) for propositions identifying two diagonally opposed error cells in the 2×2 table. Apparently the averaging inherent in the estimator for propositions with multiple error cells has a conservative effect. This suggests that we can expect a slight tendency to negative bias in the estimator of $\nabla_{\mathcal{P}}$ for complex propositions about large tables.

In summary, the results for 2×2 and 3×2 tables with reasonable marginals show very little bias in our estimator for $\nabla_{\mathcal{P}}$, even when the sample size is small.

2. The asymptotic standard error (7) closely approximates the "true" (the average over 1000 replications) standard error.

The standard error (computed about the true value of $\nabla_{\mathcal{P}}$) of the sample estimates is closely approximated by the asymptotic standard error. Again important deviations occur only for skewed marginals. Usually the asymptotic standard error slightly underestimated the true standard error, although there were specifications where the asymptotic standard error slightly exceeded the true standard error.

3. The estimated standard error is seriously biased for small samples; it generally underestimates the true standard error.

For each replication, the sample estimator of the standard error is computed by using the f_{ij} in (7). The true standard error is severely underestimated by its sample estimator when $n = 25$. For $n = 100$, however, there is substantial improvement. Only highly skewed marginals and very large values of $\nabla_{\mathcal{P}}$ present serious problems. Glaring errors occurred when our approximation criterion $nU(1 - \nabla)$ was very small; in the worst case ($n = 25, U = .01, 1 - \nabla = .1$), the simulated standard error is .99, whereas the average estimated standard error was .07. If $nU(1 - \nabla)$ is at least 2, the bias is tolerably small.

The underestimate of the standard error in small samples suggests why the continuity corrections improved the comparison of the exact and asymptotic distributions in the previous section. Undoubtedly much of the effect of the continuity correction serves to compensate for this underestimate.

4. Hypothesis tests based on the standardized value are conservative.

Recall that the standardized value of the estimate is $\hat{\nabla}_{S2}^{+}/[\text{est. var} (\hat{\nabla}_{S2})]^{1/2}$. If no prediction errors are observed in the sample—that is, if $\sum_i \sum_j \omega_{ij} f_i = 0$—then est. var $(\hat{\nabla}_{S2}) = 0$ and the standardized value is not defined. Consequently computation of sample averages and sample standard errors for the standardized values is meaningful only if at least one error occurs in every admissible replication. For $n = 25$ this condition is met only when the marginals are relatively balanced. For those experiments where computation was possible the means of the standardized values of the estimates of $\nabla_{\mathcal{P}}$ are substantially negative, rather than zero. The bias is a consequence of the continuity correction whereby we artificially increase the rule K errors. Given this bias, hypothesis tests based on these statistics tend to be quite conservative for small samples. On the other hand, the variances of the standardized values in our experiments generally are greater than the desired value of one, although not so large as to overcome the negative bias. Since the variances were computed about zero, they would in fact be closer to one without the continuity correction.

5. Confidence intervals should be used with caution when the marginals are skewed or the sample size is small.

The factors of bias, variance approximation, variance estimation, and skewness of distributions combine to determine the accuracy of an approximate confidence interval for a true $\nabla_{\mathcal{GP}}$ value. Since the previous results indicate that most of these factors (particularly the variance calculations) can be troublesome with skewed marginals and with the quantity $nU(1 - \nabla)$ much less than 5, it is not surprising that approximate confidence probabilities can be seriously in error in such a situation. For instance, when $n = 100$, $U = .01$ and $1 - \nabla_{\mathcal{GP}} = .7$ (with a single error cell in a 2×2 table) only 688 of 1000 nominal 90% confidence intervals covered the true $\nabla_{\mathcal{GP}}$; of course had the approximation been good about 900 intervals would have covered the true value. Generally it appears that when $nU(1 - \nabla)$ is 5.0 or smaller, the confidence interval tends to be much too narrow; some nominal 95% intervals contained the true $\nabla_{\mathcal{GP}}$ value only in 60% of the replications. Conversely, when $nU(1 - \nabla)$ is fairly large the intervals are slightly too wide and therefore conservative; a nominal 90% interval can have a 94 or 95% coverage probability.

In summary, all our results point in the same direction. The essential condition for the applicability of the S2 normal approximation is the textbook condition for the applicability of the normal approximation to the binomial—in our notation, $nU(1 - \nabla)$ should be at least 5 (and at most $n - 5$). When this condition is met, barring extreme skewness of marginal probabilities or remarkably large values of true $\nabla_{\mathcal{GP}}$, a researcher may use the approximation serenely.

Even the caveats that necessarily accompany such a statement are limited. The calculations are based on S1 and S2 sampling and on unweighted errors. Although we do not have direct evidence about the quality of the approximation, the same principles should apply under S3 sampling or with weighted errors.

6.5 SAMPLING DISTRIBUTIONS WHEN CERTAIN EVENTS ARE EXCLUDED

In Section 4.5 a modification of ∇ was introduced to correct for the presence of irrelevant or logically impossible cells. In sampling condition S2, ∇' or ∇'' can be estimated by replacing population probabilities by sample fractions. Since the correction factors in ∇' and ∇'' involve unknown population probabilities, the sampling distribution of estimated ∇ differs from the ordinary one.[11]

In the special case of $P'_{ij} = P_{ij}$, the changes can be stated readily. (The

---

[11]The variance derivation for ∇' was originally performed by Richard J. Morrison.

general case is just a more complex application of the same methods.) As before, the delta method applies, yielding an asymptotic normal distribution with mean ∇' or ∇'' as appropriate; the asymptotic variance again takes the following form:

$$\text{Var}(\hat{\nabla}) = (n-1)^{-1}\left[\sum_i \sum_j a_{ij}^2 P_{ij} - \left(\sum_i \sum_j a_{ij} P_{ij}\right)^2\right] \qquad (18)$$

For $\hat{\nabla}'$, the a_{ij} are (as derived in Appendix 6.5):

$$a_{ij} = U^{-1}\left\{ K\left[\sum_{j'}\bar{\phi}_{ij'}P_{.j'} + \sum_{i'}\bar{\phi}_{i'j}P_{i'.}\right] + \bar{H}\omega_{ij} - (1-\nabla)(\pi_{i.} + \pi_{.j})\right\} \qquad (19)$$

where $\bar{H} = 1 - H, \bar{\phi}_{ij} = 1 - \phi_{ij}$; H is defined in Section 4.5.1 and $\pi_{i.}, \pi_{.j}, K$ and U are defined in (5). For $\hat{\nabla}''$, the a_{ij} are given by

$$a_{ij} = (U'')^{-1}\left\{ \omega_{ij}\right.$$

$$\left. - \frac{K}{U''}\left[U''_{.j}(P_{.j})^{-1} + \sum_{j'}P_{.j'}\left(\bar{H}_{.j'}\right)^{-1}\left[\omega_{ij'} - \pi_{.j'}\bar{\phi}_{ij'}\left(\bar{H}_{.j'}\right)^{-1}\right]\right]\right\} \qquad (20)$$

where now

$$U''_{.j} = \frac{\sum_i \omega_{ij}P_{i.}P_{.j}}{\sum_i \bar{\phi}_{ij}P_{i.}}$$

$$\bar{H}_{.j} = \sum_i \bar{\phi}_{ij}P_{i.} = 1 - H_{.j}$$

$$U'' = \sum_j U''_{.j}$$

With these expressions confidence intervals and hypothesis tests can be carried out as before. Once again, one may replace population probabilities by sample fractions in the variance expressions without affecting the large-sample distribution. In assessing the quality of the small-sample approximation, the uncorrected rule U error rate should be used in calculating the indicator quantity $nU(1 - \nabla)$. The effect of the correction is to increase U, and the corrected U value may not adequately indicate the "effective" sample size.

6.6   SAMPLING DISTRIBUTIONS FOR DEGREE-2 ANALYSIS

The results in this chapter have all pertained to degree-1 predictions. In such predictions the ∇ value can be determined directly from the cross tabulation. With degree-2 predictions a slight additional complexity arises, because the data must first be converted to extensive or condensed ordinal form. However the delta method for the large-sample distribution still applies. The results stated here and derived in Appendix 6.5 are based on sampling condition S2, a single random sample from the entire population with no known population probabilities.

For predictions evaluated in the extensive form

$$\nabla_{\mathcal{P}} = 1 - \frac{\sum_g \sum_i \sum_h \sum_j \omega_{gihj} P_{gihj}}{\sum_g \sum_i \sum_h \sum_j \omega_{gihj} P_{gi..} P_{..hj}} \tag{21}$$

where the derived probabilities P_{gihj} are determined from the population probabilities by

$$P_{gihj} = P_{gh} P_{ij}$$

Under sampling condition S2, ∇ is estimated by replacing P_{ij} by the sample estimate f_{ij}, yielding

$$\hat{\nabla}_{\mathcal{P}} = 1 - \frac{\sum_g \sum_h \sum_i \sum_j \omega_{gihj} f_{gh} f_{ij}}{\sum_g \sum_h \sum_i \sum_j \omega_{gihj} f_{g.} f_{.h} f_{i.} f_{.j}} \tag{22}$$

Once again it follows from the delta method that the sampling distribution of $\nabla_{\mathcal{P}}$ can be approximated for large samples by a normal distribution with mean $\nabla_{\mathcal{P}}$ and variance

$$\mathrm{Var}\left(\hat{\nabla}_{\mathcal{P}}\right) = (n-1)^{-1}\left[\sum_i \sum_j a_{ij}^2 P_{ij} - \left(\sum_i \sum_j a_{ij} P_{ij}\right)^2\right]$$

where the a_{ij} are now

$$a_{ij} = 2 U_2^{-1} \left[\Upsilon_{ij} - \frac{K_2}{U_2} (\Upsilon_{i.} + \Upsilon_{.j}) \right] \tag{23}$$

with

$$K_2 = \sum_g \sum_h \sum_i \sum_j \omega_{gihj} P_{gihj}$$

$$U_2 = \sum_g \sum_h \sum_i \sum_j \omega_{gihj} P_{g.} P_{.h} P_{i.} P_{.j}$$

$$\Upsilon_{ij} = \sum_g \sum_h \omega_{gihj} P_{gh}$$

$$\Upsilon_{i.} = \sum_g \sum_h \sum_j \omega_{gihj} P_{g.} P_{.h} P_{.j}$$

$$\Upsilon_{.j} = \sum_g \sum_h \sum_i \omega_{gihj} P_{g.} P_{.h} P_{i.}$$

As usual, the P's appearing in the quantities in the variance may be replaced by sample f's without altering the large-sample distribution. Confidence intervals and hypothesis tests for *a priori* ∇ values are calculated just as before.

The procedure for the condensed ordinal form is virtually the same. Extending the notation of Chapter 5,

$$\hat{P}(\mathbf{C}) = 2 \sum_{g=1}^{R-1} \sum_{h=1}^{C-1} \sum_{i=g+1}^{R} \sum_{j=h+1}^{C} f_{gh} f_{ij}$$

$\hat{P}(\mathbf{D})$ and the various $\hat{P}(\mathbf{T})$ are similarly defined by substituting sample proportions for degree-1 probabilities. Also,

$$\hat{\nabla}_{\mathscr{P}} = 1 - \frac{\hat{K}}{\hat{U}}$$

where

$$\hat{K} = \omega_{\mathbf{C}} \hat{P}(\mathbf{C}) + \omega_{\mathbf{D}} \hat{P}(\mathbf{D}) + \omega_{\mathbf{T}_{\overline{Y}X}} \hat{P}(\mathbf{T}_{\overline{Y}X})$$

$$+ \omega_{\mathbf{T}_{Y\overline{X}}} \hat{P}(\mathbf{T}_{Y\overline{X}}) + \omega_{\mathbf{T}_{YX}} \hat{P}(\mathbf{T}_{YX})$$

and

$$
\hat{U} = \frac{1}{2} \left[\omega_{\mathbf{C}} + \omega_{\mathbf{D}} \right] \hat{P}\left(\overline{\mathbf{T}}_Y\right) P\left(\overline{\mathbf{T}}_X\right) + \omega_{\mathbf{T}_{\overline{Y}x}} \hat{P}\left(\overline{\mathbf{T}}_Y\right) \hat{P}\left(\mathbf{T}_X\right)
$$

$$
+ \omega_{\mathbf{T}_{Y\overline{x}}} \hat{P}\left(\mathbf{T}_Y\right) \hat{P}\left(\overline{\mathbf{T}}_X\right) + \omega_{\mathbf{T}_{Yx}} \hat{P}\left(\mathbf{T}_Y\right) \hat{P}\left(\mathbf{T}_X\right)
$$

Again the asymptotic sampling distribution is normal with mean $\nabla_{\mathcal{P}}$ and variance

$$
\mathrm{Var}\left(\hat{\nabla}_{\mathcal{P}}\right) = (n-1)^{-1} \left[\sum_i \sum_j a_{ij}^2 P_{ij} - \left(\sum_i \sum_j a_{ij} P_{ij} \right)^2 \right]
$$

with a_{ij} defined by

$$
a_{ij} = 2U^{-1} \Bigg[\Big\{ \omega_{\mathbf{C}} \big[P(Y>i, X>j) + P(Y<i, X<j) \big]
$$

$$
+ \omega_{\mathbf{D}} \big[P(Y>i, X<j) + P(Y<i, X>j) \big]
$$

$$
+ \omega_{\mathbf{T}_{\overline{Y}x}} \big[P(Y>i, X=j) + P(Y<i, X=j) \big]
$$

$$
+ \omega_{\mathbf{T}_{Y\overline{x}}} \big[P(Y=i, X>j) + P(Y=i, X<j) \big]
$$

$$
+ \omega_{\mathbf{T}_{Yx}} P(Y=i, X=j) \Big\} \tag{24}
$$

$$
- (1 - \nabla_{\mathcal{P}}) \Big\{ \frac{1}{2} \big[\omega_{\mathbf{C}} + \omega_{\mathbf{D}} \big] \big[P\left(\overline{\mathbf{T}}_Y\right) P(X \neq j)
$$

$$
+ P\left(\overline{\mathbf{T}}_X\right) P(Y \neq i) \big]
$$

$$
+ \omega_{\mathbf{T}_{\overline{Y}x}} \big[P\left(\overline{\mathbf{T}}_Y\right) P(X=j) + P\left(\mathbf{T}_X\right) P(Y \neq i) \big]
$$

$$
+ \omega_{\mathbf{T}_{Y\overline{x}}} \big[P\left(\mathbf{T}_Y\right) P(X \neq j) + P\left(\overline{\mathbf{T}}_X\right) P(Y=i) \big]
$$

$$
+ \omega_{\mathbf{T}_{Yx}} \big[P\left(\mathbf{T}_Y\right) P(X=j) + P\left(\mathbf{T}_X\right) P(Y=i) \big] \Big\} \Bigg]
$$

where $P(Y>i) = \sum_{i'>i} P_{i'}$, $P(X>j) = \sum_{j'>j} P_{.j'}$, and so on. Note that the summations in the condensed ordinal form variance refer to the original $Y \times X$ cross classification.

6.7 INFERENCE FOR *EX POST* PREDICTIONS

Our previous development of the sampling theory for estimates of $\nabla_{\mathcal{P}}$ explicitly assumes that the predictive rule $\mathcal{P}$ is specified *a priori*, independently of the actual results of the data. We believe that the most effective use of prediction analysis methods requires thoughtful prior consideration of research alternatives. Although, ideally, there should be a prior specification of a proposition $\mathcal{P}$ (or a few alternatives), many propositions will be found *ex post facto*, after the empirical data have been collected, tabulated, and examined.

As briefly indicated in the introduction to this chapter, the statistical theory required for such "postdictions" differs from the more standard methods just outlined. To see why, suppose that a researcher has to compare the means of a certain quantity under a wide variety of treatments, say reduction of blood pressure under 15 different drugs. An *ex post* method would be to compare only the largest and smallest sample means. An erroneous way to do it would be to run a t test, taking the difference of the largest and smallest means, dividing by the standard error, and comparing the result with a tabled t-value to assess significance. This procedure is wrong for two reasons. First, the fact that we are subtracting the smallest mean from the largest one implies that the expected difference will be greater than zero (substantially so with 15 possible means and the usual sampling variability), even if the true population means are identical. As the mean value of t is zero, a t test cannot possibly be appropriate. Second, this procedure uses a one-tailed test. If the true means are equal, in a replication of the experiment the difference in means under these two treatments might be negative. So the statistical analysis should be two-tailed, recognizing both positive and negative differences. The two errors combine to yield spurious results, which are due purely to chance but which are claimed as nonchance (statistically significant) by the misuse of standard, *a priori* methods in an *ex post* problem.

There are several ways in which *ex post* prediction logic statements may be formed. *In extremis* one can hold only that the data are "not nonrelated" and test the null hypothesis that there exists no statement in the prediction logic with an associated $\mathcal{U}$ matrix for which $\nabla_{\mathcal{P}}$ differs from zero—in either a positive or negative direction. We address this problem in Section 6.7.1. Alternatively the investigator may have prespecified a certain set of alternative predictions. For example, in a 3×3 table the investigator might specify the two statements shown in Table 3.3 for the inflation example. He may then test the null hypothesis that $\nabla_{\mathcal{P}} = 0$ for both of the specified statements versus the alternative that at least one $\nabla_{\mathcal{P}} \neq 0$, again in either a positive or negative direction. Section 6.7.2 analyzes this more general case. Third, one may use the data to find a rule that is "best" using sample implementations of the optimality criteria discussed in Section 4.4. We discuss this possibility briefly in Section 6.7.3. In each case the statistical theory is closely related to the chi-square test. The results are not entirely satisfactory; all tests are two-tailed (whereas the entire point of prediction analysis is to specify propositions with *positive* ∇ values) and we have no way of finding confidence intervals for an *ex post* ∇ value. The available procedures only act as screening devices to indicate whether the positive ∇ value for an apparently interesting *ex post* proposition is possibly attributable to chance.

6.7.1 The Relation of the Chi-Square Test to Prediction Analysis

Since the "postdiction" methods are based on the chi-square test of association, the reader might now choose to review the discussion of that test in Section 2.3.

The relation between the chi-square procedure and *ex post* prediction analysis involves correspondences of null hypotheses and hypothesis testing methods. The null hypothesis of the chi-square procedure is statistical independence, or $P_{ij} = P_{i.}P_{.j}$ for every cell (i,j). As demonstrated in Section 3.2.6, the hypothesis that a *particular* ∇_{φ} is zero does *not* imply statistical independence. However if *every* ∇_{φ} is zero, statistical independence must hold. To see this consider the prediction $\mathcal{P}_{ij}$, which has cell (i,j) as the only error event, and the associated ∇ value

$$\nabla_{ij} = 1 - \frac{P_{ij}}{P_{i.}P_{.j}}$$

If ∇_{ij} is zero, $P_{ij} = P_{i.}P_{.j}$. If this is true for every ∇_{ij}, then, first, every ∇_{φ} is zero [by equation (16) of Chapter 3] and, second, the independence definition is satisfied. Conversely statistical independence necessarily means that every ∇_{φ} is zero. Thus the hypothesis of statistical independence is equivalent to the hypothesis that every ∇_{φ} is zero.

The hypothesis test for an *ex post* ∇ value can be outlined as follows. One takes the estimated $\hat{\nabla}$, then divides by its estimated standard error, and squares the result. This value is compared to a quantity from a chi-square table, not a normal table.

The variance formula (for condition S2) given in Section 6.2 can be simplified under the hypothesis of statistical independence. As shown in

Appendix 6.6 the variance is

$$\mathrm{Var}_0(\hat{\nabla}) = \left[(n-1)U^2\right]^{-1}\left(\sum_i \sum_j \omega_{ij}^2 P_{i.}P_{.j} - \sum_i \pi_{i.}^2 P_{i.} - \sum_j \pi_{.j}^2 P_{.j} + U^2\right)$$

(25)

where $U, \pi_{i.}$, and $\pi_{.j}$ are defined in (5). This variance can be estimated by replacing $P_{i.}$ by $f_{i.}$ and $P_{.j}$ by $f_{.j}$ in these definitions.

The test statistic for an *ex post* ∇ is then

$$\hat{Z}^2 = \frac{\hat{\nabla}^2}{\mathrm{est.\ var}_0(\hat{\nabla})}$$

(26)

The hypothesis of independence (all ∇'s zero) is rejected if

$$\hat{Z}^2 > \frac{n-1}{n}\chi^2(\alpha)$$

(27)

where $\chi^2(\alpha)$ is the tabled chi-squared value with $(R-1)(C-1)$ degrees of freedom and the desired significance level α. The theorem, proved in Appendix 6.7, which justifies this procedure is the following:

The largest possible $\hat{Z}^2$ value for a given sample, Z^{2*}, satisfies

(28)

$$Z^{2*} = \frac{n-1}{n}\chi^2$$

where χ^2 is the ordinary chi-square statistic given by equation (1) of Chapter 2. Note that χ^2 is a statistic computed from the data whereas $\chi^2(\alpha)$ corresponds to values published in a table of the chi-square distribution. From this it follows that under the null hypothesis

$$\mathrm{Prob}\left[\hat{Z}^2 > \frac{n-1}{n}\chi^2(\alpha)\right] \leqslant \mathrm{Prob}\left[Z^{2*} > \frac{n-1}{n}\chi^2(\alpha)\right]$$

(29)

$$= \mathrm{Prob}\left[\chi^2 > \chi^2(\alpha)\right] = \alpha$$

where the first step follows from the fact that $\hat{Z}^2 \leqslant Z^{2*}$, the second from (28), and the last from the fact that the chi-square statistic has (to a good approximation) a chi-square distribution. So the *ex post* test described in this section is conservative.

Another way to use (29) is to consider the relation between the chi-square test and the *ex post* ∇ test. The analysis shows that

1. If the chi-square statistic is not significant, $\hat{\nabla}_{\mathscr{P}}$ is not significant for any proposition selected *ex post*.
2. If the chi-square statistic is significant, then there is at least one weighted sum of the $\hat{\nabla}_{ij}$'s (such as the one corresponding to Z^{2*}) that is also significant.

The first statement says that if the chi-square test does not assert that a relation exists, neither will an *ex post* del test. The second statement says that if the chi-square test asserts that a relation exists, then one can find a set of error weights that can lead to an "*ex post* significant" result. However these error weights are determined by the data and may have no particular meaning in context; furthermore some of them may be negative. As argued in Section 4.1, negative error weights seem inappropriate in an error reduction context. Therefore it cannot be guaranteed that one can always find *ex post* a meaningful proposition and set of error weights to yield a statistically significant $\hat{\nabla}$ even when the chi-square test asserts that a relation exists. From this perspective consider the many empirical studies that interpret data *ex post* on the basis of a significant chi-square test. Even if chi-square is "significant," (29) shows that the interpretation may not be.[12]

For an example let us consider the results for the MacRae data given in Table 2.3. The value of χ^2 for the data is $10.89 < 12.6$, the tabulated chi-square value for 6 degrees of freedom and a 5% significance level. Accordingly, no *ex post* prediction should yield a significant ∇. In particular, the two propositions tested in Table 6.2 (and found significant on an *a priori* basis) yield $\hat{Z}^2$ values of 8.280 and 6.300, respectively. Comparing these with $12.46 = [(n-1)/n]\chi^2(.05)$ [for six degrees of freedom], we see that neither approaches significance at the .05 level. Here is an example of the value of *a priori* specification of a theory and a warning about the deceptiveness of *ex post* reasoning. Without an *a priori* theory an investigator could not conclude that a nonchance relation existed; given the *a priori* theory, the data would provide substantial, statistically significant confirmation. Conversely if the proposition had been developed *ex post* from

---

[12]Readers who are familiar with the Scheffé (1959, pp. 67–72) [see also Hays (1972, Ch. 12, 14)] multiple comparison method may recognize the parallel between this result and the theorem connecting the "maximal comparison" of means with an F-ratio. The problem of uninterpretable weights for *ex post* optimal ∇ has a parallel in the difficulty of interpreting the maximal comparison. When we are interested only in $\nabla_{\mathscr{P}}$ with unweighted errors, a method analogous to Tukey's studentized range test (also described in Scheffé, pp. 73–75) would be preferable. We have no such method.

the data and tested as if it had been specified in advance, spurious statistical significance would have resulted.

The use of this *ex post* test and the chi-square distribution is based on a large-sample approximation. In view of the equivalence of the chi-square statistic and the largest squared standardized $\hat{V}$, the chi-square rule of thumb that "expected errors" *in each cell* should total at least five ought to be a reasonable requirement for a good approximation. Where many error cells are combined *ex post*, this rule is probably too restrictive. The additive nature of $\hat{V}$ suggests again that the *total* number of "expected errors" ought to exceed five. [Alternatively Lancaster (1969, p. 175) suggests that there should be at least one expected error in each cell for large tables.]

Other methodological problems of *ex post* selection of propositions are more serious than those of sample size. First, since the proposition is, almost by definition, selected from the data to yield a large $\hat{V}$, the estimate of true ∇ is biased upward. In particular, even if independence holds so that any true ∇ is zero, it is almost certain that some sample $\hat{V}$ values will be positive. These $\hat{V}$-values will be the ones that are selected; hence the *ex post* selection automatically overestimates the true value, at least when all true ∇ values are zero. There seems to be no easy way of correcting or measuring this bias. We return to this problem in Section 6.7.3.

As shown by the MacRae example, a potentially serious error occurs if the *a priori* test is used when the prediction is selected from the data. As the relation of the *ex post* test to chi-square indicates, there are two aspects to this error:

1. A one-tailed test is used when a two-tailed test is perhaps more plausible. In such cases the nominal one-tailed significance level is half what it should be, thus claiming more statistical significance than is justified.
2. A normal density or its squared version, a chi-square distribution with *one* degree of freedom, is used instead of a chi-square density with $(R-1)(C-1)$ degrees of freedom. Since the latter density places much more probability on large numerical values, the significance probability can be drastically altered. For instance, an observed value of 10 would be extremely improbable (highly significant) with 1 degree of freedom, but right at the expected value for 10 degrees of freedom. This problem does not arise in 2×2 tables, but the one- versus two-tailed problem remains.

The essence of this discussion is that the researcher can gain a great deal in statistical precision by selecting a prediction in advance. Such a course may not always be possible. Sometimes one can specify a limited number of predictions, more than one but much less than 2^{RC}. The next section deals with testing the statistical significance of the collection of $\hat{\nabla}$'s obtained for these predictions.

6.7.2   The Chi-Square Test for a Prespecified Set of Alternative Predictions

In going from *a priori* to *ex post* specification of predictions we went from one prediction to all possible predictions. In this subsection we deal with a chi-square test for the intermediate case of a prespecified set of a limited number of alternative predictions. The derivations of this subsection are more complicated and the results are not used again. Those who wish to move on can rely conservatively on the chi-square test outlined in the previous subsection.

The essential result of this subsection is that with a limited number of predictions being considered one can use a chi-square test with fewer degrees of freedom. The advantage is that a smaller standardized value is required to obtain statistical significance, thus improving the power of the test to detect nonzero ∇ values when they exist. The fewer the predictions being considered, the smaller is the number of degrees of freedom and the greater is the power.

The exact number of degrees of freedom is not, unfortunately, the number of predictions considered, but rather the number of linearly independent ∇ values. The definition of linear independence is best illustrated by considering linear dependence. A set of ∇ values, say $\nabla_1, \ldots, \nabla_T$, is *linearly dependent* if one of them can be written as a weighted sum (linear combination) of the others. The weights, as in our previous weighted-average interpretations, may involve the marginal probabilities and the various error weights. If ∇_T can be written in such a way, say as

$$\nabla_T = c_1 \nabla_1 + \cdots + c_{T-1} \nabla_{T-1}$$

then the hypothesis that $\nabla_1 = 0, \nabla_2 = 0, \ldots, \nabla_{T-1} = 0$ implies that $\nabla_T = 0$ also. Hence ∇_T does not represent an extra dimension, or degree of freedom, in the data. Note that the c values may be functions of unknown population marginal probabilities; regardless of what these values might be, $\nabla_1 = \cdots = \nabla_{T-1} = 0$ still implies that $\nabla_T = 0$. No assumption of "known marginals" is implied or required.[13] For example, in a 2×2 table, consider

---

[13]We emphasize that this statement is a consequence of the fact that all the ∇'s are zero. If some of the ∇'s are nonzero, knowledge of the marginals is required to compute ∇_T. Consequently this section only applies to the null hypothesis that all the ∇ values are zero.

the ∇ values for

1. The single error cell $(1,2)$.
2. The single error cell $(2,1)$.
3. The two equally weighted error cells $(1,2)$ and $(2,1)$.

By equation (3) of Chapter 3 these three ∇ values are linearly dependent, since ∇_3 is a weighted sum of ∇_1 and ∇_2. In fact since it can be proved with a little algebra that

$$\nabla_2 = \frac{P_{1.}P_{.2}}{P_{2.}P_{.1}} \nabla_1$$

the set ∇_1, ∇_2 is itself linearly dependent. So there is only one linearly independent element in the set and the appropriate degrees of freedom for the *ex post* test is one. This is hardly surprising as the standard chi-square test has only one degree of freedom in a 2×2 table; generally there are at most $(R-1)(C-1)$ linearly independent ∇ values in an $R \times C$ table. This fact suggests that the advantage in restricting the number of predictions occurs primarily in rather large tables.

For $R \times C$ tables the determination of the appropriate number of linearly independent ∇ values can be tricky. The general guide was suggested earlier: if knowledge that a set of ∇ values is zero implies that a different ∇ is zero, that ∇ must be linearly dependent on the others. In particular if the unweighted error set for a prediction is the union of disjoint, unweighted error sets for other rules, then the ∇ for the "union" rule is zero when each of the component ∇'s is 0, and the "union" ∇ is linearly dependent on the others. In the preceding 2×2 example the principle indicates that ∇_3 is a linear combination of ∇_1 and ∇_2. A more subtle form of linear combination occurs in the dependence of ∇_2 on ∇_1 or vice versa. In a general $R \times C$ cross tabulation suppose that whenever $\mathcal{P}_a$ identifies unweighted error cells in a column, $\mathcal{P}_b$ identifies the other cells in that column as error cells; further suppose that if $\mathcal{P}_a$ identifies no error cells in a column, $\mathcal{P}_b$ does likewise. Then it follows that if $\nabla_a = 0$, $\nabla_b = 0$ also, so that ∇_b and ∇_a are linearly dependent. Now suppose that $\mathcal{P}_c$ identifies as errors in a row those cells that are nonerrors for $\mathcal{P}_b$ in that row. With $\nabla_b = 0$, ∇_c must be zero, and ∇_c is linearly dependent on ∇_b and in turn on ∇_a. In the 2×2 example ∇_a is ∇_1, ∇_c is ∇_2, and ∇_b goes with the prediction that identifies cell $(2,2)$ as the only error cell. This example indicates that determination of linear dependence can be complex. A general but more technical approach is given in Appendix 6.8.

Once the number of linearly independent ∇'s is specified, the test is straightforward. The null hypothesis of statistical independence is tested against the limited alternative that at least one of the alternative ∇ values is nonzero. (Again the independence hypothesis implies that all these ∇ values are zero, but it is logically possible that all might be zero without independence holding. Thus the null and alternative hypotheses do not exhaust all possibilities.) One forms the same squared standardized value as in the previous subsection. The only change is that one compares this quantity to the chi-squared value with only the specified degrees of freedom. The mathematical result that justifies this test is very similar to that of the preceding subsection, and is sketched in Appendix 6.8.

This test is quite conservative, perhaps excessively so, even if the correct number of degrees of freedom is specified. It is a two-sided test. The appropriate alternate (research) hypothesis should almost always be that one or more of the selected ∇ values is positive. It would be desirable to have a one-sided test; as yet we do not.[14]

6.7.3   Selection of an "Optimal" Proposition

The problem of *ex post* selection of an optimal proposition was discussed in the context of populations in Section 4.4. The actual procedures in this section simply replace population values by sample estimates. Thus our primary focus will be on the statistical properties of such methods.

Recall that in Section 4.4 we found "optimal" rules by ranking the single-error-cell predictions $\mathcal{P}_{ij}$ in order of decreasing ∇_{ij}, or equivalently in order of increasing error ratios $P_{ij}/(P_{i.}P_{.j})$. One accumulates error cells until either (1) a specified minimum value of U is reached or (2) the value of ∇ falls below a specified minimum level, depending on the notion of optimality being used.

The process in the basic S2 sampling case is identical, with the true P_{ij} being replaced by sample estimates f_{ij}. Otherwise definitions and concepts of Section 4.4 apply (including the cautions against an uncritical reading of the word *optimal*). The new issues that arise are statistical ones of potential bias and erroneous error probabilities.

In Section 6.7.1 we indicate that almost any *ex post* selection of a proposition tends to give an upward bias to the estimate of $\nabla_{\mathcal{P}}$, at least for true $\nabla_{\mathcal{P}} = 0$. Since the conceptual bias built in to specifying error cells just to obtain large $\hat{\nabla}$ values is present whether true ∇ is zero or not, it is not surprising that there is a mathematical bias present in general.

---

[14]Alternatively one could use a simple Bonferroni method for a simultaneous test of T estimated ∇ values. One would simply run each test at significance level α/T, thus guaranteeing that the error probability does not exceed α. This procedure is likely to be excessively conservative for even modest-sized T.

Bartlett (1974, Chapter 3) examined this bias for his RE selection procedure described in Section 4.4.6. In applying RE selection to multivariate analysis, as described in the next chapter, Bartlett performed Monte Carlo studies with typical sample sizes of 100 to 1600. He found a rather complex pattern, depending on the true ∇ level, the size of the table, the skewness of marginal probabilities, and the method used for selecting a proposition. With true ∇ values all equal to zero, modest-sized (for instance, 4×4) tables, samples of 100 and a selection method that simply maximized RE he found substantial biases for $n = 100$ (average $\hat{\nabla} = .325$) and $n = 200$ (average $\hat{\nabla} = .227$). The bias was relatively small for $n = 900$ (average $\hat{\nabla} = .097$), though positive. In general, increased table size had some effect; the bias increased as R and C were increased. The bias decreased as marginal probabilities became more skewed. (As Bartlett points out, this is not surprising; when one marginal probability becomes zero, the ultimately skewed case, bias is reduced because table size is reduced.) In one study Bartlett set his rule to select error cells only when estimated $\nabla_{ij} \geqslant .4$. Naturally when such a rule could be carried out very large $\hat{\nabla}$ values were obtained; however in 7 of 10 cases no cells could be identified as error cells under this rule. A method that picks out few error cells automatically yields large $\hat{\nabla}$ values. In summary, the *ex post* selection of a proposition can yield spuriously large $\hat{\nabla}$ values when n is 200 or less, particularly for large tables and few error cells.

When true ∇ is positive, the picture is more encouraging. If $n = 100$ and true ∇ is large (on the order of .7) the bias is small and sometimes negative. Further the *ex post* selection found the correct error cells in a very large fraction of trials. When $n = 900$ the bias was negligible for $\nabla > .2$. His trials typically were made with 3×8 to 4×12 tables and the biases, with $\nabla \approx .7$, seemed not to be much affected by size of table or skewness of marginals. The selection rules he used were not so restrictive as to eliminate many real error cells, as $\hat{\nabla}_{ij} \geqslant .4$ would have been. Such restrictive rules would have yielded some positive bias, but presumably not a large one. Thus a researcher can be reasonably confident, even with $n = 100$, that if a strong relation exists, it will be found and assessed accurately.

However a researcher who has obtained a moderate to large estimated ∇ by *ex post* methods faces something of a dilemma. As just stated, if the true ∇ is large, then a reasonably large estimated ∇ results, but if the true ∇ is small, then the selection bias previously shown may yield a moderately large estimated ∇. How can the researcher assess which case is more likely?

A partial answer is the chi-square test of Section 6.7.1, which tests whether the estimated ∇ is compatible with complete statistical independence. The chi-square test refers to all possible linear combinations of estimated ∇ values, whereas the problem here is to test a single $\hat{\nabla}$ value selected *ex post*. Wasserman (1973) examined the result of applying the chi-square test to a U^*-optimal prediction rule. He found that the test was very conservative for 3×5 tables and extremely so for 5×7 tables. In 5×7 tables he found that the true hypothesis of independence was rejected at the .05 level only 2 times in 2250 trials, a 0.1% rejection rate, as against a nominal 5% rate. The actual significance probability did not depend in any evident way on the minimal value of U specified; apparently the large selection bias present with small U values is roughly cancelled by the large associated variance. His results indicate that a $\hat{\nabla}$ value found *"ex post significant"* by the chi-square test is very likely to be nonzero (although the selection bias may cause an overestimate of the true value); on the other hand, a value that falls slightly short of significance by the chi-square test might be significant under a more exact, less conservative test. Thus the chi-square test is not a perfect answer to the problem of assessing an estimated ∇ value for an *ex post* proposition.

A good argument can be made that no procedure can be entirely satisfactory for an *ex post* assessment. There are so many uncertainties in such an approach, so many possible biases in selection of variables and prediction rules, that no *ex post* analysis can be considered established unless it is confirmed by later studies. The prediction analysis framework works well in this context, for the prediction rule developed *ex post* in one analysis can be applied *a priori* in a replication.

6.8 EFFECT OF CLASSIFICATION ERROR

In this chapter we have examined methods for making inferences about population ∇ values in the face of sampling variability. In so doing we have ignored some other sources of error, which might generally be called sampling bias and classification error. *Sampling bias* refers to data-gathering methods that systematically overrepresent certain categories and underrepresent others. Very little can be said about the effects of such biases. If the bias overrepresents a particular row or column without distorting the conditional probabilities in that row or column, a correction similar to our S3 method could be applied. However the true marginal probabilities would rarely be known, and there is no obvious direction of bias. Thus, as usual, the effects of sampling bias are very difficult to assess.

Classification or *measurement error* tends to provide a more coherent pattern. There is a fairly well-known phenomenon that random classification or measurement error tends to reduce the perceived degree of relation between variables. To assess whether this pattern holds for ∇ we must specify a random (as opposed to biased) measurement error pattern. There are many possible measurement error patterns. We have investigated one simple but basic case. Specifically assume that

1. Misclassifications of X-state and of Y-state occur independently, so that there is no bias toward specific cells.
2. The probability that a case is misclassified into state x_j given that its true state is $x_{j'}, j \neq j'$ is ζ_j regardless of the "originating" state $x_{j'}$. Similarly the probability that apparent Y is y_i given that the true Y is $y_{i'}$ equals θ_i for all $y_{i'}$ states.
3. The sums of conditional misclassification probabilities are less than 1: $\sum_{j=1}^{C} \zeta_j < 1$, $\sum_{i=1}^{R} \theta_i < 1$. In particular this will be true if $\zeta_j < 1/C$ and $\theta_i < 1/R$, so that the probability of misclassification to any one other state is less than the probability of correct classification. For instance, the probability of misclassification on a dichotomous variable must be less than $1/2$. This is certainly not an excessively restrictive assumption.

Several possible patterns of conditional misclassification probabilities are shown in Table 6.3. The simplest pattern is that shown in (A); here $\zeta_1 = \zeta_2 = \zeta_3 = .2$ and the sum is .6. A more complex pattern is shown in (B); $\zeta_1 = .1$, $\zeta_2 = \zeta_3 = .2$, and the sum is .5. The pattern in (C) violates condition 2 in that the "incoming" probability of x_1, for example, depends on the original state, x_2 or x_3. The pattern in (D) has $\zeta_1 = \zeta_2 = \zeta_3 = .4$ (so that the sum is too large).

Misclassification obeying 1—3 results in the same reduction in strength of observed relation that has been observed in correlation studies. If we let the true probabilities be denoted by P_{ij}, we have established in Appendix 6.9 that the apparent value of ∇, call it ∇^*, is

$$\nabla^* = 1 - \frac{\left(\sum_i \sum_j \omega_{ij} P_{ij} \right) \cdot A + B}{\left(\sum_i \sum_j \omega_{ij} P_{i.} P_{.j} \right) \cdot A + B} = 1 - \frac{KA + B}{UA + B}$$

where A and B are positive constants.

Now multiplying both the numerator and denominator of K/U by A will not affect the fraction. Adding the positive value B to both will bring the fraction toward one, and therefore move ∇ toward zero. (A baseball player who starts with K hits in U at bats and then gets B consecutive hits obviously improves his batting average.) It follows that the apparent ∇

Table 6.3 Conditional misclassification patterns

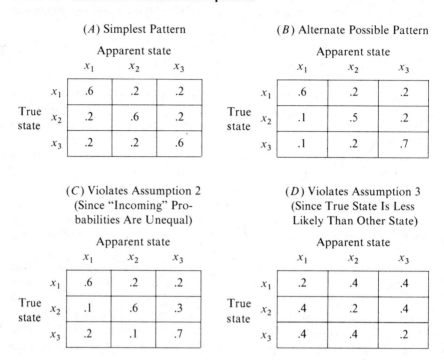

(A) Simplest Pattern

Apparent state

		x_1	x_2	x_3
True state	x_1	.6	.2	.2
	x_2	.2	.6	.2
	x_3	.2	.2	.6

(B) Alternate Possible Pattern

Apparent state

		x_1	x_2	x_3
True state	x_1	.6	.2	.2
	x_2	.1	.5	.2
	x_3	.1	.2	.7

(C) Violates Assumption 2 (Since "Incoming" Probabilities Are Unequal)

Apparent state

		x_1	x_2	x_3
True state	x_1	.6	.2	.2
	x_2	.1	.6	.3
	x_3	.2	.1	.7

(D) Violates Assumption 3 (Since True State Is Less Likely Than Other State)

Apparent state

		x_1	x_2	x_3
True state	x_1	.2	.4	.4
	x_2	.4	.2	.4
	x_3	.4	.4	.2

falls between zero and the true ∇. In correlation studies random misclassification biases the measure toward zero; random classification errors also conceal relations as measured by ∇.

APPENDIX 6.1   PROPERTIES OF ESTIMATORS

The estimators presented in Section 6.2 were selected more for simplicity than for statistical optimality. In this appendix the optimality properties (or lack thereof) are considered.

The condition S1 estimator is the most objectionable from the perspective of optimal inference theory. Since the estimated marginal fractions are unlikely to equal the known true probabilities, the estimator is likely to be inefficient. This conjecture is reinforced by the fact that the condition S1 variance can exceed the condition S2 variance, although the researcher has more information in condition S1. The only virtues of $\hat{V}_{S1}$ are simplicity, unbiasedness, and consistency. Unbiasedness follows from the linearity of $\hat{V}_{S1}$ in the unknown f_{ij}'s and the unbiasedness of the f_{ij}'s for estimating P_{ij}.

Consistency follows from unbiasedness and

$$\text{Var}(\hat{\nabla}_{S1}) = \frac{K(1-K)}{nU^2}$$

As $n \to \infty$, $\text{var}(\hat{\nabla}_{S1}) \to 0$, and $\hat{\nabla}_{S1} \to E(\hat{\nabla}_{S1}) = \nabla$. As mentioned in the text, if sampling situation S1 arises in a research project where statistical efficiency is important, fitting to the known marginals via the Deming-Stephan algorithm would be desirable.

The estimators in conditions S2 and S3 are maximum likelihood in the sense that each is a function only of the maximum likelihood estimators (f_{ij} and $f_{ij}/f_{.j}$, respectively). This and the Fisher regularity conditions, as presented in, for instance, Mood, Graybill, and Boes (1974), guarantee that these estimators are asymptotically unbiased and efficient. [Sufficiency and the Rao-Blackwell (Wilks, 1962, p. 357) approach to efficiency seem infeasible for these estimators.]

The absence of statistical optimality properties should be viewed in the perspective of the intent of this work, which is to provide investigators with means for evaluating predictions with cross-tabulated data. Failure to obtain the last few percentage points of inferential efficiency is not a heavy price to pay at this stage.

Nonetheless some improvements ought to be possible. In particular, *ex post* estimation of ∇ values introduces a blatant bias, as indicated in Section 6.7. More sophisticated methods [perhaps based on a James-Stein (1961) shrinking toward $\nabla = 0$] might well reduce this bias. Furthermore there has been no application of Bayesian methods for ∇ inference. The calculation of posterior probability intervals for one or more $\nabla_{\mathcal{P}}$ values based on a Dirichlet prior and multinomial sampling is not a trivial task; with modest data and strong prior information the effort might well be worthwhile.

APPENDIX 6.2   SAMPLING VARIANCES

As the condition S1 variance is based on the well-known binomial distribution it need not be discussed.

The sampling theory for sampling conditions S2 and S3 is based on the delta method. In effect the actual estimator is replaced by a linear approximation using Taylor's series methods. With large n and therefore small variances, the linear form closely approximates the actual estimate. In this appendix we sketch the routine computations involved.

First, the theory of Taylor's series, as presented in any calculus text, indicates that the coefficient of $f_{ij} = n_{ij}/n$ in the linear approximation to $\hat{\nabla}$

is the first partial derivative of $\hat{V}$ with respect to f_{ij} evaluated at an appropriate centering point. Since the method is to be applied for large n, which implies that f_{ij} is near P_{ij}, the centering point has each $f_{ij} = P_{ij}$. The derivative of $\hat{V}_{S2}$ with respect to f_{ij} at $f_{ij} = P_{ij}$ is indicated, as a convenient shorthand, by $\partial \nabla / \partial P_{ij}$. By the expression for the derivative of a ratio

$$\frac{\partial \nabla}{\partial P_{ij}} = -U^{-2}\left(U\omega_{ij} - K\frac{\partial U}{\partial P_{ij}}\right)$$

where K and U are the rule K and rule U error rates. To calculate the derivative of U note that we may rewrite $U = \sum_{i'}\sum_{j'}\omega_{i'j'}P_{i'.}P_{.j'}$ and that $\partial P_{i'.}/\partial P_{ij} = 1$ if $i = i'$ and 0 if $i \neq i'$ and that $\partial P_{.j'}/\partial P_{ij} = 1$ if $j = j'$ and 0 if $j \neq j'$. Using the formula for the derivative of a product and some care with subscripts yields

$$\frac{\partial U}{\partial P_{ij}} = \sum_{j'}\omega_{ij'}P_{.j'}\times 1 + \sum_{i'}\omega_{i'j}P_{i'.}\times 1 = \pi_{i.} + \pi_{.j}$$

as defined in Chapter 6.
   Substitution yields

$$\frac{\partial \nabla}{\partial P_{ij}} = -U^{-1}\left[\omega_{ij} - \frac{K}{U}(\pi_{i.} + \pi_{.j})\right] \equiv -a_{ij}$$

The Taylor's series approximation, then, is

$$\nabla_{\mathcal{P}} \cong c + \sum_{i}\sum_{j} -a_{ij}P_{ij},$$

where c is a constant and therefore irrelevant to variance calculations.
   Second, the calculation of the variance of a linear function of the f_{ij} values requires the variances and covariances of these values. Under sampling condition S2 the distribution of the numbers falling in the various cells is multinomial. So by a standard calculation for the multinomial distribution, as in Mood, Graybill, and Boes (1974, 196–197), for example,

$$\mathrm{Var}(f_{ij}) = \mathrm{Var}\left(\frac{n_{ij}}{n}\right) = \frac{P_{ij}(1 - P_{ij})}{n}$$

$$\mathrm{Cov}(f_{ij}, f_{i'j'}) = \frac{-P_{ij}P_{i'j'}}{n} \qquad \text{unless } i = i' \text{ and } j = j'$$

Recall that $\mathrm{cov}(f_{ij}, f_{ij}) = \mathrm{var}(f_{ij})$.

Third, the variance of a linear function $c - \sum_i \sum_j a_{ij} f_{ij}$ may be written as

$$\sum_i \sum_j \sum_{i'} \sum_{j'} a_{ij} \mathrm{Cov}(f_{ij}, f_{i'j'}) a_{i'j'}$$

as in D. Morrison (1967, p. 79). In matrix terms this is $\mathbf{A}' \boldsymbol{\Sigma} \mathbf{A}$ where $\boldsymbol{\Sigma}$ is the $RC \times RC$ covariance matrix of the f_{ij}'s. Because the covariances have a special form here the expression can be simplified as follows:

$$\sum_i \sum_j \sum_{i'} \sum_{j'} a_{ij} \mathrm{Cov}(f_{ij}, f_{i'j'}) a_{i'j'} = \frac{\displaystyle\sum_i \sum_j a_{ij}^2 P_{ij}}{n} - \frac{\displaystyle\sum_i \sum_j \sum_{i'} \sum_{j'} a_{ij} a_{i'j'} P_{ij} P_{i'j'}}{n}$$

$$= \frac{\displaystyle\sum_i \sum_j a_{ij}^2 P_{ij}}{n} - \frac{\left(\displaystyle\sum_i \sum_j a_{ij} P_{ij}\right)^2}{n}$$

The a_{ij} values actually used in the Chapter 6 expression for the asymptotic variance are the negatives of the partial derivatives; since the a_{ij} values occur only in squared terms [or since $\mathrm{var}(\hat{V}) = \mathrm{var}(1 - \hat{V})$], multiplying by -1 has no effect. Substituting $a_{ij} = -\partial V / \partial P_{ij}$ yields the formula in Chapter 6 for sampling condition S2. Actually, we use $n - 1$ instead of n in the denominator because it seems to improve the small-sample approximation, as discussed in Section 6.4.

There is an apparent technical problem with this derivation in that we have not explicitly considered the restriction $\sum_i \sum_j P_{ij} = 1.0$. It is not necessary to do so since this restriction is reflected in the covariances. Technically we have allowed the f_{ij}'s to vary in a full RC-dimensional space; the fact that $\sum_i \sum_j f_{ij} = 1$ implies that the covariance matrix is singular so that the f_{ij}'s are in a lower-dimensional space. The application of the delta method does *not* require a nonsingular covariance matrix. Although possible, it is awkward to recompute derivatives while taking the restriction on the sum into account and the result would be the same anyway. Therefore in this and all other derivations the restriction is ignored.

This type of computation occurs repeatedly in the appendixes. For simplicity *we consistently* let a_{ij} be the negative of the partial derivative and *ignore summation restrictions*.

The computation for sampling condition S3 is similar. The negative of the first partial derivative of $\hat{V}_{S3}$ at $f_{ij} = P_{ij}/\eta_{.j}$ becomes

$$a_{ij} = U_{S3}^{-1} \eta_{.j} \left[\omega_{ij} - (1 - \nabla_{\mathscr{P}}) \pi_{i.} \right]$$

where

$$U_{S3} = \sum_i \sum_j \omega_{ij} \left[\sum_{j'} P(y_i|x_{j'}) P_{.j'} \right] P_{.j}$$

Since in condition S3 separate and independent multinomial samples are taken from each $X = x_j$ subpopulation with respective sample sizes $n_{.j} = nf_{.j}$,

$$\text{Cov}(f_{ij}, f_{i'j'}) = 0 \qquad \text{if } j \neq j' \qquad (i \text{ and } i' \text{ may be equal or not equal})$$

$$\text{Cov}(f_{ij}, f_{i'j}) = -f_{.j}^2 P(y_i|x_j) P(y_{i'}|x_j) / n_{.j}$$

$$= -\frac{f_{.j}(P_{ij}P_{i'j})}{(nP_{.j}^2)} \qquad \text{if } i \neq i'$$

$$\text{Var}(f_{ij}) = \frac{f_{.j}^2 P(y_i|x_j)(1 - P(y_i|x_j))}{n_{.j}}$$

$$= \frac{f_{.j}P_{ij}}{(nP_{.j})} - \frac{f_{.j}P_{ij}^2}{(nP_{.j}^2)}$$

Replacing these values in $A'\Sigma A$ and simplifying gives the expression for $\text{var}(\hat{V}_{S3})$ of Chapter 6.

APPENDIX 6.3 COMPUTATIONAL FORMULA FOR THE ASYMPTOTIC VARIANCE

In Chapter 6 the condition S2 variance was given in the form:

$$(n-1)^{-1} \left[\sum_i \sum_j a_{ij}^2 P_{ij} - \left(\sum_i \sum_j a_{ij} P_{ij} \right)^2 \right]$$

We used this form because it recurs in all the other S2 variance expressions. This form could thus be the basis of a general computer program, with the a_{ij} for various sampling conditions computed by subroutines.

For hand calculation, however, it should be noted that using

$$\sum_i P_{i.}\pi_{i.} = \sum_i \sum_j \pi_{i.}P_{ij} = \sum_j \pi_{.j}P_{.j} = \sum_i \sum_j \pi_{.j}P_{ij} = U$$

and

$$a_{ij} = \frac{\left[\omega_{ij} - B\left(\pi_{i.} + \pi_{.j}\right)\right]}{U}$$

we find

$$\sum_i \sum_j a_{ij}P_{ij} = B\left\{1 - \left[\frac{\sum_i \sum_j (\pi_{i.} + \pi_{.j})P_{ij}}{U}\right]\right\}$$

$$= B\left[1 - \frac{(U+U)}{U}\right] = -B$$

Hence

$$\left(\sum_i \sum_j a_{ij}P_{ij}\right)^2 = B^2$$

APPENDIX 6.4   COMPARISON OF TWO ESTIMATED DEL VALUES FOR THE SAME DATA

The delta method extends readily to the problem of comparing two estimated ∇ values. In this appendix the method is illustrated for sampling condition S2; the ideas can readily be extended to other sampling schemes.

Since the delta method approximates the estimated ∇'s as linear functions of the f_{ij}'s,

$$\hat{\nabla}_{\mathscr{P}(1)} = c_1 - \sum_i \sum_j a_{ij}^{(1)} f_{ij}$$

$$\hat{\nabla}_{\mathscr{P}(2)} = c_2 - \sum_i \sum_j a_{ij}^{(2)} f_{ij}$$

the difference is also approximately linear

$$\hat{\nabla}_{\mathcal{P}^{(1)}} - \hat{\nabla}_{\mathcal{P}^{(2)}} = c_1 - c_2 - \sum_i \sum_j (a_{ij}^{(1)} - a_{ij}^{(2)}) f_{ij}$$

Since the asymptotic expected estimates equal the respective true values, the asymptotic expected difference is the difference in true values. Furthermore with the variances and covariances of Appendix 6.2, it follows that asymptotically

$$\text{Var}(\hat{\nabla}_{\mathcal{P}_1} - \hat{\nabla}_{\mathcal{P}_2}) = (n-1)^{-1} \left[\sum_i \sum_j (a_{ij}^{(1)} - a_{ij}^{(2)})^2 P_{ij} - \left(\sum_i \sum_j (a_{ij}^{(1)} - a_{ij}^{(2)}) P_{ij} \right)^2 \right]$$

All that is needed is to replace a_{ij} by $a_{ij}^{(1)} - a_{ij}^{(2)}$ in the typical variance calculation, be it for standard degree-1 propositions, propositions with holes, or degree-2 propositions.

We have not investigated the problem of comparing three or more estimated ∇ values. A solution similar to a one-way analysis of variance should be feasible.

APPENDIX 6.5 VARIANCES FOR DEGREE-2 PREDICTIONS AND FOR PREDICTIONS WITH IRRELEVANT EVENTS

The calculation of the condition S2 distribution of $\hat{\nabla}$ both for degree-2 predictions and for the situation where certain events are to be considered as irrelevant or logically impossible follows the standard form:

$$\text{Var}(\hat{\nabla}) = (n-1)^{-1} \left[\sum_i \sum_j a_{ij}^2 P_{ij} - \left(\sum_i \sum_j a_{ij} P_{ij} \right)^2 \right]$$

with a_{ij} the negative of the partial derivative of $\hat{\nabla}$ with respect to f_{ij} at $f_{ij} = P_{ij}$. The only remaining problem to be sketched in the appendix is the computation of partial derivatives. There are four distinct ∇ values, which are considered in order: the extensive form for pairs, the condensed ordinal form, the global correction for irrelevant events, and the set-by-set correction for irrelevant events.

For the extensive form for pairs of events the general ∇ value is

$$\nabla_{\mathcal{P}} = 1 - \frac{\sum_g \sum_i \sum_h \sum_j \omega_{gihj} P_{gihj}}{\sum_g \sum_i \sum_h \sum_j \omega_{gihj} P_{gi..} P_{..hj}} = 1 - \frac{K_2}{U_2}$$

where $\omega_{gihj} = \omega_{igjh}$ since reversing pairs of subscripts merely reverses the arbitrary order in which pairs are taken and $P_{gihj} = P_{gh}P_{ij}$ so that

$$P_{gi..} = P_{g.}P_{i.} \quad \text{and} \quad P_{..hj} = P_{.h}P_{.j}$$

With the same shorthand as in Appendix 6.2, it follows that

$$a_{ij} = -\frac{\partial \nabla_{\mathcal{G}}}{\partial P_{ij}} = (U_2)^{-2}\left(U_2\frac{\partial K_2}{\partial P_{ij}} - K_2\frac{\partial U_2}{\partial P_{ij}}\right)$$

Again the problem is keeping track of subscripts. The most convenient bookkeeping method is to attach primes to the differentiating variable and calculate

$$a_{i'j'} = (U_2)^{-2}\left(U_2\frac{\partial K_2}{\partial P_{i'j'}} - K_2\frac{\partial U_2}{\partial P_{i'j'}}\right)$$

Now

$$\frac{\partial K_2}{\partial P_{i'j'}} = \sum_g\sum_h\sum_i\sum_j \omega_{gihj}\left(P_{gh}\frac{\partial P_{ij}}{\partial P_{i'j'}} + P_{ij}\frac{\partial P_{gh}}{\partial P_{i'j'}}\right)$$

Further

$$\frac{\partial P_{ij}}{\partial P_{i'j'}} = 1 \quad \text{if } i = i' \text{ and } j = j' \text{ and } 0 \text{ otherwise}$$

whereas

$$\frac{\partial P_{gh}}{\partial P_{i'j'}} = 1 \quad \text{if } g = i' \text{ and } h = j' \text{ and } 0 \text{ otherwise}$$

So

$$\frac{\partial K_2}{\partial P_{i'j'}} = \sum_g\sum_h \omega_{gi'hj'}P_{gh} + \sum_i\sum_j \omega_{i'ij'j}P_{ij}$$

$$= 2\sum_i\sum_j \omega_{i'ij'j}P_{ij} = 2\Upsilon_{i'j'}$$

by the symmetry in ω subscripts noted above and the definition of Υ in Section 6.6.

Similarly

$$\frac{\partial U_2}{\partial P_{i'j'}} = \sum_g \sum_i \sum_h \sum_j \omega_{gihj} \left(P_{g.} P_{.h} P_{i.} \frac{\partial P_{.j}}{\partial P_{i'j'}} + P_{g.} P_{i.} P_{.j} \frac{\partial P_{.h}}{\partial P_{i'j'}} \right.$$

$$\left. + P_{g.} P_{.h} P_{.j} \frac{\partial P_{i.}}{\partial P_{i'j'}} + P_{.h} P_{i.} P_{.j} \frac{\partial P_{g.}}{\partial P_{i'j'}} \right)$$

$$= \sum_g \sum_h \sum_i \omega_{gihj'} P_{g.} P_{.h} P_{i.}$$

$$+ \sum_g \sum_i \sum_j \omega_{gij'j} P_{g.} P_{i.} P_{.j}$$

$$+ \sum_g \sum_h \sum_j \omega_{gi'hj} P_{g.} P_{.h} P_{.j}$$

$$+ \sum_h \sum_i \sum_j \omega_{i'ihj} P_{.h} P_{i.} P_{.j}$$

$$= 2 \Upsilon_{i'.} + 2 \Upsilon_{.j'}$$

by the same argument.

Thus, dropping primes on subscripts,

$$a_{ij} = 2 U_2^{-1} \left[\Upsilon_{ij} - \frac{K_2}{U_2} (\Upsilon_{i.} + \Upsilon_{.j}) \right]$$

as claimed in (23) of Chapter 6.

The calculation for the condensed ordinal form is tedious and lengthy. A typical calculation is the derivative of the probability of concordance;

$$P(\mathbf{C}) = P(X_\mathrm{I} < X_\mathrm{II} \quad \text{and} \quad Y_\mathrm{I} < Y_\mathrm{II}) + P(X_\mathrm{I} > X_\mathrm{II} \quad \text{and} \quad Y_\mathrm{I} > Y_\mathrm{II})$$

By radial symmetry

$$P(X_\mathrm{I} < X_\mathrm{II} \quad \text{and} \quad Y_\mathrm{I} < Y_\mathrm{II}) = \frac{1}{2} P(\mathbf{C}) = \sum_{g=1}^{R-1} \sum_{h=1}^{C-1} \sum_{i=g+1}^{R} \sum_{j=h+1}^{C} P_{gh} P_{ij}$$

Careful subscript checking and the rule for differentiating products yields

$$\frac{\partial P(X_\mathrm{I} < X_\mathrm{II} \quad \text{and} \quad Y_\mathrm{I} < Y_\mathrm{II})}{\partial P_{i'j'}} = \sum_{i=i'+1}^{R} \sum_{j=j'+1}^{C} P_{ij} + \sum_{g=1}^{i'-1} \sum_{h=1}^{j'-1} P_{gh}$$

$$= P(Y > i' \quad \text{and} \quad X > j') + P(Y < i' \quad \text{and} \quad X < j')$$

Note that impossible summations such as $\Sigma_{g=1}^{0}$ are defined as zero. (The reader who wishes to check this calculation might find it helpful to take specific values for i' and j', perhaps $i'=2$, $j'=3$.) From this result it is plausible, and in fact true, that

$$\frac{\partial P(X_{\mathrm{I}}>X_{\mathrm{II}} \quad \text{and} \quad Y_{\mathrm{I}}>Y_{\mathrm{II}})}{\partial P_{i'j'}} = \frac{\partial P(X_{\mathrm{I}}<X_{\mathrm{II}} \quad \text{and} \quad Y_{\mathrm{I}}<Y_{\mathrm{II}})}{\partial P_{i'j'}}$$

$$= \frac{1}{2}\frac{\partial P(\mathbf{C})}{\partial P_{i'j'}}$$

Similar calculations for all the derivatives show

$$\frac{\partial P(\mathbf{C})}{\partial P_{ij}} = 2\big[P(Y>i,X>j) + P(Y<i,X<j) \big]$$

$$\frac{\partial P(\mathbf{D})}{\partial P_{ij}} = 2\big[P(Y>i,X<j) + P(Y<i,X>j) \big]$$

$$\frac{\partial P(\mathbf{T}_{\bar{Y}X})}{\partial P_{ij}} = 2\big[P(Y>i,X=j) + P(Y<i,X=j) \big]$$

$$\frac{\partial P(\mathbf{T}_{Y\bar{X}})}{\partial P_{ij}} = 2\big[P(Y=i,X>j) + P(Y=i,X<j) \big]$$

$$\frac{\partial P(\mathbf{T}_{YX})}{\partial P_{ij}} = 2P_{ij} = 2\big[P(Y=i,X=j) \big]$$

$$\frac{\partial P(\mathbf{T}_{X})}{\partial P_{ij}} = 2P_{.j} = 2P(X=j)$$

$$\frac{\partial P(\mathbf{T}_{Y})}{\partial P_{ij}} = 2P_{i.} = 2P(Y=i)$$

$$\frac{\partial P(\bar{\mathbf{T}}_{X})}{\partial P_{ij}} = 2\big[P(X\neq j) \big]$$

$$\frac{\partial P(\bar{\mathbf{T}}_{Y})}{\partial P_{ij}} = 2\big[P(Y\neq i) \big]$$

This list, plus the standard formulas for the derivative of a quotient and a sum, yield the values of a_{ij} and therefore the variance.

The derivatives for irrelevant events proceed in a very similar way. For the global correction

$$\nabla' = 1 - \frac{K}{U/\overline{H}} = 1 - \frac{K\overline{H}}{U}$$

Hence

$$a_{ij} = -\frac{\partial \nabla'}{\partial P_{ij}} = U^{-1}\left(K\frac{\partial \overline{H}}{\partial P_{ij}} + \overline{H}\frac{\partial K}{\partial P_{ij}} - \frac{K\overline{H}}{U}\frac{\partial U}{\partial P_{ij}}\right)$$

$$= \frac{K}{U}\left[\sum_{j'}\overline{\phi}_{ij'}P_{.j'} + \sum_{i'}\overline{\phi}_{i'j}P_{i'.}\right] + \frac{\overline{H}}{U}\omega_{ij} - \frac{K\overline{H}}{U^2}\left[\sum_{j'}\omega_{ij'}P_{.j'} + \sum_{i'}\omega_{i'j}P_{i'.}\right]$$

where the calculations of $\partial \overline{H}/\partial P$, $\partial K/\partial P$, and $\partial U/\partial P$ are very similar to those of Appendix 6.2. This yields the expression for $\text{var}(\hat{\nabla})$ in Section 6.5 immediately.

Finally the second, set-by-set correction for irrelevant events results in

$$\nabla'' = 1 - \frac{K}{\sum_j U_{.j}/\overline{H}_{.j}} = 1 - \frac{K}{U''}$$

For this value the formula for the derivative of a quotient must be applied both to ∇'' and to U''. The result is as shown in equation (20) of Chapter 6.

APPENDIX 6.6 VARIANCES UNDER INDEPENDENCE

The typical null hypothesis for *a priori* analysis will be that $\nabla_{\mathcal{P}}=0$. As we have noted several times this hypothesis alone does not imply statistical independence. In *ex post* analysis, however, the natural null hypothesis is that *all* $\nabla_{\mathcal{P}}=0$; this does imply independence. The independence condition that $P_{ij} = P_{i.}P_{.j}$ implies a modified form of variance expression.

Under sampling condition S2 the asymptotic variance is, by Appendixes 6.2 and 6.3,

$$\left[(n-1)U^2\right]^{-1}\left[\left[\sum_i \sum_j (\omega_{ij} - (1-\nabla)(\pi_{i.} + \pi_{.j}))^2 P_{ij}\right] - K^2\right]$$

Expanding the square and using independence and $\nabla = 0$ yields

$$\left[(n-1)U^2\right]^{-1}\left[\left[\sum_i \sum_j \omega_{ij}^2 P_{i.}P_{.j} - 2\sum_i \sum_j \omega_{ij}(\pi_{i.}+\pi_{.j})P_{i.}P_{.j}\right.\right.$$

$$\left.\left.+ \sum_i \sum_j (\pi_{i.}+\pi_{.j})^2 P_{i.}P_{.j}\right] - K^2\right]$$

But $\sum_i \sum_j \omega_{ij}\pi_{i.}P_{i.}P_{.j} = \sum_i \pi_{i.}P_{i.}\sum_j \omega_{ij}P_{.j} = \sum_i \pi_{i.}^2 P_{i.}$ since $\pi_{i.} = \sum_j \omega_{ij}P_{.j}$; similarly $\sum_i \sum_j \omega_{ij}\pi_{.j}P_{i.}P_{.j} = \sum_j \pi_{.j}^2 P_{.j}$. Furthermore

$$\sum_i \sum_j (\pi_{i.}+\pi_{.j})^2 P_{i.}P_{.j} = \sum_i \sum_j \pi_{i.}^2 P_{i.}P_{.j} + 2\sum_i \sum_j \pi_{i.}\pi_{.j}P_{i.}P_{.j}$$

$$+ \sum_i \sum_j \pi_{.j}^2 P_{i.}P_{.j}$$

$$= \sum_i \pi_{i.}^2 P_{i.} + 2U^2 + \sum_j \pi_{.j}^2 P_{.j}$$

using $\sum_i P_{i.} = \sum_j P_{.j} = 1$ and $\sum_i \pi_{i.}P_{i.} = \sum_j \pi_{.j}P_{.j} = U$. Finally under independence $K = U$. Substitution yields the final expression for the variance

$$\mathrm{Var}_0(\hat{\nabla}_{S2}) = \left[(n-1)U^2\right]^{-1}\left(\sum_i \sum_j \omega_{ij}^2 P_{i.}P_{.j} - \sum_i \pi_{i.}^2 P_{i.} - \sum_j \pi_{.j}^2 P_{.j} + U^2\right)$$

In sampling condition S3 if the sample sizes are proportional to the $X = x_j$ probabilities, so that $f_{.j} = P_{.j}$, the same expression holds for the variance. The algebra is very similar and need not be included.

APPENDIX 6.7   PROOF OF CHI-SQUARE BOUND FOR STANDARDIZED SQUARED DEL VALUE

In Section 6.7.1 it is stated that in sampling condition S2 the largest possible value for $\hat{\nabla}^2/\mathrm{est.\ var}_0 (\hat{\nabla})$ over all possible sets of ω_{ij} weights is

$$\left(\frac{n-1}{n}\right)\sum_i \sum_j \frac{(nf_{ij} - nf_{i.}f_{.j})^2}{nf_{i.}f_{.j}} = \left(\frac{n-1}{n}\right)x^2$$

In investigating the largest possible standardized del value it is convenient to use weighted averages of $\hat{\nabla}_{ij}$ values, where $\hat{\nabla}_{ij} = 1 - f_{ij}/(f_{i.}f_{.j})$ is the estimated ∇ value for the proposition that identifies cell (i,j) as the only error cell. As in equation (16) of Chapter 3, any $\hat{\nabla}$ may be written as a weighted average of these values:

$$\hat{V} = 1 - \frac{\displaystyle\sum_i \sum_j \omega_{ij} f_{ij}}{\displaystyle\sum_i \sum_j \omega_{ij} f_{i.} f_{.j}} = \sum_i \sum_j b_{ij} \hat{V}_{ij}$$

where $b_{ij} = \omega_{ij} f_{i.} f_{.j} / (\sum_i \sum_j \omega_{ij} f_{i.} f_{.j})$. To find the estimated variance of an arbitrary weighted average $\sum_i \sum_j b_{ij} \hat{V}_{ij}$, we use

$$\text{est. var}_0(\hat{V}_{ij}) = \frac{(n-1)^{-1}(1-f_{i.})(1-f_{.j})}{(f_{i.} f_{.j})}$$

$$\text{est. cov}_0(\hat{V}_{ij}, \hat{V}_{i'j}) = \frac{-(n-1)^{-1}(1-f_{.j})}{f_{.j}} \qquad i \neq i'$$

$$\text{est. cov}_0(\hat{V}_{ij}, \hat{V}_{ij'}) = \frac{-(n-1)^{-1}(1-f_{i.})}{f_{i.}} \qquad j \neq j'$$

$$\text{est. cov}_0(\hat{V}_{ij}, \hat{V}_{i'j'}) = (n-1)^{-1} \qquad i \neq i', j \neq j'$$

The variance result follows from Appendix 6.6, replacing $P_{i.}$ and $P_{.j}$ by $f_{i.}$ and $f_{.j}$ to obtain the estimated variance. The expressions for the covariances follow by an application of the delta method to the joint sampling distribution of two $\hat{V}$ values. Alternatively these results can be derived immediately from the variance and covariance expressions in Lancaster (1969, p. 227). Routine but long algebra yields

$$\text{est. var}_0\left(\sum_i \sum_j b_{ij} \hat{V}_{ij}\right) = (n-1)^{-1}\left[b_{..}^2 + \sum_i \sum_j \frac{b_{ij}^2}{f_{i.} f_{.j}} - \sum_i \frac{b_{i.}^2}{f_{i.}} - \sum_j \frac{b_{.j}^2}{f_{.j}}\right]$$

where

$$b_{i.} = \sum_j b_{ij}, \qquad b_{.j} = \sum_i b_{ij}, \qquad \text{and } b_{..} = \sum_i \sum_j b_{ij}$$

The problem can therefore be restated as that of maximizing

$$\frac{\left(\displaystyle\sum_i \sum_j b_{ij} \hat{V}_{ij}\right)^2}{\text{est. var}_0\left(\displaystyle\sum_i \sum_j b_{ij} \hat{V}_{ij}\right)}$$

An indeterminacy arises in that multiplying all the b_{ij} by a nonzero constant leaves the ratio unchanged. To handle this problem it is convenient to require that the estimated variance equal one. This merely fixes the weights in a manner differing from the usual arbitrary convention, which has the largest weight equal to one. The problem can now be stated as that of maximizing $(\Sigma_i \Sigma_j b_{ij} \hat{\nabla}_{ij})^2$ or equivalently maximizing $\Sigma_i \Sigma_j b_{ij} \hat{\nabla}_{ij}$, under the constraint that est. $\text{var}_0(\Sigma\Sigma b_{ij} \hat{\nabla}_{ij}) = 1$. Differentiation of the Lagrangian

$$L = \sum_i \sum_j b_{ij} \hat{\nabla}_{ij} - \lambda \left[\text{est. var}_0 \left(\sum_i \sum_j b_{ij} \hat{\nabla}_{ij} \right) - 1 \right]$$

with respect to the b_{ij} and to λ yields

$$\hat{\nabla}_{ij} - 2\lambda (b_{..} + b_{ij}/(f_{i.}f_{.j}) - b_{i.}/f_{i.} - b_{.j}/f_{.j}) = 0$$

$$\text{est. var}_0 \left(\sum_i \sum_j b_{ij} \hat{\nabla}_{ij} \right) = 1$$

Letting $b_{ij}^* = (1/2\lambda)(f_{i.}f_{.j} - f_{ij})$ and then choosing λ to scale the b_{ij}^* so that est. $\text{var}_0(\Sigma_i \Sigma_j b_{ij}^* \hat{\nabla}_{ij}) = 1$ solves these equations;

$$b_{i.}^* = (1/2\lambda) \left[\sum_j (f_{i.}f_{.j} - f_{ij}) \right] = (1/2\lambda)(f_{i.} - f_{i.}) = 0$$

similarly, $b_{.j}^* = b_{..}^* = 0$, and

$$2\lambda b_{ij}^* / f_{i.}f_{.j} = (f_{i.}f_{.j} - f_{ij})/(f_{i.}f_{.j}) = \hat{\nabla}_{ij}$$

With this choice of b_{ij}^*

$$\text{est. var}_0 \left(\sum_i \sum_j b_{ij}^* \hat{\nabla}_{ij} \right) = (n-1)^{-1} \sum_i \sum_j \frac{\left\{ \frac{1}{2\lambda}(f_{i.}f_{.j} - f_{ij}) \right\}^2}{f_{i.}f_{.j}}$$

$$= (n-1)^{-1}(2\lambda)^{-2} \sum_i \sum_j \frac{(f_{i.}f_{.j} - f_{ij})^2}{f_{i.}f_{.j}}$$

Therefore the maximum value of the squared standardized weighted average

$$\frac{\left(\sum_i \sum_j b_{ij} \hat{\nabla}_{ij} \right)^2}{\text{est. var}_0 \left(\sum_i \sum_j b_{ij} \hat{\nabla}_{ij} \right)}$$

is

$$
\frac{\left[\displaystyle\sum_i \sum_j \frac{1}{2\lambda}(f_{i.}f_{.j}-f_{ij})\left[(f_{i.}f_{.j}-f_{ij})/(f_{i.}f_{.j})\right]\right]^2}{(n-1)^{-1}(2\lambda)^{-2}\displaystyle\sum_i \sum_j (f_{i.}f_{.j}-f_{ij})^2/(f_{i.}f_{.j})}
$$

$$
=(n-1)\sum_i \sum_j \frac{(f_{ij}-f_{i.}f_{.j})^2}{f_{i.}f_{.j}}
$$

$$
=\frac{(n-1)}{n}\sum_i \sum_j \frac{(nf_{ij}-nf_{i.}f_{.j})^2}{nf_{i.}f_{.j}}=\left(\frac{n-1}{n}\right)\chi^2
$$

as desired.

This bound cannot be exactly attained by a weighted ∇ value. The ω_{ij} values implied by the b_{ij}^* are proportional to $1-f_{ij}/(f_{i.}f_{.j})$ so that some of them are negative. Furthermore

$$
\hat{U}=\sum_i \sum_j \omega_{ij}f_{i.}f_{.j}=0
$$

which leaves $\hat{\nabla}$ undefined. It is possible to come arbitrarily close to the bound by modifying any one of the ω_{ij} to make $\hat{U}$ slightly different from zero.

APPENDIX 6.8 CHI-SQUARE RESULTS FOR A RESTRICTED NUMBER OF PREDICTIONS

A method for testing the null hypothesis of statistical independence against a restricted alternative that at least one of T specified ∇ values is not zero was stated in Section 6.7.2. In this appendix the mathematics underlying that test are sketched. In addition, the determination of appropriate degrees of freedom is indicated.

The key result is once again provided by a multivariate extension of the delta method. If the specified ∇ values are $\nabla_1,\ldots,\nabla_T$, then for large samples, the joint density of $\hat{\nabla}_1,\ldots,\hat{\nabla}_T$ is approximately T-variate normal. Under the null hypothesis all expected values are zero. The variance of any $\hat{\nabla}$ is given in Appendix 6.7. The covariance of any pair, say $\hat{\nabla}_1$ and $\hat{\nabla}_2$, is

$$
(n-1)^{-1}\left[\sum_i \sum_j a_{ij}^{(1)}a_{ij}^{(2)}P_{ij}-\left(\sum_i \sum_j a_{ij}^{(1)}P_{ij}\right)\left(\sum_i \sum_j a_{ij}^{(2)}P_{ij}\right)\right]
$$

which reduces to

$$\left[(n-1)U^{(1)}U^{(2)}\right]^{-1}\left[\sum_i \sum_j \omega_{ij}^{(1)}\omega_{ij}^{(2)}P_{i.}P_{.j} + U^{(1)}U^{(2)}\right.$$

$$\left. -\sum_i \pi_{i.}^{(1)}\pi_{i.}^{(2)}P_{i.} - \sum_j \pi_{.j}^{(1)}\pi_{.j}^{(2)}P_{.j}\right]$$

under independence. (The superscripts refer, of course, to the respective prediction rules.) The covariance matrix Σ can be estimated by replacing $P_{i.}$ and $P_{.j}$ by $f_{i.}$ and $f_{.j}$.

If Σ is nonsingular, since the vector of $\hat{\nabla}$'s is approximately normal with mean zero, it follows that, in vector notation,

$$\left(\hat{\nabla}_1 \dots \hat{\nabla}_T\right)\Sigma^{-1}\left(\hat{\nabla}_1 \dots \hat{\nabla}_T\right)'$$

is approximately chi-square with T degrees of freedom. Estimating Σ does not change the large-sample distribution. (Just as there seems to be no improvement using the t distribution rather than the normal in the single ∇ case, we do not expect to improve the approximation by using an F distribution instead of the chi-square.) However there is no guarantee that Σ will have full rank and therefore be nonsingular. The rank ν of Σ is, by a standard multivariate analysis theorem [again see D. Morrison (1967, p. 93)], the number of linearly independent $\hat{\nabla}$ values. This fact provides a general way of finding ν, which will be the degrees of freedom for the chi-square test of Section 6.7.2. One calculates the rank of Σ as the number of linearly independent columns in it, if necessary by a numerical method such as Gauss-Jordan elimination. The $\hat{\nabla}$ values corresponding to the linearly independent columns may be renumbered as $\hat{\nabla}_1, \dots, \hat{\nabla}_\nu$, where $\hat{\nabla}_1$ is to be isolated as the value of most interest *ex post* to the researcher. (It may safely be assumed that $\hat{\nabla}_1$, to be interesting, is not identically zero, so that it may be included among the linearly independent $\hat{\nabla}$ values.) If Σ is the covariance matrix of these $\hat{\nabla}$'s alone, it must be of full rank and have an inverse. Therefore

$$\left(\hat{\nabla}_1 \dots \hat{\nabla}_\nu\right)\Sigma_\nu^{-1}\left(\hat{\nabla}_1 \dots \hat{\nabla}_\nu\right)' = \chi_\nu^2$$

has approximately a chi-square density with ν degrees of freedom for large samples, even if the entries of Σ_ν are estimated from the data.

The test in Section 6.7.2 was not based on this quadratic form, but rather on $(\hat{\nabla}_1)^2/\text{est. var}_0(\hat{\nabla}_1)$. Of course an alternative, and statistically more powerful, test would reject the hypothesis of independence if χ_ν^2 exceeded a tabulated chi-square point. Such a test would be harder computationally and, more to the point, would not isolate a specific $\hat{\nabla}$ and prediction rule as an *ex post* alternative to independence. The test proposed in Section 6.7.2 is conservative in the sense that

$$\hat{Z}_1^2 = \frac{\hat{V}_1^2}{\text{est. var}_0(\hat{V}_1)} \leqslant \chi_\nu^2$$

This can be seen by a geometric argument. The quadratic form χ_ν^2 represents the total squared (Mahalanobis) distance of the vector $(\hat{V}_1, \dots, \hat{V}_\nu)$ from the hypothesized vector $(0, \dots, 0)$, whereas $\hat{Z}_1^2$ represents the squared (Mahalanobis) distance from the zero vector along a single axis. In ν-dimensional space, χ_ν^2 is the squared length of the hypotenuse of a right triangle with $\hat{Z}_1^2$ as the squared length of one leg. By the Pythagorean theorem, $\hat{Z}_1^2$ must be less than (or in the degenerate case equal to) χ_ν^2. A more formal argument can be made in terms of an orthogonalization of Σ; this is a standard trick of multivariate statistics and need not be developed here.

APPENDIX 6.9 PROBABILITIES UNDER MISCLASSIFICATION

In this appendix the algebra of the misclassification result of Section 6.8 is carried out. The calculations are fairly direct results of elementary probability theory. A good example is the calculation of the probability that the apparent Y-state is y_i.

$$P(\text{apparent } Y = y_i) = P(\text{actual } Y \text{ is } y_i \text{ and classification correct})$$

$$+ \sum_{i' \neq i} P(\text{actual } Y \text{ is } y_{i'} \text{ and classification is } y_i)$$

$$= P_{i.}\left[1 - \sum_{i' \neq i} P(\text{classified } Y = y_{i'} | \text{actual } Y = y_i)\right]$$

$$+ \sum_{i' \neq i} P_{i'.} P(\text{classified } Y = y_i | \text{actual } Y = y_{i'})$$

$$= P_{i.}\left(1 - \sum_{i' \neq i} \theta_{i'}\right) + \sum_{i' \neq i} P_{i'.}\theta_i$$

$$= P_{i.}\left(1 - \sum_{i' \neq i} \theta_{i'}\right) + (1 - P_{i.})\theta_i$$

$$= P_{i.}\left(1 - \sum_{i'=1}^{R} \theta_{i'}\right) + \theta_i$$

Similarly

$$P\,(\text{apparent } X = x_j) = P_{.j}\left(1 - \sum_{j'=1}^{C} \zeta_{j'}\right) + \zeta_j$$

Furthermore the assumed independence of misclassification probabilities and a similar calculation shows that $P\,(\text{apparent } X = x_j$ and apparent $Y = y_i) =$

$$\text{apparent } P_{ij} = P_{ij}\left(1 - \sum_{i' \neq i} \theta_{i'}\right)\left(1 - \sum_{j' \neq j} \zeta_{j'}\right)$$

$$+ \sum_{i' \neq i} P_{i'j}\theta_i\left(1 - \sum_{j' \neq j} \zeta_{j'}\right)$$

$$+ \sum_{j' \neq j} P_{ij'}\left(1 - \sum_{i' \neq i} \theta_{i'}\right)\zeta_j$$

$$+ \sum_{i' \neq i}\sum_{j' \neq j} P_{i'j'}\theta_i\zeta_j$$

$$= P_{ij}\left[\left(1 - \sum_{i' \neq i} \theta_{i'}\right)\left(1 - \sum_{j' \neq j} \zeta_{j'}\right)\right]$$

$$+ (P_{.j} - P_{ij})\theta_i\left(1 - \sum_{j' \neq j} \zeta_{j'}\right)$$

$$+ (P_{i.} - P_{ij})\left(1 - \sum_{i' \neq i} \theta_{i'}\right)\zeta_j$$

$$+ (1 - P_{i.} - P_{.j} + P_{ij})\theta_i\zeta_j$$

$$= P_{ij}\left[\left(1 - \sum_{i'=1}^{R} \theta_{i'}\right)\left(1 - \sum_{j'=1}^{C} \zeta_{j'}\right)\right]$$

$$+ P_{i.}\left(1 - \sum_{i'=1}^{R} \theta_{i'}\right)\zeta_j$$

$$+ P_{.j}\left(1 - \sum_{j'=1}^{C} \zeta_{j'}\right)\theta_i + \theta_i\zeta_j$$

Therefore by substitution and algebraic expansion

$$\nabla_{\text{apparent}} = 1 - \frac{\displaystyle\sum_i \sum_j \omega_{ij}(\text{apparent } P_{ij})}{\displaystyle\sum_i \sum_j \omega_{ij}(\text{apparent } P_{i.})(\text{apparent } P_{.j})}$$

$$= 1 - \frac{\left(\displaystyle\sum_i \sum_j \omega_{ij} P_{ij}\right)A + B}{\left(\displaystyle\sum_i \sum_j \omega_{ij} P_{i.} P_{.j}\right)A + B}$$

where

$$A = \left(1 - \sum_{i'=1}^{R} \theta_{i.}\right)\left(1 - \sum_{j'=1}^{C} \zeta_{j'}\right)$$

$$B = \sum_i \sum_j \omega_{ij}\left[P_{i.}\left(1 - \sum_{i'=1}^{R} \theta_{i'}\right)\zeta_j + P_{.j}\left(1 - \sum_{j'=1}^{C} \zeta_{j'}\right)\theta_i + \theta_i\zeta_j\right]$$

The condition that $\sum_i \theta_i$ and $\sum_j \zeta_j$ are less than one evidently guarantees that both A and B are positive, as claimed in Section 6.8.

7

MULTIVARIATE METHODS

Our extensive treatment of bivariate prediction has laid the groundwork for a more general treatment of multivariate prediction analysis. Specifically this chapter addresses the problems of a researcher who has several possible predictor variables and a single dependent variable. The researcher wants to relate prediction success to the combined and separate effects of the various predictor variables.

First, in Section 7.1 we discuss how to state predictions based on several variables; in Section 7.2 we present a measure of the overall success of such predictions. Then in Section 7.3 we indicate some possible ways of assessing the partial contribution of each of the predictor variables in a multivariate prediction of dependent variable states. The interrelationship of the multivariate, bivariate, and partial PRE measures is given in the basic accounting structure considered in Sections 7.4 and 7.5. In Section 7.6 there is a brief digression exploring the relation of prediction analysis methods to correlational ideas. In Section 7.7 the effects of sampling error are considered, whereas Section 7.8 addresses one possible line of attack on the formidable problems of *ex post* analysis.

This chapter is a beginning. Much remains to be done, in terms of understanding the more subtle properties of the various measures introduced here. Yet the examination of these measures and their ramifications should prove of great value in analyzing complex multivariate propositions which cannot be treated adequately with current methods. This chapter therefore is an exploration. There is no pretense of complete coverage; at best some new paths may be broken.

7.1 STATING MULTIVARIATE PREDICTIONS

The first issue in the analysis of predictions involving several variables is of course how one states such predictions. In this section we indicate a simple format and briefly discuss the relation of multivariate statements to component bivariate predictions; in the next section we turn to the evaluation of prediction success. To keep notational problems within reasonable bounds this section, and indeed the whole chapter, is written in terms of three variables. The extension to more variables is only briefly indicated, for the basic concepts, measures, and accounting structure directly parallel trivariate analysis. As in the bivariate case we denote the dependent variable as Y, with states $y_1, \ldots, y_i, \ldots, y_R$. The two independent variables (one of which might, in certain problems, be regarded as an intervening variable) are X with states $x_1, \ldots, x_j, \ldots, x_C$ and W with states $w_1, \ldots, w_k, \ldots, w_S$. The subscript S is the number of strata in a three-way cross classification. The two states of any dichotomous variable may alternatively be designated by, for example, y and $\bar{y}$ instead of y_1 and y_2. Relating either just X-states or just W-states to Y-states is accomplished by bivariate prediction logic.

7.1.1 Logical Combinations

How do we combine the two variables for a multivariate prediction of Y-states? The process is very simple. We predict some set of Y-states for every possible pair of X- and W-states. The pair (x_j, w_k) constitutes a *logical combination* $(X = x_j \ \& \ W = w_k)$. The set of all such combinations is the Cartesian product $X \times W$. This composite is in effect a new variable V with $v_1 = (x_1, w_1), \ldots, v_{CS} = (x_C, w_S)$.

For example, consider the venerable American political statement, "When the Democrats control the House of Representatives, Southern, rural Congressmen tend to be committee chairmen." The dependent variable would have categories c (for *c*hairman) and $\bar{c}$, whereas the predictor would have s (for *S*outhern) and $\bar{s}$, and r (for *r*ural) and $\bar{r}$ categories, respectively. The composite variable V would have four categories, $v_1 = (s, r)$, $v_2 = (s, \bar{r})$, $v_3 = (\bar{s}, r)$, and $v_4 = (\bar{s}, \bar{r})$. The numbering is arbitrary, of course. With this notation the prediction is simple: $v_1 \rightsquigarrow c$.

7.1.2 Prediction Statements

With the composite variable V we are back in the two-variable case, and the basic prediction logic can be used. In the trivariate case the general expression is as follows:

$$\mathcal{P} : \left\{ \mathcal{P}_{jk} \right\} = \left\{ (x_j, w_k) \rightsquigarrow \mathcal{S}(x_j, w_k) \right\} \tag{1}$$

where
$$\mathcal{S}(x_j, w_k) \subseteq \{y_1, y_2, \ldots, y_i, \ldots, y_R\}$$

which merely says that for each (x_j, w_k) pair a set $\mathcal{S}(x_j, w_k)$ of Y-states is predicted. With a fourth variable Z with states $\{z_m\}$, we would write $(x_j, w_k, z_m) \rightsquigarrow \mathcal{S}(x_j, w_k, z_m)$, and so on.

As in the bivariate case, any statements that are equivalent in elementary formal logic are equivalent in the prediction logic. For example, $[(w_1$ or $w_2)$ & $x_1] \rightsquigarrow y_1$ is equivalent to $(w_1$ & $x_1) \rightsquigarrow y_1$, $(w_2$ & $x_1) \rightsquigarrow y_1$, $(w_1$ & $x_2) \rightsquigarrow (y_1, y_2, y_3, \ldots), \ldots, (w_S$ & $x_C) \rightsquigarrow (y_1, y_2, y_3, \ldots)$.

7.1.3   Error Sets

As in the bivariate case we will use the error set $\mathcal{E}_{\mathcal{P}}$ as an alternate, and often more convenient, way of specifying a proposition $\mathcal{P}$. The concept of error weights also extends directly; ω_{ijk} is greater than zero if $Y = y_i$ is an error for the proposition when $W = w_k$ and $X = x_j$, and zero otherwise. Thus weighted errors and mixed strategy predictions can be used, just as in the bivariate case. To simplify exposition, however, the chapter deals with unweighted errors.

For example, if $R = 3$, $C = 2$, and $S = 2$ for the proposition

$$\mathcal{P} : x_1 \text{ \& } w_1 \rightsquigarrow y_1$$
$$x_1 \text{ \& } w_2 \rightsquigarrow (y_1, y_2 \text{ or } y_3)$$
$$x_2 \text{ \& } w_1 \rightsquigarrow y_2$$
$$x_2 \text{ \& } w_2 \rightsquigarrow (y_1, y_2 \text{ or } y_3)$$

then
$$\omega_{211} = \omega_{311} = \omega_{121} = \omega_{321} = 1$$

and all other ω's are zero;

$$\mathcal{E}_{\mathcal{P}} = \{(y_2, x_1, w_1), (y_3, x_1, w_1), (y_1, x_2, w_1), (y_3, x_2, w_1)\}$$

as illustrated in Table 7.1.

Table 7.1   Trivariate error set and hypothetical probabilities[a]

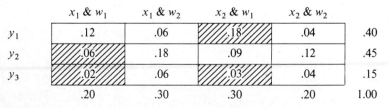

	x_1 & w_1	x_1 & w_2	x_2 & w_1	x_2 & w_2	
y_1	.12	.06	.18	.04	.40
y_2	.06	.18	.09	.12	.45
y_3	.02	.06	.03	.04	.15
	.20	.30	.30	.20	1.00

[a]Shading denotes error cells.

7.1.4   Computation

The foregoing formulation of the prediction language of course immediately implies a computational cost in data analysis. To evaluate multivariate predictions we must cross-tabulate all the variables; in the three-variable case we must calculate the proportions P_{ijk} (or the sample analogues f_{ijk}). In multiple regression, by comparison, it is only necessary to make pairwise comparisons of the variables, since all multiple and partial correlations can be calculated in terms of two-variable correlations.

The complete cross tabulation is required if one wishes to obtain the essential feature of prediction logic: its capacity to make different predictions for different categories of the independent variables. Even in the simplest structure, that of such statements as $(x_1 \, \& \, w_1) \rightsquigarrow y_1$, it is necessary to obtain information regarding the complete multivariate probability structure.

The cost and pain of extensive computation are being increasingly eased by advancing computer technology. On the other hand, there is a tradition in the social sciences not to cross-tabulate extensively because of the small number of cases to be expected in each cell. The *a priori* theorist can disregard this wisdom, even if dealing with only moderate-sized samples. As we have already seen in the bivariate case, it is the total precision of predictions and not individual cell entries that controls applications in statistical inference. Therefore we urge that data considerations not deter one from formulating multivariate predictions.

7.1.5   The Relation of Propositions for Component Variables to Propositions for the Composite Variable V

The preceding section made a simple extension of the bivariate prediction logic to represent multivariate statements as a bivariate relation between the dependent variable and the composite variable V. Not all complex models arrive at their V directly; rather they are built up gradually out of a number of component predictions. Frequently the researcher wishes to evaluate one or more of these components against the more complex model. It is appropriate therefore to enrich the prediction logic language by specifying means of combining error sets from various propositions; conversely we ask which component theories qualify as natural predecessors of a more complex theory.

To address these questions we illustrate with three dichotomous variables; the general case is a direct extension. There are four possible bivariate predictions between one state of one dichotomy and the states of another; for example, $x \rightsquigarrow \varnothing$, $x \rightsquigarrow y$, $x \rightsquigarrow \bar{y}$, and $x \rightsquigarrow [y$ or $\bar{y}]$. Since variable X has two states this leads to $4 \times 4 = 16$ possible $\mathcal{P}$'s, corresponding to the 2^{RC} possible patterns of error cells in the underlying 2×2 table. Note that each error cell [say $(\bar{y}, x)$] of a 2×2 bivariate table corresponds to two error cells $[(\bar{y}, x, w), (\bar{y}, x, \bar{w})]$ in $Y \times V$. Hereafter in this chapter when we refer to the error cells of a bivariate proposition, we refer to their representation in $Y \times V$.

Three types of bivariate predictions are considered: Y is predicted by X, Y is predicted by W, and X is predicted by W. This is without loss of generality since, for example, any statement where Y predicts X has a logically equivalent statement where X predicts Y. We thus regard Y as the dependent variable, with X an intervening variable between W and Y. This is illustrated in Figure 7.1.

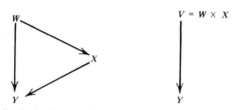

Figure 7.1   Generalized trivariate prediction. The arrows designate prediction statements.

Since there are 16 possible forms of each of the three bivariate predictions relating dichotomies, they can be grouped in $16^3 = 4096$ possible combinations; however some of these are obviously equivalent, since $Y \times V$ contains only $2^{RCS} = 2^8 = 256$ possible trivariate error sets. These 256 possible patterns of error cells are precisely what we wish to derive by appropriate combination of the bivariate propositions.

The multivariate proposition $\mathscr{P}_{YV} = \mathscr{P}_{YXW}$ might be stated in the form of a connected set of bivariate statements. To illustrate, in the committee chairman example the composite $v_1 \leadsto c$ could also have been stated: $s \leadsto c$ or $r \leadsto c$. A more complex example is represented by the combination

$$\mathscr{P}_{YXW} : \mathscr{P}_{YW} \text{ or } (\mathscr{P}_{XW} \& \mathscr{P}_{YX})$$

What is the relation between the error structures of the component bivariate predictions and that of $\mathscr{P}_{YXW}$? Conversely can all trivariate prediction rules be decomposed into bivariate predictions via some simple rules? It would be convenient to have such methods to be able to evaluate the contribution of each predictor variable to the success of the trivariate prediction. Given the complexity of possible error patterns, we cannot classify all combinations into a few categories. Furthermore it may well be that an investigator will specify a multivariate prediction directly, without building from bivariate propositions. Nonetheless some simple illustrations of combination procedures should be helpful.

7.1.6 Combining Propositions

The two simplest ways of combining propositions in symbolic logic are represented by the connectives & and *or*—or formally, *conjunction* and *disjunction*. We need to examine the analogues of such combinations in the prediction logic. The simplest example is the combination "$w \leadsto y$ & $x \leadsto y$." Suppose we call the component predictions $\mathcal{P}_{YW}$ and $\mathcal{P}_{YX}$. The error event for $\mathcal{P}_{YW}$ is $\bar{y}w$; for $\mathcal{P}_{YX}$ it is $\bar{y}x$. In multivariate terms $\bar{y}xw$ and $\bar{y}\bar{x}w$ are errors for $\mathcal{P}_{YW}$ and $\bar{y}xw$ and $\bar{y}x\bar{w}$ are errors for $\mathcal{P}_{YX}$. The conjunction & is falsified in formal logic if either component is false; thus in prediction logic "$\mathcal{P}_{YW}$ & $\mathcal{P}_{YX}$" is in error for any case that falls in either error set. The error cells for "$w \leadsto y$ & $x \leadsto y$" are therefore $\bar{y}xw$, $\bar{y}x\bar{w}$, and $\bar{y}\bar{x}w$. In general, the error cases for "$\mathcal{P}_{YW}$ & $\mathcal{P}_{YX}$" are those that violate either bivariate component. On the other hand, "$\mathcal{P}_{YW}$ or $\mathcal{P}_{YX}$" is falsified in formal logic if *both* component statements are false. By analogy, the error events for "$\mathcal{P}_{YW}$ or $\mathcal{P}_{YX}$" are those that are errors for both $\mathcal{P}_{YW}$ and $\mathcal{P}_{YX}$; in the simplest case "$w \leadsto y$ or $x \leadsto y$" is in error only when $\bar{y}xw$ occurs and is logically equivalent to $(x \& w) \leadsto y$. In general, $\mathcal{P}_{YX}$ & $\mathcal{P}_{YW}$ is a bolder statement than $\mathcal{P}_{YX}$ *or* $\mathcal{P}_{YW}$ and has a larger error set, hence greater precision.

More complex logical combinations also can be useful. For example, we could treat as errors those events that are errors for at least one *but not both* of two bivariate propositions. Combining $x \leadsto y$ and $w \leadsto y$ this way makes $\bar{y}x\bar{w}$ and $\bar{y}xw$ the trivariate error cells. This procedure has the special property that $\bar{y}$ is an error (in $\mathcal{P}_{YW}$) when w occurs and an error (in $\mathcal{P}_{YX}$) when x occurs, but not an error (in $\mathcal{P}_{YXW}$) when *both* occur. The predictive value of w is considered to be suppressed when x occurs and vice versa; for example, two drugs might cure a disease when taken alone but cancel each other when taken together.

We have been considering logical operations which treat $\mathcal{P}_{YX}$ and $\mathcal{P}_{YX}$ operations which treat $\mathcal{P}_{YX}$ and $\mathcal{P}_{YW}$ symmetrically. There is no idea that the X prediction, say, is applied first. There is an alternative, asymmetric means of combining $\mathcal{P}_{YX}$ with $\mathcal{P}_{YW}$, called *contingent refinement*, which is useful when the "first" bivariate rule makes some one-to-many predictions. As a first illustration, suppose $X = \{x_1, x_2, x_3\}$, $Y = \{y_1, y_2\}$, and

$$\mathcal{P}_{YX} : x_1 \leadsto y_1$$
$$x_2 \leadsto (y_1 \text{ or } y_2)$$
$$x_3 \leadsto y_2$$

whereas $W = \{w_1, w_2\}$ and

$$\mathcal{P}_{YW} : w_1 \rightsquigarrow y_1$$
$$w_2 \rightsquigarrow y_2$$

The $\mathcal{P}_{YX}$ rule is imprecise for $X = x_2$; we refine the prediction for $X = x_2$ by using $\mathcal{P}_{YW}$ but leave the $X = x_1$ and $X = x_3$ cases to $\mathcal{P}_{YX}$. The refined prediction is

$$\mathcal{P}_{YXW} : x_1 \rightsquigarrow y_1$$
$$x_2 \,\&\, w_1 \rightsquigarrow y_1$$
$$x_2 \,\&\, w_2 \rightsquigarrow y_2$$
$$x_3 \rightsquigarrow y_2$$

The error cell structure is illustrated for Table 7.2. In fact this kind of refinement is the basis for the Rosenthal prediction of Section 7.5.

Table 7.2   Illustration of contingent refinement[a]

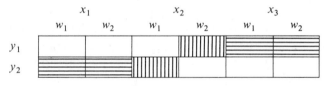

[a]Horizontal shading indicates error cells for $\mathcal{P}_{YX}$. Vertical shading indicates error cells for $\mathcal{P}_{YW}$, applied only when $X = x_2$.

A second example illustrates the general refinement process. Suppose $X = \{x_1, x_2, x_3\}$, $Y = \{y_1, y_2, y_3\}$, and

$$\mathcal{P}_{YX} : x_1 \rightsquigarrow y_1$$
$$x_2 \rightsquigarrow (y_1 \text{ or } y_2)$$
$$x_3 \rightsquigarrow (y_1, y_2 \text{ or } y_3)$$

whereas $W = \{w_1, w_2\}$ and

$$\mathcal{P}_{YW} : w_1 \rightsquigarrow (y_1 \text{ or } y_3)$$
$$w_2 \rightsquigarrow y_2$$

For $X = x_1$, $\mathcal{P}_{YX}$ predicts a single state, so no refinement is needed. For $X = x_3$, the $\mathcal{P}_{YW}$ rule completely determines the error cells, since there are no $\mathcal{P}_{YX}$ errors under x_3. The interesting situation is $X = x_2$. The $\mathcal{P}_{YW}$ rule can be used to differentiate between y_1 and y_2, so that x_2 & $w_1 \leadsto y_1$ and x_2 & $w_2 \leadsto y_2$. The $\mathcal{P}_{YW}$ rule also predicts y_3 when $W = w_1$ whereas $\mathcal{P}_{YX}$ has y_3 as an error for $X = x_2$. Since the refinement process assumes that $\mathcal{P}_{YX}$ is the dominant rule, the conflict when $X = x_2$ and $W = w_1$ is resolved by making y_3 an error. The error cell structure is illustrated in Table 7.3.

Table 7.3   Illustration of contingent refinement[a]

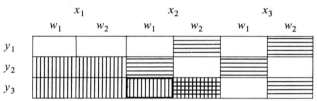

[a]Vertical shading indicates error cells for $\mathcal{P}_{YX}$. Horizontal shading indicates error cells for $\mathcal{P}_{YW}$, applied only when $X = x_2$ or x_3. Heavy border indicates conflict of $\mathcal{P}_{YX}$ error and $\mathcal{P}_{YW}$ prediction.

The contingent refinement process thus refines the prediction by introducing a more precise and differentiated prediction. The resultant $\mathcal{P}_{YXW}$ rule depends on which bivariate rule is regarded as dominant or first; had $\mathcal{P}_{YW}$ been regarded as dominant in either example, the error structure would have been different.

The various types of logical operations on propositions provide considerable flexibility for investigators who wish to combine bivariate predictions. Not all trivariate rules can be derived from these procedures; some of them must be regarded as essentially trivariate. For example, trivariate propositions such as "If $X = x$ then $w \leadsto y$ and if $X = \bar{x}$ then $w \leadsto \bar{y}$" specify different conditions under which alternative bivariate propositions relating the same two variables are asserted to apply.

7.2   MEASURING MULTIVARIATE PREDICTION SUCCESS

Once a researcher has specified a multivariate prediction, the next step is to assess its success with a set of cross-classified data. In this section we develop a ∇ measure for this purpose; since the trivariate problem is reduced to a bivariate problem via the composite V variable, the trivariate ∇ is equivalent to bivariate ∇ for a prediction $\mathcal{P}_{YV}$. As before, we first define population ∇; the issues of sample data are dealt with in Section 7.7. Some notation is required. We denote the population proportion (or probability) of $Y = y_i$, $X = x_j$, and $W = w_k$ by P_{ijk}. As before we use dots to

Table 7.4 Notation for the general $R \times CS$ cross classification

	$x_1 \& w_1$	$x_1 \& w_2$	$\ldots$	$x_1 \& w_k$	$\ldots$	$x_1 \& w_S$	$x_2 \& w_1$	$\ldots$	$x_j \& w_k$	$\ldots$	$x_C \& w_1$	$x_C \& w_2$	$\ldots$	$x_C \& w_k$	$\ldots$	$x_C \& w_S$	
y_1	P_{111}	P_{112}		P_{11k}		P_{11S}	P_{121}		P_{1jk}		P_{1C1}	P_{1C2}		P_{1Ck}		P_{1CS}	$P_{1..}$
y_2	P_{211}	P_{212}		P_{21k}		P_{21S}	P_{221}		P_{2jk}		P_{2C1}	P_{2C2}		P_{2Ck}		P_{2CS}	$P_{2..}$
$\ldots$																	$\ldots$
y_i	P_{i11}	P_{i12}		P_{i1k}		P_{i1S}	P_{i21}		P_{ijk}		P_{iC1}	P_{iC2}		P_{iCk}		P_{iCS}	$P_{i..}$
$\ldots$																	$\ldots$
y_R	P_{R11}	P_{R12}		P_{R1k}		P_{R1S}	P_{R21}		P_{Rjk}		P_{RC1}	P_{RC2}		P_{RCk}		P_{RCS}	$P_{R..}$
	$P_{.11}$	$P_{.12}$	$\ldots$	$P_{.1k}$		$P_{.1S}$	$P_{.21}$	$\ldots$	$P_{.jk}$	$\ldots$	$P_{.C1}$	$P_{.C2}$	$\ldots$	$P_{.Ck}$	$\ldots$	$P_{.CS}$	1.0

indicate which variables have been "summed out"; for instance, the probability that $X = x_j$ and $W = w_k$ is

$$P_{.jk} = \sum_{i=1}^{R} P_{ijk}$$

and the probability that $Y = y_i$ is

$$P_{i..} = \sum_{j=1}^{C} \sum_{k=1}^{S} P_{ijk} = \sum_{j=1}^{C} P_{ij.}$$

The general probability table is shown in Table 7.4. Note that the only change from the bivariate probability setup is in the subscript notation. Therefore the multivariate ∇ differs only notationally from bivariate ∇. Also, parallel to statement (10) of Chapter 3, a proposition $\mathcal{P}_{YXW}$ is not admissible if and only if for every cell belonging to $\mathcal{E}_{\mathcal{P}}$ either $P_{i.}$ or $P_{.jk} = 0$.

To illustrate, let us again examine the proposition that Southern rural congressmen tend to be committee chairmen. Table 7.5 shows the Wolfinger and Heifetz (1965) data for the U.S. House of Representatives as of January, 1964. The data are for subcommittees, which are far more numerous than full committees, and are for Democrats only. For now we treat these data as a population. The error cell is shaded and has probability 22/256. The relevant marginal probabilities are 147/256 and 73/256. Hence under the bivariate definition of ∇

$$\nabla_{(s\&r)\rightsquigarrow c} = 1 - \frac{22/256}{(147/256)(73/256)} = .475$$

Thus the prediction finds moderate support in the data.

The general definition of multivariate ∇ follows directly.

When W and X are known, under rule K we predict, for any case with state (x_j, w_k) that the Y-state is in $\mathcal{S}(x_j, w_k)$. The expected number of errors in N cases is thus

$$N \sum_i \sum_j \sum_k \omega_{ijk} P_{ijk}$$

When W and X are unknown, rule U predicts with probability $P_{.jk}$ that the Y-state will be in $\mathcal{S}(x_j, w_k)$. The expected number of rule U errors is

$$\sum_j \sum_k N P_{.jk} \sum_i \omega_{ijk} P_{i..} = N \sum_i \sum_j \sum_k \omega_{ijk} P_{i..} P_{.jk}$$

Table 7.5 Subcommittee chairmanship, Democrats, U.S. House of Representatives, by region and urban-rural, 1964

Congressman is	Constituency type[a]				
	South		North		
	Rural	Urban	Rural	Urban	
Subcommittee chairman	51	7	19	32	109
Not subcommittee chairman	22	16	34	75	147
	73	23	53	107	256

Source: Wolfinger and Heifetz (1965, Table 1, p. 343).
[a]As the table was originally published in percentage form, there may be some slight errors in the number of cases shown in each cell. Shading indicates error cell.

Then, applying the PRE model, we define the multivariate ∇ measure for the trivariate prediction logic proposition $\mathcal{P}_{YXW}$ as:

$$\nabla_{\mathcal{P}_{YXW}} = 1 - \frac{\sum_i \sum_j \sum_k \omega_{ijk} P_{ijk}}{\sum_i \sum_j \sum_k \omega_{ijk} P_{i..} P_{.jk}}$$

$$= 1 - \frac{\mathcal{P}_{YXW} \text{ error, given } X \text{ and } W}{\mathcal{P}_{YXW} \text{ error, not knowing } V(X \times W)}$$

(2)

7.2.1 Logical Equivalence, Dependent Variables, and the ∇ Measure

Consequently multivariate $\nabla_{\mathcal{P}}$ is simply the bivariate $\nabla_{\mathcal{P}}$ when the composite variable V is used to predict Y. The equivalence property of $\nabla_{\mathcal{P}}$ implies that we obtain the same value of $\nabla_{\mathcal{P}}$ when Y is used to predict V. However ∇ is not invariant when either X or W is substituted for Y as the dependent variable, as, for example, when a composite variable $V' = Y \times W$ is used to predict X.

This point is illustrated by the probability and error structure shown in

Table 7.1. With Y as dependent variable,

$$\nabla_{\mathscr{P}_{YXW}} = 1 - \frac{P_{211} + P_{311} + P_{121} + P_{321}}{P_{2..}P_{.11} + P_{3..}P_{.11} + P_{1..}P_{.21} + P_{3..}P_{.21}}$$

$$= 1 - \frac{.06 + .02 + .18 + .03}{(.45)(.2) + (.15)(.2) + (.40)(.3) + (.15)(.3)} = -.018$$

But if we regard X as the dependent variable,

$$\nabla_{\mathscr{P}_{XYW}} = 1 - \frac{P_{211} + P_{311} + P_{121} + P_{321}}{P_{.1.}P_{2.1} + P_{.1.}P_{3.1} + P_{.2.}P_{1.1} + P_{.2.}P_{3.1}} = -.055$$

and with W as the dependent variable

$$\nabla_{\mathscr{P}_{WYX}} = .049$$

Although $\mathscr{P}_{YXW}$, $\mathscr{P}_{XYW}$, and $\mathscr{P}_{WYX}$ identify the same error cells and, hence, are logically equivalent, their corresponding measures of prediction success are not equal. Indeed one of the three $\nabla_{\mathscr{P}}$ measures is positive, even though the other two are negative. Consequently, the multivariate measure fails to maintain the logical equivalence property of bivariate $\nabla_{\mathscr{P}}$.

From the previous examples it is clear that the differences in multiple ∇ values here result only from differences in the rule U error rates. As a PRE measure, ∇ is a relative quantity, comparing errors under two distinct information conditions. As in Chapter 3, the rule K error rate is unaffected by changing the labeling of variables as *dependent* or *independent*. However the symmetry in the evaluation of scope and precision, which allowed the bivariate rule U error rate to be similarly unaffected, fails in the multivariate model. Thus in multivariate analyses, it is not always true that logically equivalent propositions have identical ∇ values. It is still true that arbitrary redefinitions of the states of a given variable that do not affect the counting of prediction errors will not alter the value of ∇.

In summary, the multivariate ∇ measure is simply the bivariate ∇ applied with a composite independent variable. Once the definition of the composite variable V is given, no new principles are required. The multivariate analysis really begins when a researcher attempts to analyze the separate effects of the various predictor variables. This is the task of the next section.

7.3 MEASURING PARTIAL PREDICTION SUCCESS

There are at least two possible approaches to assessing the extra predictive value of an additional predictor variable W. One approach is similar to the $\mathcal{E}_+, \mathcal{E}_-$ analysis of Section 3.2.4. In comparing a bivariate and a trivariate prediction it is easy to restate the two propositions as bivariate propositions in the $Y \times V$ cross classification. Then the discussion of the comparison of two propositions in Chapters 3 and 4 applies directly. This approach has two drawbacks in terms of multivariate analysis: first, it requires the statement of two distinct prediction rules; second, it does not explicitly involve a distinction between the X and W predictors, since the $\mathcal{E}_+$ and $\mathcal{E}_-$ sets of error cells may not have any simple definition in terms of either predictor alone.

7.3.1 Partial Prediction Success in a Single Subpopulation

The alternative is to use only the trivariate proposition and to attempt to assess the predictive value of the added variable W, over and above the value of X alone. There is a standard statistical method for this "extra value" analysis, namely, the idea of "partialing out" or "holding constant" the value of the original predictor X. When X is restricted to have a particular value x_j, or in other words, for the cases in the $X = x_j$ subpopulation, variations in Y-state cannot be attributable to changes in X information since X is constant, but may possibly be predicted from the respective W-states. Thus assessing the success of $\mathcal{P}_{YXW}$ within fixed $X = x_j$ subpopulations indicates the extra value of W information. The basic partial prediction success measure will be derived as a ∇ value for a specific X subpopulation.

To define partial ∇ we need more notation. Probabilities within the subpopulation having $X = x_j$ are by definition conditional probabilities given $X = x_j$. Conditional probabilities are defined by[1]

$$P(y_i, w_k | x_j) = \frac{P_{ijk}}{P_{\cdot j \cdot}}$$

$$P(y_i | x_j) = \frac{P_{ij \cdot}}{P_{\cdot j \cdot}}$$

$$P(w_k | x_j) = \frac{P_{\cdot jk}}{P_{\cdot j \cdot}}$$

[1] We do not use the notation $P_{ik|j}$ since $P_{14|3}$ could mean $P(y_1, w_4 | x_3)$ or $P(y_1, x_4 | w_3)$ or any of several other possibilities.

For example, if the Wolfinger–Heifetz data are taken to be the entire population

$$P(\text{nonchairman, rural}|\text{Southern}) = \frac{P_{\bar{c}sr}}{P_{.s.}} = \frac{(22/256)}{(96/256)}$$

$$= \frac{22}{96}$$

which corresponds to the fact that 22 of the 96 Southern congressmen were nonchairmen and rural.

The basic partial ∇ measure is simply the ordinary ∇ measure defined on a fixed subpopulation, or in other words, a ∇ measure defined with conditional probabilities. For example, the partial ∇ for $(s\ \&\ r) \leadsto c$ holding *Southern* fixed is calculated for the subpopulation of 96 Southern congressmen (as opposed to the total population of 256 congressmen).

For rule K we predict *chairman* for all 73 rural congressmen and err in 22 cases. For rule U we predict *chairman* for 73 congressmen selected randomly from the Southern subpopulation. As the fraction of Southerners who are not chairmen is $38/96$, we expect $73(38/96) = 28.90$ errors. The PRE, which is the partial ∇ given *Southern*, is

$$\nabla_{(s\&r)\leadsto c|s} = 1 - \frac{22}{28.9} = .239$$

indicating a modest additional value of the *rural* information in predicting chairmanships, over and above the *Southern* information.

The general definition of partial ∇ given $X = x_j$ is very similar. There are $NP_{.j.}$ cases with $X = x_j$. Under rule K the prediction that Y falls in $\mathsf{S}(x_j, w_k)$ is made for $NP_{.jk} = NP(w_k|x_j)P_{.j.}$ observations. The total number of errors for the $\mathsf{S}(x_j, w_k)$ prediction equals [where $P(y_i|x_j, w_k) = P_{ijk}/P_{.jk}$]

$$NP(\omega_k|x_j)P_{.j.} \sum_i \omega_{ijk} P(y_i|x_j, w_k)$$

Under rule U the prediction that Y will fall in $\mathsf{S}(x_j, w_k)$ is made for $NP(w_k|x_j)P_{.j.}$ randomly chosen observations, with

$$NP(w_k|x_j)P_{.j.} \sum_i \omega_{ijk} P(y_i|x_j)$$

expected errors. With each rule we must sum over the possible w_k values to obtain the total error rate. Thus the partial ∇ given $X = x_j$ is the PRE measure

$$\nabla_{\mathscr{P}_{YXW}|x_j} = 1 - \frac{\mathscr{P}_{YXW} \text{ errors given } X = x_j \text{ and } W}{\mathscr{P}_{YXW} \text{ errors given } X = x_j \text{ but not } W}$$

$$= 1 - \frac{N \sum_i \sum_k \omega_{ijk} P(y_i|x_j,w_k) P(w_k|x_j) P_{.j.}}{N \sum_i \sum_k \omega_{ijk} P(y_i|x_j) P(w_k|x_j) P_{.j.}} \qquad (3)$$

$$= 1 - \frac{\sum_i \sum_k \omega_{ijk} P(y_i,w_k|x_j)}{\sum_i \sum_k \omega_{ijk} P(y_i|x_j) P(w_k|x_j)}$$

The numerical properties of this partial ∇ are the same as those of ordinary ∇, since the partial is merely the ordinary ∇ applied within the $X = x_j$ subpopulation. In particular it is zero when W and Y are independent within the $X = x_j$ subpopulation (which does not imply that W and Y are independent generally). Again the idea of partial ∇ given a particular X-state can be accommodated within the basic framework with minor modifications.[2]

7.3.2 Overall Partial Prediction Success

However, this version of partial ∇ generally gives a different result for each possible x_j state. It would be desirable to have an overall measure of the predictive value of W over and above the value of X to the prediction, rather than a collection of different values for different X-states. For example, with seven X-states the information that the respective partial ∇ values are .8, .1, -2, 0, .9, .4, and .2 does not summarize the extra predictive value adequately. We need a measure based on total errors rather than errors within specific X subpopulations.

The solution is a simple extension of the calculations of the subpopula-

[2]Davis (1967) has defined a partial gamma coefficient that applies these ideas to ordinal variables. His partial gamma is the ordinary gamma coefficient between two variables defined on a subpopulation of pairs of cases that have the same value of a third variable. Since ordinary γ is a special case of ordinal ∇, partial gamma is a special case of partial ∇. The prediction rule applies to those pairs tied on the third variable but not tied on either of the other two and is the usual concordance prediction. Davis did not attempt a partial coefficient for the subpopulation of nontied cases. For further developments in multivariate degree-2 analysis, see Smith (1974), Somers (1974), and Wilson (1974).

tion partial ∇. Under rule K a total of

$$NP_{.j.} \sum_i \sum_k \omega_{ijk} P(y_i, w_k|x_j)$$

errors are committed within the $X = x_j$ subpopulation; the grand total of rule K errors over all subpopulations is therefore

$$N \sum_j P_{.j.} \sum_i \sum_k \omega_{ijk} P(y_i, w_k|x_j) = N \sum_i \sum_j \sum_k \omega_{ijk} P_{ijk}$$

Note that these are just the rule K errors for multiple $\nabla_{\mathcal{P}_{YXW}}$. Under rule U the corresponding grand total is

$$N \sum_i \sum_j \sum_k \omega_{ijk} P(y_i|x_j) P(w_k|x_j) P_{.j.}$$

Since this quantity may be rewritten as

$$N \sum_i \sum_j \sum_k \omega_{ijk} P(y_i|x_j) P_{.jk}$$

these errors are also just the rule U errors for multiple $\nabla_{\mathcal{P}_{YXW}}$ *except* that the subpopulation y_i probabilities have been substituted for the overall population y_i probabilities. The overall partial is, once again, a PRE measure:

$$\nabla_{\mathcal{P}_{YXW}|X} = 1 - \frac{\mathcal{P}_{YXW} \text{ errors given } X \text{ and } W}{\mathcal{P}_{YXW} \text{ errors given } X \text{ but not } W}$$

$$= 1 - \frac{\displaystyle\sum_i \sum_j \sum_k \omega_{ijk} P_{ijk}}{\displaystyle\sum_i \sum_j \sum_k \omega_{ijk} P(y_i|x_j) P(w_k|x_j) P_{.j.}} \tag{4}$$

$$= 1 - \frac{\displaystyle\sum_i \sum_j \sum_k \omega_{ijk} P_{ijk}}{\displaystyle\sum_i \sum_j \sum_k \omega_{ijk} P(y_i|x_j) P_{.jk}}$$

The subpopulation partials are denoted $\nabla_{\mathcal{P}|x_j}$ and the overall partial is called $\nabla_{\mathcal{P}|X}$.

The overall partial, as an average, may hide some relations that would be revealed in the component partials. In particular, if some of the component partials were near 1.0 whereas others were substantially negative, the overall partial might average out to zero. This would indicate that, *overall*, W information has no predictive value for this proposition; but the large component partials imply that in a revised prediction, W is valuable in those X subpopulations.

7.3.3 Example: Choice in Coalition Formation

The data in Table 7.6, taken from Riker and Zavoina (1970), serve to illustrate this "overall" partial ∇ coefficient. The dependent variable is the action of a specified player in a given coalition formation trial; the subject may vote for or against another player. The independent variables are the action of the subject on the previous trial (for or against) and the results of the previous trial (won or lost). One Riker-Zavoina prediction (actually their null model) was that any action that led to a win on the previous trial would be repeated and that a subject who voted "for" and lost on the previous trial would change and vote "against"; the error cells for this prediction are shaded in Table 7.6. For these data the multiple ∇ is .227.

Partial ∇ can be used to assess the additional predictive value of won or lost information over and above the predictive value of the previous vote. By expression (4),

$$\nabla_{\mathcal{P}_{YXW}|X}$$

$$= 1 - \frac{(31 + 22 + 36)/280}{\left(\frac{57}{142}\right)\left(\frac{94}{142}\right)\left(\frac{142}{280}\right) + \left(\frac{85}{142}\right)\left(\frac{48}{142}\right)\left(\frac{142}{280}\right) + \left(\frac{63}{138}\right)\left(\frac{86}{138}\right)\left(\frac{138}{280}\right)}$$

$$= .158$$

Table 7.6 Voting in a coalition experiment[a]

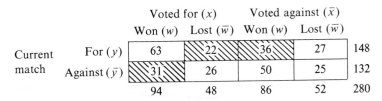

		Previous match				
		Voted for (x)		Voted against ($\bar{x}$)		
		Won (w)	Lost ($\bar{w}$)	Won (w)	Lost ($\bar{w}$)	
Current match	For (y)	63	22	36	27	148
	Against ($\bar{y}$)	31	26	50	25	132
		94	48	86	52	280

Source: Riker and Zavoina, (1970, p. 60).
[a]Shaded cells represent error cells for Riker-Zavoina learning predictions.

indicating, perhaps unexpectedly, that the result of the previous vote has only modest predictive value once the previous choice is known.

7.3.4 Weighted Average Interpretation of Partial ∇

It is not surprising, given previous weighted-average interpretations, that the overall partial ∇ is a weighted average of the subpopulation partial ∇ values:

$$
\nabla_{\mathcal{P}|X} = \frac{\sum_j P_{.j.}\left[\left(\sum_i \sum_k \omega_{ijk} P(y_i|x_j) P(w_k|x_j)\right)\right] \nabla_{\mathcal{P}|x_j}}{\sum_j P_{.j.}\left[\sum_i \sum_k \omega_{ijk} P(y_i|x_j) P(w_k|x_j)\right]}
\tag{5}
$$

As before, the weights depend on the precision of the prediction in each subpopulation. In this case the precision weights are multiplied by the probabilities of the various X subpopulations; it seems most plausible that prediction success in a subpopulation that contains many cases (and so has high probability) should be more heavily weighted than one containing few cases. Note that in the Wolfinger-Heifetz data the value of $\nabla_{(s\&r)\leadsto c|\bar{s}}$ is not defined, for there are no error cells in the $\bar{s}$ subpopulation, but by the same token the $\bar{s}$ prediction has zero precision and therefore zero weight. In that special case the overall partial ∇ must equal the partial for the only relevant subpopulation.

The Riker-Zavoina data provide a more general example. For the subpopulation previously voting "for" the partial is

$$
1 - \frac{(31+22)/142}{(57/142)(94/142)+(85/142)(48/142)} = .203
$$

whereas for the subpopulation previously voting "against" it is

$$
1 - \frac{36/138}{(63/138)(86/138)} = .083
$$

As the category previously voting "for" is slightly more probable and has a prediction of greater precision, it is weighted more heavily in the overall partial:

$$\nabla_{\mathscr{P}_{YXW}|X} = \frac{\left(\dfrac{142}{280}\right)(.468)(.203) + \left(\dfrac{138}{280}\right)(.284)(.083)}{\left(\dfrac{142}{280}\right)(.468) + \left(\dfrac{138}{280}\right)(.284)}$$

$$= .158$$

The overall partial is nearer the "voted for" (x) subpopulation partial.

7.3.5 Properties of Partial ∇

Partial ∇ is a PRE measure. A value for $\nabla_{\mathscr{P}|X}$ of .6 means that prediction error is reduced 60% by adding W-state information beyond that achieved knowing only the X-state. The average partial achieves its maximum value of one when the combined X and W information allows for perfect prediction. It is zero if W information has, on average, no predictive value; in particular it is zero if W and Y are conditionally independent given X (that is, independent in *each* $X = x_j$ subpopulation) as in Table 7.7. Note that this example shows that conditional independence does not imply absolute independence, so that a zero partial ∇ need not imply that W by itself has no value in predicting Y. The fact that partial ∇ is zero in this case implies that W contributes no information beyond that contained in X for this prediction of Y. And, as usual, the partial ∇ can be negative; this occurs when W information, in conjunction with given X data, is worse (under a given prediction $\mathscr{P}_{YXW}$) than no extra information beyond the X data. The average partial $\nabla_{\mathscr{P}|X}$ summarizes the total error reduction whereas the component partial $\nabla_{\mathscr{P}|x_j}$ values indicate the error reduction for specific subpopulations.

We now have three ∇ measures—the basic bivariate ∇, the multivariate ∇, and the average partial ∇, the latter being subdivided into components. To begin exploring relations among these measures we now turn to the accounting task of assessing the sequential effects of using first X, then W in predicting Y.

7.4 ERROR ACCOUNTING AND PREDICTION SHIFTS

To begin the process of error accounting, a final look at the Wolfinger-Heifetz data will help. The multivariate measure $\nabla_{(s\&r)\leadsto c}$ indicates the proportionate error reduction of the trivariate prediction; $\nabla_{s\leadsto c}$ measures

Table 7.7 Hypothetical probabilities illustrating conditional independence

(*A*) Trivariate Probabilities[a]

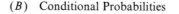

	w_1		w_2		
	x_1	x_2	x_1	x_2	
y_1	.004	.294	.036	.126	.46
y_2	.036	.126	.324	.054	.54
	.040	.420	.360	.180	1.00

[a]Shading shows error set for $\mathscr{P}_{YXW}$: $(w_1$ & $x_2) \rightsquigarrow y_1$ & $(w_2$ & $x_1) \rightsquigarrow y_2$.

(*B*) Conditional Probabilities

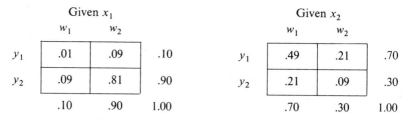

Given x_1

	w_1	w_2	
y_1	.01	.09	.10
y_2	.09	.81	.90
	.10	.90	1.00

Given x_2

	w_1	w_2	
y_1	.49	.21	.70
y_2	.21	.09	.30
	.70	.30	1.00

(*C*) Bivariate Probabilities[b]

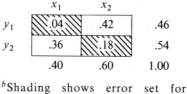

	x_1	x_2	
y_1	.04	.42	.46
y_2	.36	.18	.54
	.40	.60	1.00

[b]Shading shows error set for $\mathscr{P}_{YX}$: $x_1 \longleftrightarrow y_2$.

the success of the bivariate prediction; and $\nabla_{(s\&r) \rightsquigarrow c|s}$ assesses the additional value of "rural" information. As stated, there should be a relation among these quantities; ideally, one would hope that the value of "Southern" information would combine with the additional value of "rural" information to yield the joint prediction success of both types of information. In this section we develop that combination process and begin to investigate the complexities and apparent paradoxes that arise in the process.

The simplest situation has each variable with two states, as in the Wolfinger-Heifetz example. There the relevant multivariate, bivariate, and partial ∇ values are

$$\nabla_{(s\&r)\rightsquigarrow c} = 1 - \frac{22/256}{(147/256)(73/256)} = .475$$

$$\nabla_{s\rightsquigarrow c} = 1 - \frac{38/256}{(147/256)(96/256)} = .311$$

$$\nabla_{(s\&r)\rightsquigarrow c|s} = 1 - \frac{22/96}{(38/96)(73/96)} = .239$$

Therefore knowledge that a congressman is Southern reduces chairman-ship prediction error 31%; the additional knowledge that he is "rural" reduces the remaining 69% by another 24%. The remaining error is 76% of 69%, or 52%. The trivariate ∇ says that the combined "Southern, rural" information reduces error 48%, leaving the same residual 52% error rate. In terms of relative error rates, the suggested equation is

$$\frac{\text{Trivariate error knowing both } s \text{ and } r}{\text{Trivariate error knowing neither } s \text{ nor } r}$$

$$= \frac{\text{Bivariate error knowing } s}{\text{Bivariate error knowing neither}} \times \frac{\text{Error knowing } r \text{ (with given } s)}{\text{Error not knowing } r \text{ (with given } s)}$$

In this situation and in all situations where the trivariate error set is contained in the error set of a bivariate sufficiency prediction $(x \rightsquigarrow y)$, this simple accounting holds. (The general case does not yield such simple results, as we shall see.) Translating the verbal equation into symbols gives

$$\frac{P_{\bar{c}sr}}{P_{\bar{c}..}P_{.sr}} = \frac{P_{\bar{c}s.}}{P_{\bar{c}..}P_{.s.}} \times \frac{P(r \text{ and } \bar{c}|s)}{P(r|s)P(\bar{c}|s)}$$

which, in fact, holds true since

$$\frac{P_{\bar{c}s.}}{P_{\bar{c}..}P_{.s.}} \frac{P(r \text{ and } \bar{c}|s)}{P(r|s)P(\bar{c}|s)} = \frac{P_{\bar{c}s.}(P_{\bar{c}sr}/P_{.s.})}{P_{\bar{c}..}P_{.s.}(P_{.sr}/P_{.s.})(P_{\bar{c}s.}/P_{.s.})} = \frac{P_{\bar{c}sr}}{P_{\bar{c}..}P_{.sr}}$$

Thus error analysis is simple and direct for any single error cell.

Error analyses of more general predictions are more complex. In moving from a bivariate to a trivariate prediction, a researcher not only alters the available information by adding a variable W but also alters the prediction rule itself. Changing from a bivariate $\mathcal{P}_{YX}$ to a trivariate $\mathcal{P}_{YXW}$ may alter prediction precision as well as prediction success. These factors will be accounted for.

7.4.1 U-shift: Precision Adjustments

Since the denominator U of any ∇ measure has already been interpreted as a means of assessing precision, it is relatively easy to account for the change in precision in changing from $\mathcal{P}_{YX}$ to $\mathcal{P}_{YXW}$. We define the U-shift factor as

$$\psi^{(U)}_{\mathcal{P}_{YXW}/\mathcal{P}_{YX}} = \frac{\text{Precision of } \mathcal{P}_{YXW}}{\text{Precision of } \mathcal{P}_{YX}}$$

$$= \frac{\sum_i \sum_j \sum_k \omega_{ijk} P_{i..} P_{.jk}}{\sum_i \sum_j \omega'_{ij} P_{i..} P_{.j.}} \tag{6}$$

where ω_{ijk} is the error weight measure associated with $\mathcal{P}_{YXW}$ and ω'_{ij} is the measure associated with $\mathcal{P}_{YX}$. The U-shift is merely the ratio of respective precision measures. Although the two *a priori* theories $\mathcal{P}_{YXW}$ and $\mathcal{P}_{YX}$ imply a belief that their respective rule K errors will be less than their respective rule U errors, in general there is no *a priori* expectation regarding the comparison of rule U errors between the two theories. We therefore define the U-shift measure as an error ratio, not an error reduction. The U-shift simply assesses the relative precision of the two predictions.

7.4.2 K-Shift: Initial Bivariate Information Adjustment

To see what further accounting is necessary we must consider the purpose of the various ∇ measures. Each assesses the value of certain information. Trivariate $\nabla_{\mathcal{P}_{YXW}}$ assesses the value of both X and W information under $\mathcal{P}_{YXW}$; bivariate $\nabla_{\mathcal{P}_{YX}}$ assesses the value of X information using $\mathcal{P}_{YX}$; partial $\nabla_{\mathcal{P}_{YXW}|X}$ assesses the extra information value of W using $\mathcal{P}_{YXW}$ and is defined independently of the $\mathcal{P}_{YX}$ prediction. The U-shift accounts for changes in precision when X- (and W-) state information is not used in the prediction rules. What is missing is an assessment of the contribution to error reduction of the X information under $\mathcal{P}_{YXW}$ as compared to its contribution under $\mathcal{P}_{YX}$. We shall define a K-shift factor for this purpose. To do this we must find the error rate of $\mathcal{P}_{YXW}$ without actual information on the W-state, since the value of W information is to be reflected in partial ∇. We previously found the error rate for $\mathcal{P}_{YXW}$ using X information but not W information to be the rule U error rate for the partial $\nabla_{\mathcal{P}|X}$. This rate equals

$$\sum_i \sum_j \sum_k \omega_{ijk} P(y_i|x_j) P(w_k|x_j) P_{\cdot j}.$$

The K-shift is the ratio of this error rate to the error rate using X information with $\mathcal{P}_{YX}$, which is $\sum_i \sum_j \omega'_{ij} P_{ij}$:

$$\psi^{(K)}_{\mathcal{P}_{YXW}/\mathcal{P}_{YX}} = \frac{\sum_i \sum_j \sum_k \omega_{ijk} P(y_i|x_j) P(w_k|x_j) P_{\cdot j}}{\sum_i \sum_j \omega'_{ij} P_{ij}}$$

$$= \frac{\sum_i \sum_j \sum_k \omega_{ijk} P(y_i|x_j) P_{\cdot jk}}{\sum_i \sum_j \omega'_{ij} P_{ij}} \tag{7}$$

Again the shift measure is an error ratio, not a PRE measure, since there is no reason to assume error reduction or increase, in general.

7.4.3 K-Shift for Identical Prediction Rules

Two special cases help to illustrate the meaning of the K-shift factor. First, suppose that $\mathcal{P}_{YXW}$ and $\mathcal{P}_{YX}$ are logically equivalent. In this case the value of X information should be the same for both rules and the K-shift factor should equal one. If the two prediction rules are in fact identical, the error weights must be equal, $\omega_{ijk} = \omega'_{ij}$. Thus if (y_i, x_j) is an error for $\mathcal{P}_{YX}$, (y_i, x_j, w_k) must be an equally weighted error for $\mathcal{P}_{YXW}$, regardless of the state w_k of W. Since $\sum_k P(w_k|x_j) = 1$,

$$\sum_i \sum_j \sum_k \omega_{ijk} P(y_i|x_j) P(w_k|x_j) P_{\cdot j}.$$

$$= \sum_i \sum_j \omega'_{ij} P(y_i|x_j) P_{\cdot j} \sum_k P(w_k|x_j) = \sum_i \sum_j \omega'_{ij} P_{ij}.$$

Therefore the K-shift is one, as it should be with identical predictions; in fact it is easy to show that the U-shift also equals one in this situation, since identical predictions have equal precision.

7.4.4   *K*-Shift for Restriction of a Single Error Cell

The second special case is the restriction of a single error cell. Suppose that (y_2, x_4, w_1) is the only $\mathcal{P}_{YXW}$ error cell, for example, whereas (y_2, x_4) is the only error cell for $\mathcal{P}_{YX}$. Then nontrivial predictions are made with $\mathcal{P}_{YXW}$ only for $X = x_4$ and $W = w_1$, and only for $X = x_4$ under $\mathcal{P}_{YX}$. For $\mathcal{P}_{YX}$ a case with $X = x_4$ will be in error whenever $Y = y_2$, hence with probability $P(y_2|x_4)$. For $\mathcal{P}_{YXW}$ a case with $X = x_4$ will lead to a nontrivial prediction only with probability $P(w_1|x_4)$; of these again a fraction $P(y_2|x_4)$ will be in error. Thus the relative X-known error rate is

$$\psi^{(K)}_{\mathcal{P}_{YXW}/\mathcal{P}_{YX}} = \frac{P(w_1|x_4)P(y_2|x_4)}{P(y_2|x_4)} = P(w_1|x_4)$$

In this example it also follows, by direct calculation, that the relative precision rate is also $P(w_1|x_4)$, reflecting the fact that the error set of $\mathcal{P}_{YXW}$ is a subset of that of $\mathcal{P}_{YX}$ in the trivariate distribution. The equality of K-shift and U-shift holds generally only for any two propositions that are logically equivalent to the preceding. The basic form is $\mathcal{P}_{YX} : x \rightsquigarrow y$; $\mathcal{P}_{YXW} : (w \,\&\, x) \rightsquigarrow y$. The equality of the shifts in these situations was the cause for the relatively simple error structure of the Wolfinger–Heifetz example.

In general, the K-shift is less than one if, on average, knowledge of an X-state yields fewer errors under $\mathcal{P}_{YXW}$ than under $\mathcal{P}_{YX}$; the U-shift is less than one if completely random application of $\mathcal{P}_{YXW}$ yields fewer expected errors than random application of $\mathcal{P}_{YX}$.

7.4.5   The Accounting Equation

With these shift measures we can give a complete breakdown of the error structure of a trivariate prediction $\mathcal{P}_{YXW}$ relative to a bivariate $\mathcal{P}_{YX}$. The basic accounting equation is

$$\text{Trivariate error ratio} = \text{Bivariate error ratio} \times \text{Partial error ratio} \times \frac{K\text{-shift}}{U\text{-shift}}$$

or

$$1 - \nabla_{\mathcal{P}_{YXW}} = \left(1 - \nabla_{\mathcal{P}_{YX}}\right)\left(1 - \nabla_{\mathcal{P}_{YXW}|X}\right)\frac{\psi^{(K)}_{\mathcal{P}_{YXW}/\mathcal{P}_{YX}}}{\psi^{(U)}_{\mathcal{P}_{YXW}/\mathcal{P}_{YX}}} \tag{8}$$

Verbally this equation is

$$\frac{\mathcal{P}_{YXW}\ \text{error knowing } X \text{ and } W}{\mathcal{P}_{YXW}\ \text{error knowing neither}} = \frac{\mathcal{P}_{YX}\ \text{error knowing } X}{\mathcal{P}_{YX}\ \text{error knowing neither}}$$

$$\times\ \frac{\mathcal{P}_{YXW}\ \text{error knowing } W\ (\text{and given } X\,)}{\mathcal{P}_{YXW}\ \text{error not knowing } W\ (\text{and given } X\,)}$$

$$\times\ \frac{(\,\mathcal{P}_{YXW}\ \text{error knowing } X\,)/(\,\mathcal{P}_{YX}\ \text{error knowing } X\,)}{(\,\mathcal{P}_{YXW}\ \text{error knowing neither})/(\,\mathcal{P}_{YX}\ \text{error knowing neither})}$$

whereas formally it is

$$\frac{\displaystyle\sum_i \sum_j \sum_k \omega_{ijk} P_{ijk}}{\displaystyle\sum_i \sum_j \sum_k \omega_{ijk} P_{i..} P_{.jk}}$$

$$= \frac{\displaystyle\sum_i \sum_j \omega'_{ij} P_{ij.}}{\displaystyle\sum_i \sum_j \omega'_{ij} P_{i..} P_{.j.}} \times \frac{\displaystyle\sum_i \sum_j \sum_k \omega_{ijk} P_{ijk}}{\displaystyle\sum_i \sum_j \sum_k \omega_{ijk} P(y_i|x_j) P(w_k|x_j) P_{.j.}}$$

$$\times\ \frac{\left(\displaystyle\sum_i \sum_j \sum_k \omega_{ijk} P(y_i|x_j) P(w_k|x_j) P_{.j.}\right)\Big/\left(\displaystyle\sum_i \sum_j \omega'_{ij} P_{ij.}\right)}{\left(\displaystyle\sum_i \sum_j \sum_k \omega_{ijk} P_{i..} P_{.jk}\right)\Big/\left(\displaystyle\sum_i \sum_j \omega'_{ij} P_{i..} P_{.j.}\right)}$$

This equation gives a complete, if moderately complex, error accounting.
Fortunately the complexity of this equation does *not* lead to an explo-
sive increase in complexity for analyses involving four or more variables.
The logical combination device allows us to write X and W as a single
variable X'. If the fourth variable is denoted Z with associated subscript
m, we may write

$$1 - \nabla_{\mathcal{P}_{YXWZ}} = 1 - \nabla_{\mathcal{P}_{YX'Z}}$$

$$= \left(1 - \nabla_{\mathcal{P}_{YX'}}\right)\left(1 - \nabla_{\mathcal{P}_{YX'Z}|X'}\right) \frac{\psi^{(K)}_{\mathcal{P}_{YX'Z}/\mathcal{P}_{YX'}}}{\psi^{(U)}_{\mathcal{P}_{YX'Z}/\mathcal{P}_{YX'}}} \tag{8'}$$

where, for instance,

$$\nabla_{\mathcal{P}_{YX\cdot Z|X'}} = 1 - \frac{\displaystyle\sum_i \sum_j \sum_k \sum_m \omega_{ijkm} P_{ijkm}}{\displaystyle\sum_i \sum_j \sum_k \sum_m \omega_{ijkm} P(y_i|x_j, w_k) P(z_m|x_j, w_k) P_{\cdot jk\cdot}} \tag{4'}$$

and $1 - \nabla_{\mathcal{P}_{YX'}} = 1 - \nabla_{\mathcal{P}_{YXW}}$ is given by (2). If one combines all previous independent variables into a single composite variable before introducing another variable, the error accounting for a general multivariate problem reduces to a sequence of successive trivariate analyses. To help the reader gain intuition in trivariate analysis we present some further examples in the next section.

7.5 ERROR ACCOUNTING: EXAMPLES

In this section we present some examples to illustrate the various measures and their relationships. The intent is to show some of the kinds of error structures that can arise in the trivariate case. Two examples come from political science; the other is hypothetical and designed to give a "pure" case.

The first real example is taken from Rosenthal's (1968) analysis of electoral coalitions in the French Fourth Republic. His dependent variable is whether or not a "centrist strategy" occurs, so we may write $Y = \{cs, \overline{cs}\}$. He uses two independent variables: whether the degree of communist "danger" is high, medium, or low, so $D = \{h, m, l\}$; and whether a "rightist alternative" is available, so $A = \{ra, \overline{ra}\}$. The initial bivariate prediction is that a centrist strategy coalition occurs for "high" communist danger and does not form for "low" danger. No prediction is made in the "medium" case. Thus

$$\mathcal{P}_{YD} : h \rightsquigarrow cs, m \rightsquigarrow (cs \text{ or } \overline{cs}), l \rightsquigarrow \overline{cs}$$

Contingent refinement of this prediction uses $\{ra, \overline{ra}\}$ to predict coalitions for the "medium" communist danger case. When a rightist alternative is present the coalition is predicted; when it is not, no centrist coalition should form. Thus

$$\mathcal{P}_{YDA} : h \rightsquigarrow cs, m \text{ and } \overline{ra} \rightsquigarrow cs, m \text{ and } ra \rightsquigarrow \overline{cs}, l \rightsquigarrow \overline{cs}$$

The data are presented in Table 7.8; the double-hatched cells are errors for $\mathcal{P}_{YD}$, whereas all single- or double-hatched cells are errors for $\mathcal{P}_{YDA}$.

Table 7.8 Centrist strategies as predicted by communist "danger" and rightist alternative (1951 French legislative elections)[a,b]

Communist "danger"

	High (h)	Medium (m)		Low (l)	
District type		$\overline{ra}$	ra		
Centrist strategy (cs)	35	17	4	1	57
Not centrist strategy ($\overline{cs}$)	4	2	21	11	38
	39	19	25	12	95

Source: Rosenthal (1968, pp. 279–280).
[a]Rightist alternative—ra; no rightist alternative—$\overline{ra}$.
[b]The cross-tabulation involving rightist alternative was not available for the h and l categories on communist "danger." Since the rightist alternative variable has no predictive significance for these categories, the available data suffice for all relevant computations.

The analysis of error is interesting because the trivariate prediction is less successful than the bivariate one (though it has higher precision). From Table 7.8,

$$\nabla_{\mathscr{P}_{YD}} = 1 - \frac{1+4}{39(38/95) + 12(57/95)} = .781$$

whereas

$$\nabla_{\mathscr{P}_{YDA}} = 1 - \frac{4+2+4+1}{39(38/95) + 19(38/95) + 25(57/95) + 12(57/95)} = .758$$

The trivariate prediction has twice the precision ($U = .478$) of the bivariate proposition ($U = .240$). The presence of a rightist alternative has substantial extra predictive value, since, controlling for "danger," overall partial $\nabla = .591$. Even though partial ∇ is positive, trivariate ∇ is less than partial ∇. The reason lies in the shift factors. The U-shift is 1.99, since trivariate $\mathscr{P}$ has nearly twice the prediction precision of bivariate $\mathscr{P}$. Under rule K the bivariate error rate is $5/95 = .053$; for the trivariate prediction, the D-known but A-unknown error rate is

$$\frac{4}{95} + \frac{23}{44} \cdot \frac{19}{44} \cdot \frac{44}{95} + \frac{21}{44} \cdot \frac{25}{44} \cdot \frac{44}{95} + \frac{1}{95} = .283$$

[Note that $P(\overline{cs}|h)P(\overline{ra}|h)P(h) + P(\overline{cs}|h)P(ra|h)P(h) = P(\overline{cs}|h)P(h) =$ $(4/39)\cdot(39/95)=(4/95)$; a similar calculation yields the $1/95$ for the $D=l$ cases.] Thus the K-shift factor is very large;

$$\psi^{(K)}_{\mathscr{P}_{YDA}/\mathscr{P}_{YD}} = \frac{.283}{.053} = 5.37$$

The fact that the $\mathscr{P}_{YD}$ error cells are a subset of the $\mathscr{P}_{YDA}$ error set necessarily means that both shift factors (relative error rates) exceed one. In this case the difference in the predictions occurs when $D=m$; since this category contains a large fraction of all cases, the expected number of errors randomly applying the $\mathscr{P}_{YDA}$ rule given $D=m$ will be quite large. In contrast, the simple $\mathscr{P}_{YD}$ prediction identifies no errors at all for $D=m$. Therefore the K-shift factor is large; it inflates the $\mathscr{P}_{YDA}$ error rate. This effect carries through to yield a lower value for $\nabla_{\mathscr{P}_{YDA}}$ than $\nabla_{\mathscr{P}_{YD}}$.

For a second illustration we may reconsider the Riker–Zavoina data of Table 7.6. No bivariate prediction was specified earlier as the multiple and partial ∇ values can be computed without reference to any bivariate rule. For comparison, consider the simple persistence prediction that a subject will vote "for" (y) if and only if the previous vote was "for" (x). Then

$$\nabla_{\mathscr{P}_{YX}} = 1 - \frac{(57+63)/280}{(132/280)(142/280)+(148/280)(138/280)}$$

$$= .142$$

Since $\nabla_{\mathscr{P}_{YXW}} = .227$, the trivariate theory is more successful; the fact that $\nabla_{\mathscr{P}_{YXW|X}} = .158$ also suggests that the won-lost variable W has extra predictive value.

For a complete analysis, shift factors are required. The K-shift factor, assessing the relative error rates of the two propositions with previous vote known, is

$$\psi^{(K)} = \frac{\left(\frac{57}{142}\right)\left(\frac{94}{142}\right)\left(\frac{142}{280}\right) + \left(\frac{85}{142}\right)\left(\frac{48}{142}\right)\left(\frac{142}{280}\right) + \left(\frac{63}{138}\right)\left(\frac{86}{138}\right)\left(\frac{138}{280}\right)}{(31+26+36+27)/280}$$

$$= .881$$

The U-shift factor, assessing the relative error rates of the two propositions with no predictor values known, is

$$\psi^{(U)} = \frac{\left(\dfrac{132}{280}\right)\left(\dfrac{94}{280}\right) + \left(\dfrac{148}{280}\right)\left(\dfrac{48}{280}\right) + \left(\dfrac{148}{280}\right)\left(\dfrac{86}{280}\right)}{\left(\dfrac{132}{280}\right)\left(\dfrac{142}{280}\right) + \left(\dfrac{148}{280}\right)\left(\dfrac{138}{280}\right)}$$

$$= .823$$

The ratio of these shift factors is 1.07, indicating that the effect of a change in prediction *form* is relatively small compared to the effects of prediction *information*, as assessed by bivariate and partial ∇. The accounting equation (8) becomes

$$(1 - .227) = (1 - .142)(1 - .158)\frac{.881}{.823}$$

Another possible situation is best illustrated by a hypothetical example. The example is based on Table 7.7. Suppose that the bivariate prediction is $\mathscr{P}_{YX}:\ x_1 \longleftrightarrow y_2$ and that the trivariate prediction is $\mathscr{P}_{YWX}:(w_1,x_2)\rightsquigarrow y_1, (w_2,x_1)\rightsquigarrow y_2$. Then

$$\nabla_{\mathscr{P}_{YX}} = 1 - \frac{.04 + .18}{(.4)(.46) + (.6)(.54)} = .567$$

whereas the trivariate prediction yields

$$\nabla_{\mathscr{P}_{YXW}} = 1 - \frac{.126 + .036}{(.54)(.42) + (.46)(.36)}$$

$$= .587$$

Therefore it appears that W information has some small predictive value. But as can be seen from the conditional probability tables of Table 7.7, W and Y are independent given X, so that W has no additional predictive value. As a check,

$$\nabla_{\mathscr{P}_{YXW}|X} = 1 - \frac{.126 + .036}{(.1)(.36) + (.3)(.42)}$$

$$= 0$$

So it follows that the change from bivariate to trivariate ∇ must be entirely attributable to the change in prediction rules, as reflected in the shift factors. The K-shift is

$$\psi^{(K)} = \frac{(.1)(.36) + (.3)(.42)}{.04 + .18}$$

$$= .736$$

whereas

$$\psi^{(U)} = \frac{(.54)(.42) + (.46)(.36)}{(.46)(.4) + (.54)(.6)}$$

$$= .772$$

The shift factors explain the change from $\nabla_{\mathcal{P}_{YX}} = .567$ to $\nabla_{\mathcal{P}_{YXW}} = .587$. $\mathcal{P}_{YXW}$ has lower precision than $\mathcal{P}_{YX}$, as indicated by $\psi^{(U)} = .772$ and also by two error cells out of eight versus two error cells out of four in the bivariate table. However, knowledge of X gives an even lower error ratio, as evidenced by $\psi^{(K)} = .736$. The change in prediction success is attributable entirely to the effect of changing predictions, and not at all to the value of W information.

There is an alternative approach to multivariate analysis, related to the theory of "causal models," which is conveniently illustrated by these artificial data. Here we only present a brief sketch of this approach. [See Hildebrand, Laing, and Rosenthal (1975) for a somewhat more detailed exposition.] One possible scheme might be to predict that $w_1 \rightsquigarrow x_2$ and that $x_2 \rightsquigarrow y_1$. Both bivariate predictions are supported:

$$\nabla_{w_1 \rightsquigarrow x_2} = 1 - \frac{.04}{(.46)(.40)} = .783$$

$$\nabla_{x_2 \rightsquigarrow y_1} = 1 - \frac{.18}{(.54)(.60)} = .444$$

The X variable can be regarded as an "intervening" variable in the system. Since in formal logic $w_1 \rightarrow x_2$ and $x_2 \rightarrow y_1$ implies that $w_1 \rightarrow y_1$, it is plausible that in prediction analysis $w_1 \rightsquigarrow y_1$ should be supported, since $w_1 \rightsquigarrow x_2$ and $x_2 \rightsquigarrow y_1$ both are supported. With these data the conjecture is true:

$$\nabla_{w_1 \rightsquigarrow y_1} = 1 - \frac{.162}{(.46)(.54)} = .348$$

This example illustrates the general theorem that *if W and Y are indepen-dent given X and if $\mathcal{P}_{YX}$ and $\mathcal{P}_{XW}$ form a so-called sufficiency chain leading to $\mathcal{P}_{YW}$, then*

$$\nabla_{\mathcal{P}_{YW}} = \nabla_{\mathcal{P}_{YX}} \nabla_{\mathcal{P}_{XW}}$$

so that $\mathcal{P}_{YW}$ will be supported more weakly than $\mathcal{P}_{YX}$ and $\mathcal{P}_{XW}$. In this case we have $.348 = .783 \times .444$. The conditional independence condition is crucial; it is possible without this condition that both $w_1 \rightsquigarrow x_2$ and $x_2 \rightsquigarrow y_1$ are supported but that $w_1 \rightsquigarrow y_1$ has a negative ∇. The whole area of derived predictions and causal models has barely been touched; this example only indicates the possibility of a prediction analysis approach to the area.

These examples should, we hope, guide the researcher in the error analyses of multivariate predictions. As mentioned, the assessment of predictions has several dimensions; from the accounting process presented in this section it seems to follow that the most crucial components are prediction success, as measured by the various ∇ measures, and prediction precision, as measured by rule U error and the comparison of two propositions in these respects through the K- and U-shift factors.

No examples have been given for predictions dealing with four or more variables, since no new ideas are required. If there are three predictors—say, X, W, and Z—in a particular problem, the four-variable analysis can be reduced to a trivariate problem by combining X and W, via Cartesian product, into a composite variable X'. One can then analyze the trivariate proposition $\mathcal{P}_{YX'Z} = \mathcal{P}_{YV}$, where $V = X' \times Z = X \times W \times Z$. Generally the natural analysis is a sequential one; all "previous" variables can be combined into a composite variable, which plays the role of X in our analysis; the "new" variable then plays the role of W.

This section concludes our development of multivariate analyses with known populations. We have defined population versions of multivariate ∇, partial ∇, precision, K-shift and U-shift. Two tasks now become obvious. The analyses of this chapter need to be related to standard, correlation-based multivariate analyses, and the inferential extension from populations to samples must be made. In the next section (which is a bit more technical and can be omitted by those oriented to pragmatics as opposed to methodological genealogy) we discuss relations with standard multivariate analysis; the following section outlines methods for making statistical inferences.

7.6   RELATIONS WITH MULTIPLE REGRESSION AND CORRELATION METHODS

The multivariate prediction analysis methods described in the previous sections are, to some extent, analogues of conventional regression-correlation methods. There are important differences; in particular, the prediction shift factors are new concepts. In this section we draw analogies and explain the contrasts.

Just as multivariate prediction analysis has two basic components, a prediction rule and a prediction success measure, the conventional model has a prediction equation determined by multiple regression and an evaluation of that prediction given by the square of a correlation coefficient. The multiple regression equation yields a combined predictor variable. For instance, the equation

$$\text{Predicted } y = 1.4 + 7.2w - 2.6x$$

can be read as defining a new variable V by the linear combination $v = 1.4 + 7.2w - 2.6x$ and then predicting Y to be equal to V. The choice of the particular linear combination is made *ex post* to give "best fit" in the sense of minimizing total squared prediction error. (It is possible that the selected combination may omit one of the predictors as with $v = 1.4 + 0.0w - 2.6x$.) Thus the multiple regression equation uses linear combination, as opposed to the logical combination of prediction analysis, and derives the prediction *ex post* by least squares. Multiple regression is not commonly used in the analysis of *a priori* propositions, which are an important focus of prediction analysis.

The measure of prediction success in bivariate linear prediction is the squared correlation coefficient. In multivariate analysis the measure is the coefficient of determination R^2; as proved, for example, in Hays (1973, Ch. 16), this coefficient is the square of the correlation of predicted y with actual y. In other words, the measure of multivariate prediction success is simply the bivariate coefficient applied to Y and the composite variable V. The multivariate ∇ is defined by precisely the same process.[3]

There is also a parallel with the partial ∇ coefficient. The partial correlation coefficient is usually defined in terms of "adjusting" the variables Y and W to compensate for the effects of X. For instance, Hays (1973, Ch. 16) indicates the process of adjustment that involves predicting W and Y separately from X, and subtracting these predicted values to obtain *residual* W and Y scores. The correlation of these residual scores is the partial correlation coefficient, as usually defined. An alternative definition is more appropriate for our purposes. In this alternative model we consider the conditional probability distribution given a particular X-value. Within this subpopulation one can evaluate the value of W information for predicting Y. Under the critical assumption that the joint probability distribution is multivariate normal, it follows that (i) the predictive value of W is the same within every $X = x$ subpopulation; and (ii) the

---

[3]The coefficient of determination shares another property with ∇. In the bivariate case it is not affected by changing the labeling of independent and dependent variables; in the multivariate case it is.

predictive value of W (reduction in squared prediction error) equals the squared partial correlation coefficient. These results are proved in several multivariate analysis books, such as D. Morrison (1967, p. 91). The partial ∇ defined earlier is, in two senses, a partial analogue. As in the multivariate normal case, partial ∇ measures prediction success within a subpopulation. However there is no reason to assume that this success is constant over the various x_j subpopulations, so the overall partial ∇ is defined as a weighted average of (generally) different subpopulation values. When the multivariate normal assumption does not hold, the squared partial correlation coefficient can also be interpreted as an average error reduction, so that the analogy of partial ∇ is somewhat more general.

The aspect of multivariate ∇ analysis that has no analogue in conventional methods is the presence of prediction shift measures. There is no shift factor in regression-correlation analysis because those methods force prediction rules to be identical under certain distinct information conditions. For known X-state, the prediction analysis approach compares two distinct predictions in the K-shift measure—$\mathcal{P}_{YX}$ and $\mathcal{P}_{YXW}$ without W information; by contrast, the regression rule is always "Predict the conditional mean of Y given $X = x$, whether another variable W is to be available or not." Thus the two regression prediction rules when X is known are identical for both the bivariate and trivariate "propositions"; naturally the error rates are also the same. Similarly if X is unknown, the regression prediction rules reduce to "Predict the mean of Y"; therefore the U-shift factor is always 1.0. Thus for the standard regression-correlation model there is no change in prediction rule, hence no shift factors.

In large part this "shiftless" phenomenon is a result of the linearity of the regression-correlation model. If the multivariate normal assumption does not hold, it might be, for instance, that the best prediction rule is

$$\text{Predicted } y = 1.4 + 7.2 x \cdot w$$

In such a case the conventional approach would be to transform to a new variable $v = x \cdot w$, and ignore any possibility of multivariate analysis. Therefore the issue of prediction shift is eradicated by the assumption of a linear relation. We suspect that prediction analysis of such "nonlinear linear models" would require something like a prediction shift analysis.

In summary, the comparison of ∇ methods with regression-correlation methods reveals two basic similarities: the idea of multivariate prediction success as bivariate prediction success using a composite predictor variable, and the idea of partial prediction success as bivariate analysis conditional on a given subpopulation. There are two basic differences: the partial coefficient in ∇ analysis is not constant over subpopulations so that an average value is required, and the change in prediction has no analogue so that the shift factors are a new element in multivariate analysis.

7.7 STATISTICAL INFERENCE IN MULTIVARIATE ANALYSIS

The entire focus of this chapter has been on population measures, consistent with our strategy throughout the book. This strategy simplifies the discussion; once the ideas are mastered in a population context the problems of inference from samples can be attacked more readily. In this section we present a brief supplement to the ideas of Chapter 6 to allow for multivariate inference. As in that chapter we have to be content with large-sample, asymptotic theory.

Recall that the basic tasks of inference theory are to determine an appropriate method for estimating a population value and to assess the sampling distribution of that estimator. Here we investigate only condition S2 sampling where the data are the result of a single random sample from the whole population and where no marginal (or cell) probabilities are known in advance. Inferential methods for other combinations of sampling method and information could be derived by principles much like those of Chapter 6. Since there are so many possible combinations, rather than trying to cover them all, we will be content with the simplest possible setup.

To specify the appropriate estimate and sampling distribution, we need to extend the notation of Chapter 6. Assume that the size of the random sample is n, that the number of observations in cell (y_i, x_j, w_k) is n_{ijk}, and that the fraction in that cell is $f_{ijk} = n_{ijk}/n$. Again the dot notation indicates that the relevant subscript has been summed out, so that

$$n_{.jk} = \sum_i n_{ijk}$$

is the number of observations with $X = x_j$ and $W = w_k$, and

$$f_{.jk} = \sum_i f_{ijk}$$

is the fraction of observations with that property. Similar definitions hold for $n_{i..}$, $f_{i..}$, and so on. For conditional fractions we shall use $f(y_i, w_k | x_j)$ to denote the fraction of $X = x_j$ cases having $Y = y_i$ and $W = w_k$; thus $f(y_i, w_k | x_j) = n_{ijk}/n_{.j.} = f_{ijk}/f_{.j.}$. Similar definitions hold for $f(y_i | x_j)$ and $f(w_k | x_j)$. With this notation we can extend the ideas of Chapter 6 to cover multivariate analysis.

7.7.1 Estimates

The estimates of multivariate and partial ∇, and of the K-shift and U-shift factors, all follow by simply replacing true probabilities by sample fractions in the definitions. This procedure can be justified by the statistical principle of maximum likelihood; it is also the simplest, most natural method. By simple replacement the estimates become

$$\hat{\nabla}_{\mathscr{P}_{YXW}} = 1 - \frac{\sum_i \sum_j \sum_k \omega_{ijk} f_{ijk}}{\sum_i \sum_j \sum_k \omega_{ijk} f_{i..} f_{.jk}} = 1 - \frac{n \sum_i \sum_j \sum_k \omega_{ijk} n_{ijk}}{\sum_i \sum_j \sum_k \omega_{ijk} n_{i..} n_{.jk}} \tag{9a}$$

$$\hat{\nabla}_{\mathscr{P}_{YXW|X}} = 1 - \frac{\sum_i \sum_j \sum_k \omega_{ijk} f_{ijk}}{\sum_i \sum_j \sum_k \omega_{ijk} f(y_i|x_j) f(w_k|x_j) f_{.j.}} \tag{9b}$$

$$\hat{\psi}^{(K)}_{\mathscr{P}_{YXW}/\mathscr{P}_{YX}} = \frac{\sum_i \sum_j \sum_k \omega_{ijk} f(y_i|x_j) f(w_k|x_j) f_{.j.}}{\sum_i \sum_j \omega'_{ij} f_{ij.}} \tag{9c}$$

$$\hat{\psi}^{(U)}_{\mathscr{P}_{YXW}/\mathscr{P}_{YX}} = \frac{\sum_i \sum_j \sum_k \omega_{ijk} f_{i..} f_{.jk}}{\sum_i \sum_j \omega'_{ij} f_{i..} f_{.j.}} \tag{9d}$$

These estimates correspond exactly to the values calculated in the Wolfinger–Heifetz, Riker–Zavoina, and Rosenthal examples, if the data are regarded as a sample rather than a population.

7.7.2 Sample Variances

The complicated part of the inference procedure is the determination of sampling distributions. As in Chapter 6, the exact distributions are too complex to be helpful. Therefore we turn to asymptotic distributions; once again, the delta method allows us to conclude that the large-sample distributions all are normal. Also, the delta method indicates that the asymptotic expected value of each of the estimators is the respective population quantity. So the only new issue is the calculation of sampling variances.

In fact these calculations are not entirely new; since the trivariate and component partial ∇ values are extended cases of the bivariate ∇ treated in previous chapters, their sampling variances can be inferred from previous results. In particular, adapting equation (7) of Chapter 6 to trivariate notation, it follows that

$$\mathrm{Var}\left(\hat{\nabla}_{\mathscr{P}_{YXW}}\right) = \left[(n-1)U^2\right]^{-1}$$

$$\times\left[\left[\sum_i\sum_j\sum_k\left[\omega_{ijk} - B\left(\pi_{i..} + \pi_{.jk}\right)\right]^2 P_{ijk}\right]\right.$$

$$\left. - \left[\sum_i\sum_j\sum_k\left[\omega_{ijk} - B\left(\pi_{i..} + \pi_{.jk}\right)\right]P_{ijk}\right]^2\right] \qquad (10)$$

where

$$\pi_{i..} = \sum_j\sum_k \omega_{ijk} P_{.jk}$$

$$\pi_{.jk} = \sum_i \omega_{ijk} P_{i..}$$

$$U = \sum_i\sum_j\sum_k \omega_{ijk} P_{i..} P_{.jk}$$

$$B = 1 - \nabla_{\mathscr{P}_{YXW}}$$

Furthermore for the component partial given $X = x_j$ the estimate is exactly $\hat{\nabla}$ applied to the subpopulation. So the variance of a component partial ∇ is

$$\mathrm{Var}\left(\hat{\nabla}_{\mathscr{P}_{YXW|x_j}}\right) = \left[(n_{.j.} - 1)U_{.j.}^2\right]^{-1}$$

$$\times\left[\left[\sum_i\sum_k\left[\omega_{ijk} - B_{.j.}\left\{\pi\left(y_i|x_j\right) + \pi\left(w_k|x_j\right)\right\}\right]^2 P\left(y_i, w_k|x_j\right)\right]\right.$$

$$\left. - \left[\sum_i\sum_k\left[\omega_{ijk} - B_{.j.}\left\{\pi\left(y_i|x_j\right) + \pi\left(w_k|x_j\right)\right\}\right]P\left(y_i, w_k|x_j\right)\right]^2\right] \qquad (11)$$

for $\psi^{(K)}_{\mathcal{P}_{YXW}/\mathcal{P}_{YX}}$ they are

$$(K'_{YX})^{-1}\left[\pi(y_i|x_j)+\pi(w_k|x_j)-\pi(.|x_j)-\psi^{(K)}\omega'_{ij}\right]$$

various K, U, and π quantities not previously defined are as follows:

$$K=\sum_i\sum_j\sum_k\omega_{ijk}P_{ijk}$$

$$K'_{YX}=\sum_i\sum_j\omega'_{ij}P_{ij.}$$

$$U_X=\sum_i\sum_j\sum_k\omega_{ijk}P(y_i|x_j)P(w_k|x_j)P_{.j.}$$

$$U'_{YX}=\sum_i\sum_j\omega'_{ij}P_{i..}P_{.j.}$$

$$\pi'_{i..}=\sum_j\omega'_{ij}P_{.j.}$$

$$\pi'_{.j.}=\sum_i\omega'_{ij}P_{i..}$$

$$\pi(.|x_j)=\sum_i\sum_k\omega_{ijk}P(y_i|x_j)P(w_k|x_j)$$

Inference Procedures

se of these formulas parallels Chapter 6. The method of establishing
ence intervals is standard. Compute the estimated value; also com-
e estimated variance by "plugging in" sample fractions for popula-
robabilities in the variance formulas. The relevant numbers of
rd deviations (1.96 for 95%, etc.) are obtained from normal tables
nerally denoted as Z. The $100(1-\alpha)\%$ confidence interval is once

ted value $-Z_{\alpha/2}$ (estimated variance)$^{1/2}$

$\leqslant$ true value $\leqslant$ estimated value $+Z_{\alpha/2}$ (estimated variance)$^{1/2}$

ting purposes the natural hypotheses are that trivariate $\nabla=0$ and
rtial $\nabla=0$. These may be tested by calculating

$$\hat{Z}=\frac{\text{(estimated value)}}{\text{(estimated variance)}^{1/2}}$$

where

and

$$\pi(y_i|x_j) = \sum_k \omega_{ijk} P(w_k|x_j)$$

$$\pi(w_k|x_j) = \sum_i \omega_{ijk} P(y_i|x_j)$$

The

$$U_{.j.} = \sum_i \sum_k \omega_{ijk} P(y_i|x_j) P(w_k$$

$$B_{.j.} = 1 - \nabla_{\mathcal{P}|x_j}$$

It should be recalled that the accuracy of thes depends on large sample sizes. The rules of th directly to multivariate ∇, since one can treat th bivariate problem. In contrast, the relevant sam partial ∇ given $X = x_j$ is the number of cases number of X categories relative to the sample siz of some subsamples is small, the normal appro effective as in the bivariate case. The adequacy c component partial ∇ may be assessed by the rul with n interpreted as the number of sample cas subpopulation.

For the remaining quantities, namely, the aver measures, such considerations are less crucial, multivariate ∇, are averages over the entire c normal sampling distributions apply for large sa estimator has the true value as its asymptotic e> the variance is of the form

7.7.3

The u confi pute tion stand and g again

$$\text{Var}(\text{estimator}) = (n-1)^{-1}\left[\sum_i \sum_j \sum_k a_{ijk}^2 P_{ijk} - \right.$$

The a_{ijk} coefficients for $\hat{\nabla}_{\mathcal{P}_{YXW}|X}$, as shown in A

Estim

$$(U_X)^{-1}\left[\omega_{ijk} - K(\pi(y_i|x_j) + \pi(w_k\right.$$

for $\psi_{\mathcal{P}_{YXW}/\mathcal{P}_{YX}}^{(U)}$ they are

For te that p

$$(U'_{YX})^{-1}\left[\pi_{.jk} + \pi_{i..} - \psi^{(U)}(\right.$$

and comparing the result to a normal table. (In principle the shift factors can be tested against the natural null value of 1.0; however in many cases —those with trivariate error cells being a subset of the bivariate ones, especially—the shifts necessarily fall below 1.0, so that no test is needed.) For example, with the Rosenthal data given in Section 7.5 we may compute 95% confidence intervals as:

$$.617 \leqslant \nabla_{\mathcal{P}_{YDA}} \leqslant .898$$

$$.408 \leqslant \nabla_{\mathcal{P}_{YDA}|D} \leqslant 1.107$$

$$.589 \leqslant \nabla_{\mathcal{P}_{YD}} \leqslant .972$$

Note that the upper bound on the 95% confidence interval for the partial del exceeds the theoretical limit of one for these data. This excessive value reflects the fact that although the distribution of del must be skewed (truncated at one), the approximation is based on the normal distribution, which is symmetric. Note also that trivariate del has a tighter confidence interval than bivariate del, reflecting the trivariate proposition's greater precision.

No new principles are involved in this multivariate inference. Therefore it seems plausible that the qualitative properties of the inferences should be the same as in the bivariate case. In particular the rules of thumb and continuity corrections given in Chapter 6 should be applicable (except, as indicated, for the component partials, where subsample size becomes relevant). For the shift measures we conjecture that if all estimated error totals are at least 5, the normal approximation will be adequate. No computer studies back this guess; should the question ever become crucial, a Monte Carlo study would definitely be in order. However the consistent pattern of our previous studies lends some credence to the rules of thumb, even for the extension to the shift measures.

This section has sketched the process of multivariate inference very briefly. Since there are no really new ideas involved, the reader who is in command of Chapter 6 should find no trouble in extending an analysis to the multivariate case.

7.8 *EX POST* MULTIVARIATE ANALYSIS

The extension of *ex post* bivariate prediction analysis methods to multivariate data opens a staggering array of possibilities. With the freedom to specify not only the prediction rules to be analyzed, but also the variables that will enter the analysis, a researcher is likely to be overwhelmed with choices. The entire focus of this book has demanded the active intervention of the researcher in assessing data. To use prediction analysis methods prediction rules *must* be specified. The intent of this book has been to aid the researcher in evaluating previously given prediction rules. Yet it will not always be true that a researcher will have such rules available in advance. In fact, it may not even be possible to specify the relevant variables *a priori* at the exploratory level of data analysis. Therefore a systematic method of generating prediction rules and selecting useful predictor variables might well be helpful in multivariate research.

In this section we describe one method for selecting predictor variables and prediction rules for given data. The exposition is based on the work of Bartlett (1974).

Bartlett considered a general problem with many potential predictor variables and a specified dependent variable given in advance. No prediction rules are prespecified; part of the problem is to determine these rules *ex post*, from the data. Bartlett assumed that a reasonably large (several hundred cases, at a minimum) sample was available.

Bartlett's multivariate method is simply a sequential application of his bivariate RE optimality procedure outlined in Section 4.4.6. Of all the possible propositions one first selects the bivariate proposition that *ex post* maximizes absolute error reduction (RE). One then must attack the "then what" issue. After selecting a single "best" predictor, one has to find a "next best" predictor. Thus, having specified a bivariate prediction rule via his selection of "best X" and "best $\mathcal{P}$," one next faces the issue of "next best W." The idea for a solution is simple: split the data population into subpopulations with identical predictions given X. Thus if the prediction rule is "$Y=y_1$" for $X=x_1$, x_3, and x_6, "$Y=y_2$" for $X=x_4$, and "$Y=y_1$ or y_2" for $X=x_2$ and x_5, the population is split into three subpopulations—respectively, those with $X=x_1$, x_3, and x_6, with $X=x_4$, and with $X=x_2$ and x_5. Within such subpopulations the selection of a "next best" predictor is based on the subpopulation RE values as sketched.

Once the subpopulations have been specified, one can apply the basic algorithm separately to each subpopulation. There is no reason to assume that the "second best" predictor is the same from one X-subpopulation to the next, so the algorithm works on each subpopulation separately. Within each subpopulation, the basic algorithm finds a "next best" predictor that splits the subpopulation into "subsubpopulations" to which the algorithm may be applied once again. Thus the process can be continued as long as there are candidates for new predictor variables; the only question is how to stop this subdividing algorithm.

The stopping rule is based on two ideas. First, Bartlett noted that as the subdivision of the population continues, the value of adding new predictors will probably (though not inevitably) decrease. Thus the algorithm was designed to make it possible to stop subdividing when all remaining predictors fail to achieve a specified (absolute) error reduction. Also, as the subdivision process inevitably leaves small numbers of cases within specified subpopulations, the algorithm also stops whenever a subpopulation size reaches a specified minimal level. With these two specifications the algorithm guarantees that the *ex post* selection process cannot continue to absurd lengths. In fact, Bartlett has refined the method to avoid several other problems. His dissertation (Bartlett, 1974) is remarkably instructive for those who wish to apply prediction logic methods but have no predictive base on which to start.

Bartlett's approach is overwhelmingly *ex post*. Everything—predictor variables and prediction rules—is selected from the data. There is no intervention by the investigator. Since the predictions are selected to fit the data, the measure of success is suspect; the very process of selecting the prediction rule should inflate the measured success of that rule. However Bartlett's Monte Carlo studies, briefly summarized in Section 6.7, show that this inflation, although present, is surprisingly small. Although sampling error and the self-fulfilling prophecy (of selecting predictions to match the results) will take a toll, the degree of bias is small except in pathological cases, as noted in Chapter 6. Thus even though the algorithm is totally *ex post*, the prediction rules resulting from that algorithm seem to come fairly close to the underlying structure when such exists and their success seems to be adequately measured (if by no means perfectly) by the estimated ∇ value.

Thus we conclude (on very limited evidence) that *ex post* selection of prediction rules is reasonably reliable, given some natural precautions, and that the *a priori* ideas of prediction analysis are not so sensitive to basic assumptions that the first breath of *ex post* analysis sends them tumbling into the nearest ditch.

APPENDIX 7.1 DERIVATION OF ASYMPTOTIC VARIANCES

With simple random sampling, the entries in the cross classification have a multinomial distribution. As in the derivation in Appendix 6.2,

$$\mathrm{Var}(f_{ijk}) = n^{-1}\left[P_{ijk} - P_{ijk}^2 \right]$$

$$\mathrm{Cov}(f_{ijk}, f_{i'j'k'}) = -n^{-1}\left[P_{ijk} P_{i'j'k'} \right]$$

when $(i,j,k) \neq (i',j',k')$. As before, we replace n by $n-1$ for an improved · small-sample approximation.

The delta method replaces a given estimator by a linear function $c - \sum_i \sum_j \sum_k a_{ijk} f_{ijk}$ where $-a_{ijk}$ is the partial derivative of the estimator with respect to f_{ijk}, evaluated at $f_{ijk} = P_{ijk}$, and c need not be specified explicitly. In matrix terms, with obvious definitions, the approximation is $c - \mathbf{A'F}$. The variance of this linear function is $\mathbf{A'\Sigma A}$. With $n - 1$ replacing n, we obtain

$$\frac{\sum_i \sum_j \sum_k a_{ijk}^2 \left(P_{ijk} - P_{ijk}^2 \right)}{n-1} - \frac{\sum_i \sum_j \sum_k \sum_{i'} \sum_{j'} \sum_{k'} a_{ijk} a_{i'j'k'} P_{ijk} P_{i'j'k'}}{n-1}$$

where the sextuple sum includes all terms except main diagonals, those with $(i,j,k) = (i',j',k')$. With the P_{ijk}^2 term placed in the second sum, this simplifies to

$$(n-1)^{-1} \left[\sum_{i=1}^{R} \sum_{j=1}^{C} \sum_{k=1}^{S} a_{ijk}^2 P_{ijk} - \sum_{i=1}^{R} \sum_{j=1}^{C} \sum_{k=1}^{S} \sum_{i'=1}^{R} \sum_{j'=1}^{C} \sum_{k'=1}^{S} a_{ijk} a_{i'j'k'} P_{ijk} P_{i'j'k'} \right]$$

$$= (n-1)^{-1} \left[\sum_i \sum_j \sum_k a_{ijk}^2 P_{ijk} - \left(\sum_i \sum_j \sum_k a_{ijk} P_{ijk} \right)^2 \right]$$

as indicated in the text.

The only remaining task is the computation of partial derivatives. The computation is most difficult for $\nabla_{\mathscr{G}_{YXW|X}}$. Once this job is completed, the computations for the shift measures follow readily and will not be given here. As a convenient shorthand we write $\partial \nabla / \partial P_{ijk}$ instead of $(\partial \hat{\nabla} / \partial f_{ijk}) | f_{ijk} = P_{ijk}$. Thus by the expression for the derivative of a ratio,

$$\frac{\partial \nabla_{\mathscr{G}_{YXW|X}}}{\partial P_{ijk}} = -U_X^{-2} \left(U_X \frac{\partial K}{\partial P_{ijk}} - K \frac{\partial U_X}{\partial P_{ijk}} \right)$$

where U_X and K are defined in Section 7.7. Now

$$\frac{\partial K}{\partial P_{ijk}} = \frac{\partial \sum_i \sum_j \sum_k \omega_{ijk} P_{ijk}}{\partial P_{ijk}} = \omega_{ijk}$$

The derivative of

$$U_X = \sum_{i'} \sum_{j'} \sum_{k'} \omega_{i'j'k'} \frac{P_{i'j'.} P_{.j'k'}}{P_{.j'.}}$$

requires careful checking of subscripts. If $j \neq j'$ the derivative of $(\omega_{i'j'k'} P_{i'j'} P_{.j'k'}) / P_{.j'}$ is zero. Otherwise we again have a ratio to differentiate. The derivative of $P_{.j.}$ with respect to P_{ijk} is one; that of $P_{i'j.}$ is zero if $i \neq i'$ and one if $i = i'$; that of $P_{.jk'}$ is zero if $k \neq k'$ and one if $k = k'$. It follows that

$$\frac{\partial U_X}{\partial P_{ijk}} = \sum_{i'} \sum_{k'} \omega_{i'jk'} \frac{(\partial / \partial P_{ijk}) P_{i'j.} P_{.jk'}}{P_{.j.}}$$

$$- \sum_{i'} \sum_{k'} \omega_{i'jk'} \frac{P_{i'j.} P_{.jk'} \times 1}{P_{.j.}^2}$$

$$= \sum_{i'} \omega_{i'jk} \frac{P_{i'j.}}{P_{.j.}} + \sum_{k'} \omega_{ijk'} \frac{P_{.jk'}}{P_{.j.}}$$

$$- \sum_{i'} \sum_{k'} \omega_{i'jk'} \frac{(P_{i'j.} P_{.jk'})}{P_{.j.}^2}$$

$$= \pi(w_k | x_j) + \pi(y_i | x_j) - \pi(\cdot | x_j)$$

by the definitions of conditional probability and of the π-values given in Section 7.7.

Replacing the expressions for $\partial K / \partial P_{ijk}$ and $\partial U_X / \partial P_{ijk}$ in the expression for $\partial \nabla / \partial P_{ijk}$ yields the negative of the value for a_{ijk} given in Section 7.7.

REFERENCES

Kenneth J. Arrow. *Social Choice and Individual Values*, 2nd ed. New Haven: Yale University Press, 1963.

—— and F. H. Hahn. *General Competitive Analysis*. San Francisco: Holden-Day, 1971.

Robert J. Aumann and Michael Maschler. "The bargaining set for cooperative games." In M. Dresher, L. S. Shapley, and A. W. Tucker, Eds., *Advances in Game Theory*, Annals of Mathematics Studies, No. 52. Princeton, N. J.: Princeton University Press, 1964. Pp. 443–476.

Richard A. Bartlett. "Partition analysis of categorical data," unpublished doctoral dissertation. Philadelphia: University of Pennsylvania, 1974.

Rodolfo Benini. *Principii di Demografia*. Florence: G. Barbera, 1901. *Manuali Barbera di Scienze Giuridiche Sociali e Politiche*, No. 29. Cited by Goodman and Kruskal (1959).

Yvonne M. M. Bishop, Stephen E. Feinberg, and Paul W. Holland. *Discrete Multivariate Analysis: Theory and Practice*. Cambridge, Mass.: M.I.T. Press, 1975.

Hubert M. Blalock, Jr. *Social Statistics*. New York: McGraw-Hill, 1960.

Peter M. Blau and Richard A. Schoenherr. *The Structure of Organizations*. New York: Basic Books, 1971.

Karl Brunner and Allan Meltzer, Eds. Supplementary issue, *Journal of Monetary Economics*, forthcoming (1978).

William Buchanan. "Nominal and ordinal bivariate statistics: The practitioner's view." *American Journal of Political Science*, **18** (1974), 625–646.

Philip W. Buck. *Amateurs and Professionals in British Politics, 1918–59*. Chicago: University of Chicago Press, 1963.

Theodore Caplow. "Further development of a theory of coalitions in the triad." *American Journal of Sociology*, **64** (1959), 488–493.

Gösta Carlsson. *Social Mobility and Class Structure*. Lund: C. W. K. Gleerup, 1958.

294

J. M. Chertkoff. "A revision of Caplow's coalition theory." *Journal of Experimental Social Psychology*, **3** (1967), 172–177.

Carl F. Christ. "Econometric models, aggregate." In David L. Sills, Ed., *International Encyclopedia of the Social Sciences*, Vol. 4. New York: Macmillan, 1968. Pp. 344–350.

John A. Clausen. "Studies of the postwar plans of soldiers: A problem in prediction." In Samuel A. Stouffer et al., *Studies in Social Psychology in World War II*, Vol. 4. *Measurement and Prediction*. Princeton, N.J.: Princeton University Press, 1950. Pp. 568–622.

Jacob Cohen. "A coefficient of agreement for nominal scales." *Educational and Psychological Measurement*, **20** (1960), 37–46.

———. "Weighted kappa: nominal scale agreement with provision for scaled disagreement or partial credit." *Psychological Bulletin*, **70** (1968), 213–220.

James S. Coleman. *Introduction to Mathematical Sociology*. New York: Free Press, 1964.

Clyde Coombs. *A Theory of Data*. New York: Wiley, 1964.

Herbert L. Costner. "Criteria for measures of association." *American Sociological Review*, **30** (1965), 341–353.

James A. Davis. "A partial coefficient for Goodman and Kruskal's gamma." *Journal of the American Statistical Association*, **62** (1967), 184–193.

W. Edward Deming and Frederick F. Stephan. "On a least squares adjustment of a sampled frequency table when the expected marginal totals are known." *Annals of Mathematical Statistics*, **11** (1940*a*), 427–44.

———. "The sampling procedure of the 1940 population census." *Journal of the American Statistical Association*, **35** (1940*b*), 615–630.

Gustav Deuchler. "Über die methoden der korrelationsrechnung in der pädagogik und psychologie." *Zeitschrift für Pädagogische Psychologie und Experimentelle Pädagogik*, **15** (1914), 114–131, 145–159, 229–242. Cited by Goodman and Kruskal (1959).

Sanford M. Dornbusch and W. Richard Scott. *Evaluation and the Exercise of Authority*. San Francisco: Jossey-Bass, 1975.

J. Durbin. "Appendix note on a statistical question raised in the preceding paper." *Population Studies*, **9** (1955), 101.

Maurice Duverger. *Political Parties*. Trans. Barbara and Robert North. New York: Wiley, 1954.

A. W. F. Edwards. "The measure of association in a 2×2 table." *Journal of the Royal Statistical Society*, Series A, **126** (1963), 109–114.

Heinz Eulau. *Micro-Macro Political Analysis: Accents of Inquiry*. Chicago: Aldine, 1969.

Joseph L. Fleiss. *Statistical Methods for Rates and Proportions*. New York: Wiley, 1973.

Roy G. Francis. *The Rhetoric of Science*. Minneapolis: University of Minnesota Press, 1961.

Michele Fratianni. "Discriminating between alternative hypotheses of inflation in the Italian experience." Paper prepared for the Sixth Konstanz Seminar on Monetary Theory and Monetary Policy, June 25–27, 1975.

John E. Freund. *Mathematical Statistics*, 2nd ed. Englewood Cliffs, N.J.: Prentice-Hall, 1971.

William A. Gamson. "A theory of coalition formation." *American Sociological Review*, **26** (1961), 373–382.

D. V. Glass, Ed. *Social Mobility in Britain*. London: Routledge and Kegan Paul, 1954.

Arthur S. Goldberg. "Social determinism and rationality as bases of party identification." *American Political Science Review*, **63** (1969), 5–25.

Leo A. Goodman. "The analysis of cross-classified data: Independence, quasi-independence, and interactions in contingency tables with or without missing entries." *Journal of the American Statistical Association*, **63** (1968), 1091–1131.

———. "How to ransack social mobility tables and other kinds of cross-classification data." *American Journal of Sociology*, **75** (1969), 1–40.

———. "The analysis of multidimensional contingency tables: Stepwise procedures and direct estimation methods for building models for multiple classifications." *Technometrics*, **13** (1971), 33–62.

———and William H. Kruskal. "Measures of association for cross-classifications." *Journal of the American Statistical Association*, **49** (1954), 732–764.

———. "Measures of association for cross-classifications, II: Further discussions and references." *Journal of the American Statistical Association*, **54** (1959), 123–163.

———. "Measures of association for cross-classifications, III: Approximate sampling theory." *Journal of the American Statistical Association*, **58** (1963), 310–364.

———. "Empirical evaluation of formal theory." *Journal of Mathematical Sociology*, 3 (1974), 187–196.

Major Greenwood, Jr., and G. Udny Yule. "The statistics of anti-typhoid and anti-cholera inoculations, and the interpretation of such statistics in general." *Proceedings of the Royal Society of Medicine*, **8**, part 2 (1915), 113–194.

James F. Grizzle, C. Frank Starmer, and Gary C. Koch. "Analysis of categorical data by linear models." *Biometrics*, **25** (1969), 489–504.

Louis Guttman. "An outline of the statistical theory of prediction." Supplementary Study B-1. In Paul Horst et al., Eds., *The Prediction of Personal Adjustment*. New York: Social Science Research Council, Bulletin 48, 1941. Pp. 213–258.

Shelby J. Haberman. *The Analysis of Frequency Data*. Chicago: University of Chicago Press, 1974.

Michael T. Hannan, Jr. *Aggregation and Disaggregation in Sociology*. Lexington, Mass.: Lexington Books, 1971.

Roland K. Hawkes. "The multivariate analysis of ordinal measures." *American Journal of Sociology*, **76** (1971), 908–926.

William L. Hays. *Statistics for the Social Sciences*, 2nd. ed. New York: Holt, Rinehart & Winston, 1973.

David K. Hildebrand, James D. Laing, and Howard Rosenthal. "Prediction logic: A method for empirical evaluation of formal theory." *Journal of Mathematical Sociology*, 3 (1974a), 163–185.

———. "Prediction logic and quasi-independence in empirical evaluation of formal theory." *Journal of Mathematical Sociology*, 3 (1974b), 197–209.

———. "A prediction logic approach to causal models of qualitative variates." In David R. Heise, Ed., *Sociological Methodology, 1976*. San Francisco: Jossey-Bass, 1975. Pp. 146–175.

———. "Prediction analysis in political research." *The American Political Science Review*, **70** (1976), 509–535.

Paul G. Hoel, Sidney C. Port, and Charles J. Stone. *Introduction to Statistical Theory*. Boston: Houghton Mifflin, 1971.

George C. Homans. *Social Behavior: Its Elementary Forms*. New York: Harcourt Brace, Jovanovich, 1961.

W. James and Charles Stein. "Estimation with quadratic loss." In Jerzy Neyman, Ed., *Proc. Fourth Berkeley Symposium of Mathematical Statistics and Probability*, Vol. 1. Berkeley: University of California Press, 1961. Pp. 361–379.

Abraham Kaplan. *The Conduct of Inquiry*. San Francisco: Chandler, 1964.

John G. Kemeny, J. Laurie Snell, and Gerald L. Thompson. *Introduction to Finite Mathematics*, 3rd ed. Englewood Cliffs, N.J.: Prentice-Hall, 1974.

Jae-On Kim. "Predictive measures of ordinal association." *American Journal of Sociology*, **76** (1971), 891–907.

Wladimir Köppen. "Die aufeinanderfolge der unperiodischen witterungserscheinungen nach den grundsatzen der wahrscheinlichkeitsrechnung." *Repertorium für Meterologie*, (Petrograd: Akademiia Nauk) **2** (1870–1871), 189–238. Cited by Goodman and Kruskal, 1959.

Pieter Korteweg. "The inflation problem: A multi-country study of inflation—The Dutch case 1952–72." Rotterdam: Netherlands School of Economics, Working Paper 741 (1974). Paper presented at Eastern Economic Association Meetings, Albany, N.Y., October 25–27, 1

William H. Kruskal. "Ordinal measures of association." *Journal of the American Statistical Association*, **53** (1958), 814–861.

Solomon Kullback. *Information Theory and Statistics*. New York: Dover, 1968.

James D. Laing. "Organizational evaluation and authority: A test of a theory," unpublished doctoral dissertation. Stanford, Calif.: Stanford University, 1967.

—— and Richard J. Morrison. "Sequential games of status." *Behavioral Science*, **19** (1974), 177–196.

—— and Howard Rosenthal. "Linking form of hypothesis to type of statistic: Comment." *American Sociological Review*, **39** (1974), 277–280.

H. O. Lancaster. *The Chi-Squared Distribution*. New York: Wiley, 1969.

Stanley Lieberson. "Limitations in the application of non-parametric coefficients of correlation." *American Sociological Review*, **29** (1964), 744–746.

Seymour Martin Lipset. *Political Man*. Garden City, N.Y.: Doubleday, 1960.

Jane Loevinger. "A systematic approach to the construction and evaluation of tests of ability." *Psychological Monographs*, **61** (1947), No. 4.

R. Duncan Luce and Howard Raiffa. *Games and Decisions*. New York: Wiley, 1957.

Duncan MacRae, Jr. *Parliament, Parties, and Society in France: 1946–1958*. New York: St. Martin's Press, 1967.

John H. Madge. *The Tools of Social Science*. London: Longmans, Green, 1953.

Michael Maschler. "Playing an *n*-person game, an experiment." Princeton, N.J.: Princeton University Econometric Research Program, Research Memorandum No. 73, 1965.

Richard D. McKelvey and William Zavoina. "A statistical model for the analysis of ordinal level dependent variables." *Journal of Mathematical Sociology*, **4** (1975), 103–120.

Robert Messenger and Lewis Mandell. "A modal search technique for predictive nominal scale multivariate analysis." *Journal of the American Statistical Association*, **67** (1972), 768–772.

Alexander Mood, Franklin A. Graybill, and Duane Boes. *Introduction to the Theory of Statistics*, 3rd ed. New York: McGraw-Hill, 1974.

Raymond N. Morris. "Multiple correlation and ordinally scaled data." *Social Forces*, **48** (1970), 299–311.

Donald F. Morrison. *Multivariate Statistical Methods*. New York: McGraw-Hill, 1967.

Richard J. Morrison. "Caplow's model: A reformulation." *Journal of Mathematical Sociology*, **3** (1974a), 215–230.

———. "Choice models of coalition formation in the triad," unpublished doctoral dissertation. Pittsburgh: Carnegie-Mellon University, 1974b.

Frederick Mosteller. "Association and estimation in contingency tables." *Journal of the American Statistical Association*, **63** (1968), 1–29.

Edward N. Muller. "A test of a partial theory of potential for political violence." *American Political Science Review*, **66** (1972), 928–959.

Karl Pearson. "On a criterion that a given system of deviations from the probable in the case of a correlated system of variables is such that it can be reasonably supposed to have arisen by random sampling." *Philosophical Magazine*, **50** (1900), 157–175.

Donald C. Ploch. "Theorizing and statistics." *The American Sociologist*, **5** (1970), 143–146.

———. "Ordinal measures of association and the general linear model." In H. M. Blalock, Jr., Ed., *Measurement in the Social Sciences*. Chicago: Aldine, 1974. Pp. 343–368.

H. Edward Ransford. "Blue collar anger: Reactions to student and black protest." *American Sociological Review*, **37** (1972), 333–346.

Anatol Rapoport. *N-Person Game Theory*. Ann Arbor: University of Michigan Press, 1970.

William H. Riker. *The Theory of Political Coalitions*. New Haven: Yale University Press, 1962.

——— and William James Zavoina. "Rational behavior in politics: Evidence from a three-person game." *American Political Science Review*, **64** (1970), 48–60.

W. S. Robinson. "Ecological correlations and the behavior of individuals." *American Sociological Review*, **15** (1950), 351–357.

Natalie Rogoff. *Recent Trends in Occupational Mobility*. New York: Free Press, 1953.

Howard Rosenthal. "Voting in coalition models in election simulations." In William Coplin, Ed., *Simulation and the Study of Politics*. Chicago: Markham, 1968. Pp. 237–287.

———. "Aggregate data." In Ithiel de Sola Pool et al., Eds., *Handbook of Communication*. Chicago: Rand McNally, 1973. Pp. 915–937.

Jack Sawyer and Duncan MacRae, Jr. "Game theory and cumulative voting in Illinois: 1902–1954." *American Political Science Review*, **56** (1962), 936–946.

Henry Scheffé. *The Analysis of Variance*. New York: Wiley, 1959.

Fred T. Schreier. *Human Motivation: Probability and Meaning*. New York: Free Press, 1957.

W. Richard Scott, Sanford M. Dornbusch, Bruce C. Bushing, and James D. Laing. "Organizational evaluation and authority." *Administrative Science Quarterly*, **12** (1967), 93–117.

Robert B. Smith. "Continuities in ordinal path analysis." *Social Forces*, **53** (1974), 200–229.

Robert H. Somers. "A new asymmetric measure of association." *American Sociological Review*, **27** (1962), 799–811.

———. "On the measurement of association." *American Sociological Review*, **33** (1968), 291–292.

———. "Analysis of partial rank correlation measures based on the product moment model: Part one." *Social Forces*, **53** (1974), 229–246.

Patrick Suppes. *Introduction to Logic*. New York: Van Nostrand, Reinhold, 1957.

Henri Theil. *Statistical Decomposition Analysis*. Amsterdam: North Holland, 1972.

Warren S. Torgerson. *Theory and Methods of Scaling*. New York: Wiley, 1958.

Stanely H. Udy, Jr. *Organization of Work*. New Haven: Human Relations Area Files Press, 1967.

W. Edgar Vinacke. "Power, strategy, and the formation of coalitions in triads under four incentive conditions." Office of Naval Research, NONR 3748(02), Technical Report No. 1, 1962.

John von Neumann and Oskar Morgenstern. *Theory of Games and Economic Behavior*, 2nd ed. Princeton, N.J.: Princeton University Press, 1947.

David L. Wallace. "Asymptotic approximations to distributions." *Annals of Mathematical Statistics*, **29** (1958), 635–654.

Stanley Wasserman. "An ex-post significance test for a pattern of association based on predictive logic," unpublished M.S. thesis. Philadelphia: University of Pennsylvania, 1973.

S. S. Wilks. *Mathematical Statistics*. New York: Wiley, 1962.

Thomas P. Wilson. "A proportional-reduction-in-error interpretation for Kendall's tau-b." *Social Forces*, **47** (1968), 340–342.

———. "Critique of ordinal variables." *Social Forces*, **49** (1970), 432–444.

———. "Measures of association for bivariate ordinal hypotheses." In Hubert M. Blalock, Ed., *Measurement in the Social Sciences*. Chicago: Aldine-Atherton, 1974. Pp. 327–342.

———. "On interpreting ordinal analogies to multiple regression and path analysis." *Social Forces*, **53** (1974), 196–199.

Raymond E. Wolfinger and Joan Heifetz. "Safe seats, seniority, and power in Congress." *American Political Science Review*, **60** (1965), 337–349.

Saburo Yasuda. "A methodological inquiry into social mobility." *American Sociological Review*, **39** (1964), 16–23.

W. J. Youden. "Index for rating diagnostic tests." *Cancer*, **3** (1950), 32–35.

G. Udny Yule. "On the methods of measuring association between two attributes." *Journal of the Royal Statistical Society*, **75** (1912), 579–642. Discussion, 643–653.

AUTHOR INDEX

This index refers only to passages where an author's work is directly cited or discussed. Statistical measures associated with specific individuals, such as Yule's Q, are referenced in the subject index. The subject index also references repeated use of data from a given source, such as "Buck study."

Arrow, Kenneth J., 5n
Aumann, Robert J., 153

Bartlett, Richard A., 59n, 144-145, 229, 290-291
Benini, Rodolfo, 73
Bishop, Yvonne M. M., 48n
Blalock, Hubert M., Jr., 25
Blau, Peter M., 20-23
Boes, Duane, 46n, 233-234
Brunner, Karl, 79n
Buck, Philip W., 13, 13n, 15, 33
Busching, Bruce C., 30-31, 68-69

Caplow, Theodore, 84-88, 113-114, 117, 119-122
Carlsson, Gösta, 73
Chertkoff, Jerome M., 113-114, 118-123, 143-144
Christ, Carl F., 11n
Clausen, John A., 9
Cohen, Jacob, 74, 91n, 177n
Coleman, James S., 29
Coombs, Clyde, 4n
Costner, Herbert L., 39-40, 46n, 59-60, 72n, 175, 176n

Davis, James A., 265n
Deming, W. Edward, 131, 198

Deuchler, Gustav, 175
Dornbusch, Sanford M., 30-31, 68-69
Durbin, J., 73
Duverger, Maurice, 28

Edwards, A. W. F., 50
Eulau, Heniz, 11n

Feinberg, Stephen E., 48n
Fleiss, Joseph L., 74, 91n
Francis, Roy G., 31
Fratianni, Michele, 79n
Freund, John E., 187n, 194n

Gamson, William A., 114
Glass, D. V., 73
Goldberg, Arthur S., 18, 160, 181-183
Goodman, Leo A., 18, 39, 43, 46n, 47, 51-55, 57n, 58, 73, 125, 127-128, 169, 177n
Graybill, Franklin A., 46n, 233-234
Greenwood, Major, Jr., 124n, 129
Grizzle, James F., 48n
Guttman, Louis, 39-40, 52-53

Haberman, Shelby J., 48n
Hahn, F. H., 5n
Hannan, Michael T., Jr., 11n
Hawkes, Roland K., 173n
Hays, William L., 10n, 44n, 45, 190,

SUBJECT INDEX